Compendium of Plant Genomes

Series Editor

Chittaranjan Kole, Raja Ramanna Fellow, Government of India,
ICAR-National Research Center on Plant Biotechnology, Pusa,
New Delhi, India

Whole-genome sequencing is at the cutting edge of life sciences in the new millennium. Since the first genome sequencing of the model plant Arabidopsis thaliana in 2000, whole genomes of about 100 plant species have been sequenced and genome sequences of several other plants are in the pipeline. Research publications on these genome initiatives are scattered on dedicated web sites and in journals with all too brief descriptions. The individual volumes elucidate the background history of the national and international genome initiatives; public and private partners involved; strategies and genomic resources and tools utilized; enumeration on the sequences and their assembly; repetitive sequences; gene annotation and genome duplication. In addition, synteny with other sequences, comparison of gene families and most importantly potential of the genome sequence information for gene pool characterization and genetic improvement of crop plants are described.

Interested in editing a volume on a crop or model plant? Please contact Prof. C. Kole, Series Editor, at ckoleorg@gmail.com

More information about this series at http://www.springer.com/series/11805

Karam B. Singh · Lars G. Kamphuis · Matthew N. Nelson
Editors

The Lupin Genome

Springer

Editors
Karam B. Singh
Agriculture and Food
Commonwealth Scientific
and Industrial Research Organisation
Wembley, WA, Australia

Centre for Crop and Disease
Management
Curtin University
Bentley, WA, Australia

The UWA Institute of Agriculture
The University of Western Australia
Perth, WA, Australia

Lars G. Kamphuis
Agriculture and Food
Commonwealth Scientific and Industrial
Research Organisation
Wembley, WA, Australia

Centre for Crop and Disease
Management
Curtin University
Bentley, WA, Australia

The UWA Institute of Agriculture
The University of Western Australia
Perth, WA, Australia

Matthew N. Nelson
Agriculture and Food
Commonwealth Scientific and Industrial
Research Organisation
Wembley, WA, Australia

The UWA Institute of Agriculture
The University of Western Australia
Perth, WA, Australia

Royal Botanic Gardens Kew
Ardingly, West Sussex, UK

ISSN 2199-4781 ISSN 2199-479X (electronic)
Compendium of Plant Genomes
ISBN 978-3-030-21269-8 ISBN 978-3-030-21270-4 (eBook)
https://doi.org/10.1007/978-3-030-21270-4

This Springer imprint is published by the registered company Springer Nature Switzerland AG
The registered company address is: Gewerbestrasse 11, 6330 Cham, Switzerland

This book series is dedicated to my wife Phullara, and our children Sourav, and Devleena
Chittaranjan Kole

Preface to the Series

Genome sequencing has emerged as the leading discipline in the plant sciences coinciding with the start of the new century. For much of the twentieth century, plant geneticists were only successful in delineating putative chromosomal location, function, and changes in genes indirectly through the use of a number of "markers" physically linked to them. These included visible or morphological, cytological, protein, and molecular or DNA markers. Among them, the first DNA marker, the RFLPs, introduced a revolutionary change in plant genetics and breeding in the mid-1980s, mainly because of their infinite number and thus potential to cover maximum chromosomal regions, phenotypic neutrality, absence of epistasis, and codominant nature. An array of other hybridization-based markers, PCR-based markers, and markers based on both facilitated construction of genetic linkage maps, mapping of genes controlling simply inherited traits, and even gene clusters (QTLs) controlling polygenic traits in a large number of model and crop plants. During this period, a number of new mapping populations beyond F_2 were utilized and a number of computer programs were developed for map construction, mapping of genes, and for mapping of polygenic clusters or QTLs. Molecular markers were also used in the studies of evolution and phylogenetic relationship, genetic diversity, DNA fingerprinting, and map-based cloning. Markers tightly linked to the genes were used in crop improvement employing the so-called marker-assisted selection. These strategies of molecular genetic mapping and molecular breeding made a spectacular impact during the last one and a half decades of the twentieth century. But still they remained "indirect" approaches for elucidation and utilization of plant genomes since much of the chromosomes remained unknown and the complete chemical depiction of them was yet to be unraveled.

Physical mapping of genomes was the obvious consequence that facilitated the development of the "genomic resources" including BAC and YAC libraries to develop physical maps in some plant genomes. Subsequently, integrated genetic–physical maps were also developed in many plants. This led to the concept of structural genomics. Later on, emphasis was laid on EST and transcriptome analysis to decipher the function of the active gene sequences leading to another concept defined as functional genomics. The advent of techniques of bacteriophage gene and DNA sequencing in the 1970s was extended to facilitate sequencing of these genomic resources in the last decade of the twentieth century.

As expected, sequencing of chromosomal regions would have led to too much data to store, characterize, and utilize with the-then available computer software could handle. But the development of information technology made the life of biologists easier by leading to a swift and sweet marriage of biology and informatics, and a new subject was born—bioinformatics.

Thus, the evolution of the concepts, strategies, and tools of sequencing and bioinformatics reinforced the subject of genomics—structural and functional. Today, genome sequencing has traveled much beyond biology and involves biophysics, biochemistry, and bioinformatics!

Thanks to the efforts of both public and private agencies, genome sequencing strategies are evolving very fast, leading to cheaper, quicker, and automated techniques right from clone-by-clone and whole-genome shotgun approaches to a succession of second-generation sequencing methods. The development of software of different generations facilitated this genome sequencing. At the same time, newer concepts and strategies were emerging to handle sequencing of the complex genomes, particularly the polyploids.

It became a reality to chemically—and so directly—define plant genomes, popularly called whole-genome sequencing or simply genome sequencing.

The history of plant genome sequencing will always cite the sequencing of the genome of the model plant Arabidopsis thaliana in 2000 that was followed by sequencing the genome of the crop and model plant rice in 2002. Since then, the number of sequenced genomes of higher plants has been increasing exponentially, mainly due to the development of cheaper and quicker genomic techniques and, most importantly, the development of collaborative platforms such as national and international consortia involving partners from public and/or private agencies.

As I write this preface for the first volume of the new series "Compendium of Plant Genomes," a net search tells me that complete or nearly complete whole-genome sequencing of 45 crop plants, eight crop and model plants, eight model plants, 15 crop progenitors and relatives, and 3 basal plants is accomplished, the majority of which are in the public domain. This means that we nowadays know many of our model and crop plants chemically, i.e., directly, and we may depict them and utilize them precisely better than ever. Genome sequencing has covered all groups of crop plants. Hence, information on the precise depiction of plant genomes and the scope of their utilization are growing rapidly every day. However, the information is scattered in research articles and review papers in journals and dedicated Web pages of the consortia and databases. There is no compilation of plant genomes and the opportunity of using the information in sequence-assisted breeding or further genomic studies. This is the underlying rationale for starting this book series, with each volume dedicated to a particular plant.

Plant genome science has emerged as an important subject in academia, and the present compendium of plant genomes will be highly useful both to students and teaching faculties. Most importantly, research scientists involved in genomics research will have access to systematic deliberations on the plant genomes of their interest. Elucidation of plant genomes is of interest not only for the geneticists and breeders, but also for practitioners of an array of plant science disciplines, such as taxonomy, evolution, cytology, physiology, pathology, entomology, nematology, crop production, biochemistry, and obviously bioinformatics. It must be mentioned that information regarding each plant genome is ever-growing. The contents of the volumes of this compendium are, therefore, focusing on the basic aspects of the genomes and their utility. They include information on the academic and/or economic importance of the plants, description of their genomes from a molecular genetic and cytogenetic point of view, and the genomic resources developed. Detailed deliberations focus on the background history of the national and international genome initiatives, public and private partners involved, strategies and genomic resources and tools utilized, enumeration on the sequences and their assembly, repetitive sequences, gene annotation, and genome duplication. In addition, synteny with other sequences, comparison of gene families, and, most importantly, the potential of the genome sequence information for gene pool characterization through genotyping by sequencing (GBS) and genetic improvement of crop plants have been described. As expected, there is a lot of variation of these topics in the volumes based on the information available on the crop, model, or reference plants.

I must confess that as the series editor, it has been a daunting task for me to work on such a huge and broad knowledge base that spans so many diverse plant species. However, pioneering scientists with lifetime experience and expertise on the particular crops did excellent jobs editing the respective volumes. I myself have been a small science worker on plant genomes since the mid-1980s and that provided me the opportunity to personally know several stalwarts of plant genomics from all over the globe. Most, if not all, of the volume editors are my longtime friends and colleagues. It has been highly comfortable and enriching for me to work with them on this book series. To be honest, while working on this series I have been and will remain a student first, a science worker second, and a series editor last. And I must express my gratitude to the volume editors and the chapter authors for providing me the opportunity to work with them on this compendium.

I also wish to mention here my thanks and gratitude to the Springer staff particularly, Dr. Christina Eckey and Dr. Jutta Lindenborn for the earlier set of volumes and presently Ing. Zuzana Bernhart for all their timely help and support.

I always had to set aside additional hours to edit books beside my professional and personal commitments—hours I could and should have given to my wife, Phullara, and our kids, Sourav, and Devleena. I must mention that they not only allowed me the freedom to take away those hours from them but also offered their support in the editing job itself. I am really not sure whether my dedication of this compendium to them will suffice to do justice to their sacrifices for the interest of science and the science community.

New Delhi, India Chittaranjan Kole

Preface to the Volume

Introduction to the Lupin Genome

A major development over the last couple of decades has been the development and influence of genomic approaches to help advance our understanding of many crop species; knowledge that in turn can greatly benefit efforts to improve these crops. This book belongs to a book series by Springer called the 'Compendium of Plant Genomes' that describes genomic and related resources in many different crops. It focuses on lupins which are grain legume crops. Legumes, which belong to the Fabaceae (or Leguminosae) family, are widely distributed and form the third-largest land plant family in terms of number of species. From an agricultural point of view, they can occur as both grain crops, also known as pulses, and as fodder crops. Lupins are important ecological 'engineers', able to colonise extremely impoverished soils as well as thrive on low nutrient soils due to their ability to fix atmospheric nitrogen in symbiosis with bacteria and take up phosphorus efficiently from soils.

Lupins belong to the genus *Lupinus* in the Genistoid clade of legumes, which diverged early in Papillionoid legume evolution (Lavin et al. 2005). Lupins are receiving considerable interest recently not only for their value for sustainable farming as a break crop but also as a potential 'superfood' for fighting major health issues around diabetes and obesity. The genus *Lupinus* encompasses around 275 species that are widely distributed geographically, primarily in the Mediterranean region and North and South America, and can be found in a wide range of habitats (Drummond et al. 2012). Only a few lupin species have been domesticated and today the most widely cultivated species are *L. angustifolius* and *L. albus*, while *L. luteus* and *L. mutabilis* are niche crops. Although production has fluctuated over the last 20 years, over a million tonnes are produced every year. In 2017, the largest producers were Australia (1,031,425 t), Poland (168,678 t) and the Russian Federation (161,680 tonnes) (FAO 2017).

This volume on lupin genomics focuses primarily on narrow-leafed lupin (*L. angustifolius* or NLL), which is the main lupin crop primarily grown in Australia. Its genome has been recently sequenced with a focus on the gene-rich space and this has helped lead to the development of new breeding tools for the improvement of this and related lupin crops. This book describes these developments and also has chapters that detail the genomic and related genetic and cytogenetic resources that have been developed for NLL and how they are being used to help advance both fundamental and applied research on NLL in areas ranging from its domestication to syntenic relationships between NLL and other legume crops. Additional chapters report on genomic efforts being undertaken in other lupin crops. A brief outline of the book follows:

Chapter 1 by Dr. Wallace Cowling entitled 'Genetic diversity in narrow-leafed lupin breeding after the domestication bottleneck', and helps set the scene well for the following chapters. Narrow-leafed lupin was not fully domesticated until the 1950s in Australia and Dr. Cowling describes in detail the breeding efforts that led to this achievement and the following efforts to improve the crop. However, the breeding efforts to date have resulted in genome diversity being much lower in domesticated accessions compared with wild relatives, representing a severe domestication bottleneck. Dr. Cowling suggests methods for improving genetic diversity and the potential for long-term genetic gain, including the use of genomic information now available for this crop.

In Chap. 2, entitled 'Ecophysiology and phenology: genetic resources for genetic/genomic improvement of narrow-leafed lupin', Dr. Candy Taylor and colleagues describe the extensive genetic resources available in lupins with a focus on narrow-leafed lupin phenology. They describe how there are around 33,000 accessions of various lupin species that have been accumulated by more than 20 substantially sized and independent gene banks across the globe. They demonstrate how valuable these collections are as resources to breeding programmes to introduce new variation for traits, by focusing on examples related to phenology and in particular flowering time, and how these have benefited crop adaptation in narrow-leafed lupin.

In Chap. 3, entitled 'Overview of genomic resources available for lupins with a focus on narrow-leafed lupin (*Lupinus angustifolius*)', Dr. Karam B. Singh and colleagues provide an overview of the genomic resources available for narrow-leafed lupin with a focus on the current reference genome which underpins many of the other resources. They also describe how the narrow-leafed lupin reference genome has provided valuable insight into narrow-leafed lupin evolution and important information on some of its key plant-microbe interactions. The chapter also touches on some of the genomic resources that are in the pipeline in narrow-leafed and some other lupin species and describes the lupin genome portal, a web-based resource that houses genomic and related information for narrow-leafed lupin.

In Chap. 4, entitled 'Cytomolecular insight into lupin genomes' Dr. Karolina Susek and Dr. Barbara Naganowska summarise a large body of work that has been conducted using cytogenetic approaches in lupins, where again the focus has been on narrow-leafed lupin, which has served as a model for other lupin species. They describe cytogenetic efforts to estimate genome size, identify the number of chromosomes and integrative genetic and cytogenetic mapping in narrow-leafed lupin and discuss how insight into chromosome rearrangements has led to a hypothetical model of lupin karyotype evolution.

Chapter 5, by Dr. Lars G. Kamphuis and colleagues entitled 'Transcriptome resources paving the way for lupin crop improvement' describes the transcriptomic datasets that have been generated for lupin species from expressed sequence tags (EST) libraries through to more recent next generation RNA sequencing libraries. These datasets have been used to generate gene-based molecular markers in lupins, assist with the annotation of the narrow-leafed lupin genome and looked into specific global gene expression studies in different tissue types to address specific research questions around, for example, alkaloid biosynthesis, cluster root formation and organ abscission.

Chapter 6 by Dr. Michał Książkiewicz and Dr. Huaan Yang is entitled 'Molecular marker resources supporting the Australian lupin breeding programme' and provides a detailed overview of the different types of molecular markers that have been used in the Australian and European narrow-leafed lupin breeding programmes. It describes the implementation and accuracy of current molecular markers for domestication traits such as flowering time, seed permeability, pod shattering, alkaloid content, flower colour and disease resistance such as anthracnose caused by *Colletotrichum lupini* and phomopsis stem blight caused by *Diaporthe toxica* and concludes with the opportunities that next generation sequencing has to offer to provide additional molecular markers linked to important traits for lupin crop improvement.

Chapter 7 by Dr. Steven Cannon is entitled 'Chromosomal structure, history, and genomic synteny relationships in *Lupinus*' and capitalises on the genome sequence of narrow-leafed lupin and utilises it to infer the evolutionary history of narrow-leafed lupin relative to other legume species. Using synteny analyses the chapter demonstrates that the ancestor of all lupin species underwent a whole-genome triplication and that chromosome breakages, fusions and independent duplications subsequently led to various chromosome counts in lupin species. It also presents a detailed overview of the online resources generated to view the NLL genome and compare and contrast these to other legumes in various synteny viewers.

The next chapter (Chap. 8) by Dr. Matthew N. Nelson and colleagues entitled 'How have narrow-leafed lupin genomic resources enhanced our understanding of lupin domestication?' focuses on the domestication of lupin species pre- and post- the genomic revolution. It highlights how advances in genetic and genomic technologies have increased our understanding of lupin domestication and how it has led to the identification of key genes that control particular domestication traits such as flowering time and alkaloid content. It also highlights that the domestication process of lupins has led to a significant reduction in genetic diversity in both the Australian and European breeding programs.

Dr. Candy Taylor and colleagues explore the molecular control of time to flowering in narrow-leafed lupin in Chap. 9, which is entitled 'Genomic applications and resources to dissect flowering time control in narrow-leafed lupin'. They describe how modification of phenology was fundamental to the successful adaptation of narrow-leafed lupin to its key growing environments in southern Australia and northern Europe. They go on to recount recent advances in our understanding of the molecular mechanisms underlying these phenology changes, most notably the central role of a *Flowering Locus T* homologue in narrow-leafed lupin.

Chapter 10 by Dr. Paolo Annicchiaricho and colleagues is entitled 'Genetic and genomic resources in white lupin and the application of genomic selection'. Genotyping-by-sequencing technology has enabled cost-effective, accurate and high-density genotyping of white lupin. Two genomic selection approaches were compared and both were able to predict yield, architecture and phenology traits at moderate to high accuracy. The authors then discuss how genomic technology can be applied more broadly to other lupin crops.

In Chap. 11, Dr. Muhammad Munir Iqbal and colleagues review recent advances in 'Genomics of yellow lupin (*Lupinus luteus* L.)'. As a niche crop, yellow lupin has attracted little breeding effort or investment in genomic resources. Recently, the first genetic map for this species was released as well as transcriptomic resources. A genome sequencing project is underway for yellow lupin. The authors discuss how these rapidly developing tools can be used to help plant breeders overcome restraints holding back yellow lupin as a more widely adapted crop.

Finally, in Chap. 12, Dr. Abdelkader Aïnouche and colleagues conducted a detailed analysis of the 'The repetitive content in lupin genomes'. Focusing on four closely related species with striking differences in chromosome number and genome size, they found transposable elements accounted for most of the genome size variation, while many tandem repeats were unique to each species. The authors argue for a centralised resource to house the growing information on the repeat compartment of lupin genomes.

Our hope is that this book will provide a valuable resource to lupin/legume researchers and breeders to understand lupin genomes and a guide on how best to use rapidly developing genomic resources to understand and improve these fascinating legume species.

Wembley, Bentley, Perth, Australia Karam B. Singh
karam.singh@csiro.au
Wembley, Bentley, Perth, Australia Lars G. Kamphuis
lars.kamphuis@csiro.au
Wembley, Perth, Australia Matthew N. Nelson
matthew.nelson@csiro.au

References

Drummond CS, Eastwood RJ, Miotto STS, Hughes CE (2012) Multiple continental radiations and correlates of diversification in *Lupinus* (Leguminosae): testing for key innovation with incomplete taxon sampling. Syst Biol 61:443–460

FAO (2017) FAOSTAT: crop production data. http://www.fao.org/faostat/en/#data/QC: Food and Agriculture Organisation of the United Nations

Lavin M, Herendeen PS, Wojciechowski MF (2005) Evolutionary rates analysis of Leguminosae implicates a rapid diversification of lineages during the tertiary. Syst Biol 54:575–594

Contents

Genetic Diversity in Narrow-Leafed Lupin Breeding After the Domestication Bottleneck

1

Wallace A. Cowling

Abstract

Narrow-leafed lupins (*Lupinus angustifolius* L.) were fully domesticated as a valuable grain legume crop in Australia during the mid-twentieth century. Pedigree records are available for 31 released varieties and 93 common ancestors from 1967 to 2016, which provides a rare opportunity to study genetic diversity and population inbreeding in a crop following a domestication bottleneck. From the 1930s–1960s, partially domesticated germplasm was exchanged among lupin breeders in eastern and western Europe, Australia, and USA. Mutants of two founder parents contributed to the first fully domesticated narrow-leafed lupin variety "Uniwhite" in 1967. Four Phases of breeding are proposed after domestication in the Australian lupin breeding program: Foundation (1967–1987), First Diversification (1987–1998), Exploitation (1998–2007), and Second Diversification (2007–2016) Phases. Foundation Phase varieties had only two or three founder parents following the domestication bottleneck and high average coefficient of coancestry ($f = 0.45$). The First Diversification Phase varieties were derived from crosses with wild lupin ecotypes, and varieties in this Phase had lower average coefficient of coancestry ($f = 0.27$). Population coancestry increased in varieties of the Exploitation Phase ($f = 0.39$). The rate of inbreeding (ΔF) between the First Diversification and Exploitation Phase (10 years) was 0.09 per cycle, which equates to 9% loss of alleles per cycle due to random drift and low-effective population size ($N_e = 5.4$), assuming two 5-year cycles. New genetic diversity was introduced in the Second Diversification Phase varieties ($f = 0.24$) following more crossing with wild lupins. Genetic progress in Australian lupin breeding so far has been substantial with improvements in grain yield and disease resistance, but narrow genetic diversity will limit future genetic progress. The pedigree of the latest varieties includes 39.1% from three founder varieties in the domestication bottleneck and 48.3% from 9 wild ecotypes that survived 50 years of selection. In terms of conservation genetics, the Australian lupin breeding program is a critically endangered population, and subject to excessive random drift. Migration of genetic diversity from wild lupins or exchange with international breeding programs will improve long-term genetic gain and effectiveness of genomic selection.

The original version of this chapter was revised: Figure 1.1 has been updated with part figure. The correction to this chapter is available at https://doi.org/10.1007/978-3-030-21270-4_13

W. A. Cowling (✉)
The UWA Institute of Agriculture, The University of Western Australia, 35 Stirling Highway, Perth, WA 6009, Australia
e-mail: wallace.cowling@uwa.edu.au

K. B. Singh et al. (eds.), *The Lupin Genome*, Compendium of Plant Genomes,
https://doi.org/10.1007/978-3-030-21270-4_1

1.1 Introduction

Narrow-leafed lupins (*Lupinus angustifolius* L.) provide a rare opportunity to study the impact of a recent domestication bottleneck on reducing genetic diversity and its subsequent recovery in a self-pollinating crop. Sweet narrow-leafed lupins were fully domesticated in the mid-twentieth century, following the discovery of domestication genes in different parts of the world and the exchange of germplasm among breeders in eastern and western Europe, USA, and Australia (Gladstones 1970). Pedigree records are available for 31 varieties released in Australia from 1967 to 2016 (Cowling 1999; IP Australia 2019), and 93 common ancestors. For more information on the history and attributes of narrow-leafed lupin breeding globally, readers are directed to several extensive reviews on the subject (Clements et al. 2005; Cowling and Gladstones 2000; Cowling et al. 1998b; Gladstones 1970, 1998; Święcicki et al. 2015).

Genome diversity is much lower in domesticated narrow-leafed lupins compared with their wild relatives, and wild and landrace *L. angustifolius* ecotypes provide a wealth of genetic and phenological diversity for potential use by lupin breeders (Berger et al. 2012; Mousavi Derazmahalleh et al. 2018; Cowling et al. 1998a). Wild narrow-leafed lupins have contributed to improved grain yield and disease resistance in sweet domesticated varieties (Cowling and Gladstones 2000; Stefanova and Buirchell 2010). The progeny of several wild × domesticated lupin crosses were fully fertile and released as improved varieties in the Australian lupin breeding program in the 1980s (Cowling 1999). The best lines from this round of crossing were recombined to produce high-performing varieties released in the 2000s (Stefanova and Buirchell 2010).

This chapter investigates genetic diversity and population inbreeding in the Australian lupin breeding program over 50 years from 1967 to 2016 based on pedigrees, and suggests methods for improving genetic diversity and the potential for long-term genetic gain, including the use of genomic and pedigree information and optimal contributions selection to achieve these goals.

1.2 Analysis of Genetic Diversity and Population Inbreeding

Pedigree records exist for 31 varieties released from 1967 to 2016 including 93 common ancestors in the pedigree (Cowling 1999; IP Australia 2019, Dr. Bevan Buirchell *pers. comm.*). This information was used to develop a pedigree including founder lines, varieties, and presumed or known common ancestors (Table 1.1). The number of generations of selfing in each line ("fgen") was used to calculate the level of inbreeding in each line. The value of fgen was assumed to be "0" for F_1 progeny, "5" for released varieties, and "10" for landraces or wild ecotypes (Table 1.1).

These records were used to construct a numerator relationship matrix (A-matrix) using ASREML software (VSN International, UK), and pedigrees were plotted in a pedigree chart (Fig. 1.1). The most significant feature of the lupin pedigree is the relatively small number of individuals which contribute to variety development over 50 years (total 124), compared with animal breeding where thousands of animals in the pedigree typically contribute to future performance (Goddard and Hayes 2009).

1.2.1 Coefficient of Coancestry and Inbreeding Coefficient

From pedigree records (Table 1.1), the coefficient of coancestry or kinship coefficient (*f*) between each pair of lines was calculated as ½ the numerator relationship value (*a*-value), which is the proportion of additive genetic variance that two individuals have in common. The coancestry of two individuals is "the probability that two gametes taken at random, one from each, carry alleles that are identical by descent" (Falconer and Mackay 1996), or put another way, the chance that a randomly chosen allele in two potential crossing parents is the same allele as in the common ancestor. In *L. angustifolius*, commercial varieties are homozygous at most loci, and therefore identity by descent represents the

Table 1.1 Pedigrees of Australian narrow-leafed lupin varieties during four Phases of variety release: Phase 1 (Foundation Phase, 1967–1987), Phase 2 (First Diversification Phase, 1987–1998), Phase 3 (Exploitation Phase, 1998–2007), and Phase 4 (Second Diversification Phase, 2007–2016). Key contributing ancestors are shown together with released varieties, indicated by date of release. Where parents are not known, the symbol "0" appears. "fgen" is the number of generations of selfing in the line or variety. "Var. no." is the number of the line or variety in temporal order of the pedigree. Wild ecotypes from the Australian Lupin Collection are indicated by the suffix "w", e.g. P22750w. Where numbers were not located in the records, these are replaced with "xx", e.g. 62Axx1 is a line derived from a cross made in 1962.

Name of line or variety	Female parent	Male parent	fgen	Var. no	Phase of release
New Zealand Blue	0	0	10	V1	
Germany-iuc	0	0	10	V2	
Landrace-moll	0	0	10	V3	
Borre 1947	Germany-iuc	Landrace-moll	10	V4	
New Zealand Blue-le	New Zealand Blue	New Zealand Blue	5	V5	
New Zealand Blue-ta	New Zealand Blue	New Zealand Blue	5	V6	
New Zealand Blue-leuc	New Zealand Blue	New Zealand Blue	5	V7	
Borre-Ku	Borre 1947	Borre 1947	5	V8	
Borre-efl	Borre 1947	Borre 1947	5	V9	
62Axx1	New Zealand Blue-leuc	Borre 1947	2	V10	
62Axx2	New Zealand Blue-le	New Zealand Blue-ta	2	V11	
64Axx2	62Axx1	New Zealand Blue-ta	0	V12	
64Axx1	64Axx2	62Axx2	0	V13	
Rancher	0	0	5	V14	
66A001	64Axx1	Rancher	0	V15	
66Axx2	64Axx1	Borre-Ku	0	V16	
Uniwhite 1967	64Axx2	64Axx2	5	V17	Phase1
P20722w	0	0	10	V18	
P20723w	0	0	10	V19	
AB12	66Axx2	66Axx2	2	V20	
Borre-efl/Uw	Borre-efl	Uniwhite 1967	0	V21	
Borre-efl/Uh	Borre-efl	64Axx1	0	V22	
65G-251	0	0	5	V23	
70A61	P20722w	AB12	0	V24	
70A62	P20723w	AB12	0	V25	
71Axx1	Borre-efl/Uw	64Axx1	0	V26	
65G-251/Uh	65G-251	64Axx1	5	V27	
Pxxxx1w	0	0	10	V28	
P20639w	0	0	10	V29	
P22661w	0	0	10	V30	
72Axx1	65G-251/Uh	66A001	5	V31	
Uniharvest 1972	64Axx1	64Axx1	5	V32	Phase1
72A014	66A001	66Axx2	0	V33	
72A015	71Axx1	66A001	0	V34	
Unicrop 1973	66Axx2	66Axx2	5	V35	Phase1
Fest 1973	62Axx2	62Axx2	5	V36	Phase1

(continued)

Table 1.1 (continued)

Name of line or variety	Female parent	Male parent	fgen	Var. no	Phase of release
64A02	Uniwhite 1967	P20639w	5	V37	
73Axx1	Borre-efl/Uh	Uniharvest 1972	5	V38	
P22750w	0	0	10	V39	
P22872w	0	0	10	V40	
P22748w	0	0	10	V41	
P22721w	0	0	10	V42	
72A014-1	72A014	72A014	2	V43	
72A014-2	72A014	72A014	2	V44	
72A015-2	72A015	72A015	2	V45	
74Axx1	72Axx1	Unicrop 1973	0	V46	
74A003	74Axx1	Unicrop 1973	0	V47	
75A045	P22872w	72A014-1	0	V48	
75A054	P22721w	72A014-1	0	V49	
75A060	P22748w	72A014-1	0	V50	
75A061	P22750w	72A014-1	0	V51	
Unicrop-E	Unicrop 1973	Unicrop 1973	5	V52	
Marri 1976	66A001	66A001	5	V53	Phase1
CE2-1-1	Pxxxx1w	72A014-1	5	V54	
76A106-31	Unicrop 1973	P22661w	5	V55	
76A106-32	Unicrop 1973	P22661w	5	V56	
76A6-11-3-1-2	Marri 1976	Unicrop-E	5	V57	
79A078	70A62	70A61	0	V58	
Illyarrie 1979	72A014-1	72A014-1	3	V59	Phase1
Yandee 1980	72A014-2	72A014-2	3	V60	Phase1
Chittick 1982	72A015-2	72A015-2	3	V61	Phase1
75A061-3	75A061	75A061	2	V62	
75A054-5	75A054	75A054	2	V63	
79A078-14-10	79A078	79A078	5	V64	
84A086	75A061	CE2-1-1	0	V65	
84L528-18	CE2-1-1	76A106-31	2	V66	
84L551-13	76A106-32	76A6-11-3-1-2	2	V67	
75A054-5-8	75A054-5	75A054-5	2	V68	
75A061-3-1	75A061-3	75A061-3	2	V69	
84S019-96-2	79A078-14-10	84A086	4	V70	
P26672w	0	0	10	V71	
P22764w	0	0	10	V72	
84A086-12-17	84A086	84A086	4	V73	
84A086-73-10	84A086	84A086	4	V74	
Danja 1986	74A003	74A003	5	V75	Phase1
Wandoo 1986	73Axx1	64A02	5	V76	Phase1

(continued)

Table 1.1 (continued)

Name of line or variety	Female parent	Male parent	fgen	Var. no	Phase of release
83A025	75A061-3-1	75A054-5-8	0	V77	
Geebung 1987	73Axx1	64A02	5	V78	Phase1
Gungurru 1988	75A061	75A061	2	V79	Phase2
88L152-29	Gungurru 1988	P26672w	5	V80	
Yorrel 1989	75A045	75A045	5	V81	Phase2
Warrah 1989	75A060	75A060	5	V82	Phase2
75A045-10-8	Yorrel 1989	Yorrel 1989	1	V83	
Merrit 1991	Gungurru 1988	Gungurru 1988	2	V84	Phase2
84S019-96-2-11	84S019-96-2	84S019-96-2	1	V85	
84A086-73-10-37	84A086-73-10	84A086-73-10	0	V86	
84A041	Yorrel 1989	83A025	2	V87	
83A008-71-41(sel)	75A061-3-1	75A045-10-8	5	V88	
84S035-48-2	Yorrel 1989	84A086	3	V89	
84S035-48-4	Yorrel 1989	84A086	3	V90	
90A050	Merrit 1991	84S035-48-2	0	V91	
95L335-17-15	88L152-29	84S019-96-2-11	5	V92	
84S017	79A078-14-10	84A041	0	V93	
Myallie 1995	CE2-1-1	76A106-31	5	V94	Phase2
84S035-48-4-24	84S035-48-4	84S035-48-4	0	V95	
Wonga 1996	83A025	83A025	4	V96	Phase2
Kalya 1996	Warrah 1989	79A078-14-10	4	V97	Phase2
Tallerack 1997	84L528-18	84L551-13	5	V98	Phase2
84S017-50S-62	84S017	84S017	5	V99	
Tanjil 1998	Wonga 1996	Wonga 1996	2	V100	Phase2
Belara 1997	84S035-48-2	84S035-48-2	1	V101	Phase3
Moonah 1998	84S017	84S017	1	V102	Phase3
Quilinock 1999	84S019-96-2	84S019-96-2	1	V103	Phase3
90S085-107-33	Tanjil 1998	90A050	5	V104	
90S085-107-39	Tanjil 1998	90A050	5	V105	
91A047-58	Tanjil 1998	84A086-12-17	3	V106	
97L122-1	91A047-58	Kalya 1996	3	V107	
01LF1 bulk	90S085-107-39	0	0	V108	
01L576-108	P22764w	83A008-71-41(sel)	5	V109	
95L256-17	84A086-73-10-37	Quilinock 1999	0	V110	
97L182-5-7	84S017-50S-62	0	5	V111	
03A013R	95L335-17-15	0	0	V112	
04A010	97L182-5-7	03A013R	0	V113	
03LF1 bulk	0	95L335-17-15	0	V114	
Mandelup 2004	84A086-12-17	84S035-48-2	4	V115	Phase3

(continued)

Table 1.1 (continued)

Name of line or variety	Female parent	Male parent	fgen	Var. no	Phase of release
01A019R-23	Mandelup 2004	90S085-107-39	2	V116	
Coromup 2006	84S035-48-4-24	84A086-73-10	4	V117	Phase3
Jenabillup 2007	95L256-17	95L256-17	3	V118	Phase3
PBA Gunyidi 2012	90S085-107-39	01LF1 bulk	5	V119	Phase4
PBA Barlock 2013	97L122-1	90S085-107-33	5	V120	Phase4
06AF1 bulk	Jenabillup 2007	04A010	0	V121	
PBA Jurien 2015	03LF1 bulk	95L335-17-15	5	V122	Phase4
PBA Leeman 2016	01L576-108	Coromup 2006	5	V123	Phase4
PBA Bateman 2016	01A019R-23	06AF1 bulk	5	V124	Phase4

"end of the road" in terms of allelic variation at a locus. Progeny of a cross between two highly selfed parents that share an allele that is identical by descent will be fixed at that locus. The coefficient of coancestry between two parents is equal to the inbreeding coefficient (*F*) of their progeny.

1.2.2 Four Phases of Lupin Breeding Based on Coancestry

On the basis of average *f*-values, the 31 released varieties in the Australian lupin breeding program from 1967 to 2016 were allocated to four Phases of breeding: 11 varieties in the Foundation Phase (1967–1987), 9 varieties in the First Diversification Phase (1987–1998), 6 varieties in the Exploitation Phase (1998–2007), and 5 varieties in the Second Diversification Phase (2007–2016) (Table 1.1 and Figs. 1.1 and 1.2). The Foundation Phase is equivalent to breeding cycles one and two in Stefanova and Buirchell (2010), the First Diversification Phase to breeding cycles three and four, and the Exploitation Phase to breeding cycle five.

The average coefficient of coancestry in 11 varieties released during the Foundation Phase was high (average f = 0.45) (Fig. 1.2), as expected following a domestication bottleneck involving three main founders: "New Zealand Blue," "Borre," and "Rancher" (Fig. 1.1). "Rancher" from USA was incorporated into the breeding of several Foundation Phase varieties ("Marri," "Chittick," "Illyarrie," and "Yandee"). "65G-251" from USA forms a small proportion of the pedigree of "Danja" but did not contribute to pedigrees beyond this Phase. The high level of coancestry in the Foundation Phase varieties is almost equivalent to the mating of a noninbred individual with itself, that is, the first selfing of a cross progeny assuming no prior inbreeding (f = 0.5). On average, for two randomly mated foundation varieties in the Foundation Phase, there is a 45% chance that a randomly chosen allele is identical by descent, or put another way, the progeny of this mating will be fixed for ancestral alleles at 45% of loci by genetic drift alone. Such high levels of random drift are typical of "domestication bottlenecks" and dramatically reduce the potential for future genetic gain by selection (Falconer and Mackay 1996). The pedigrees of several varieties "Marri," "Chittick," "Wandoo," and "Danja" did not contribute beyond the Foundation Phase.

The average coefficient of coancestry decreased in the First Diversification Phase (f = 0.26) (Fig. 1.2), after intercrossing of 72A014 (the "Illyarrie" progenitor) with wild lupin ecotypes (Table 1.1 and Fig. 1.1). This level of coefficient of coancestry among varieties is equivalent to full-sib or parent-offspring mating, assuming unrelated parents.

Intercrossing of high-performing lines from the First Diversification Phase led to more population inbreeding in the Exploitation Phase, and

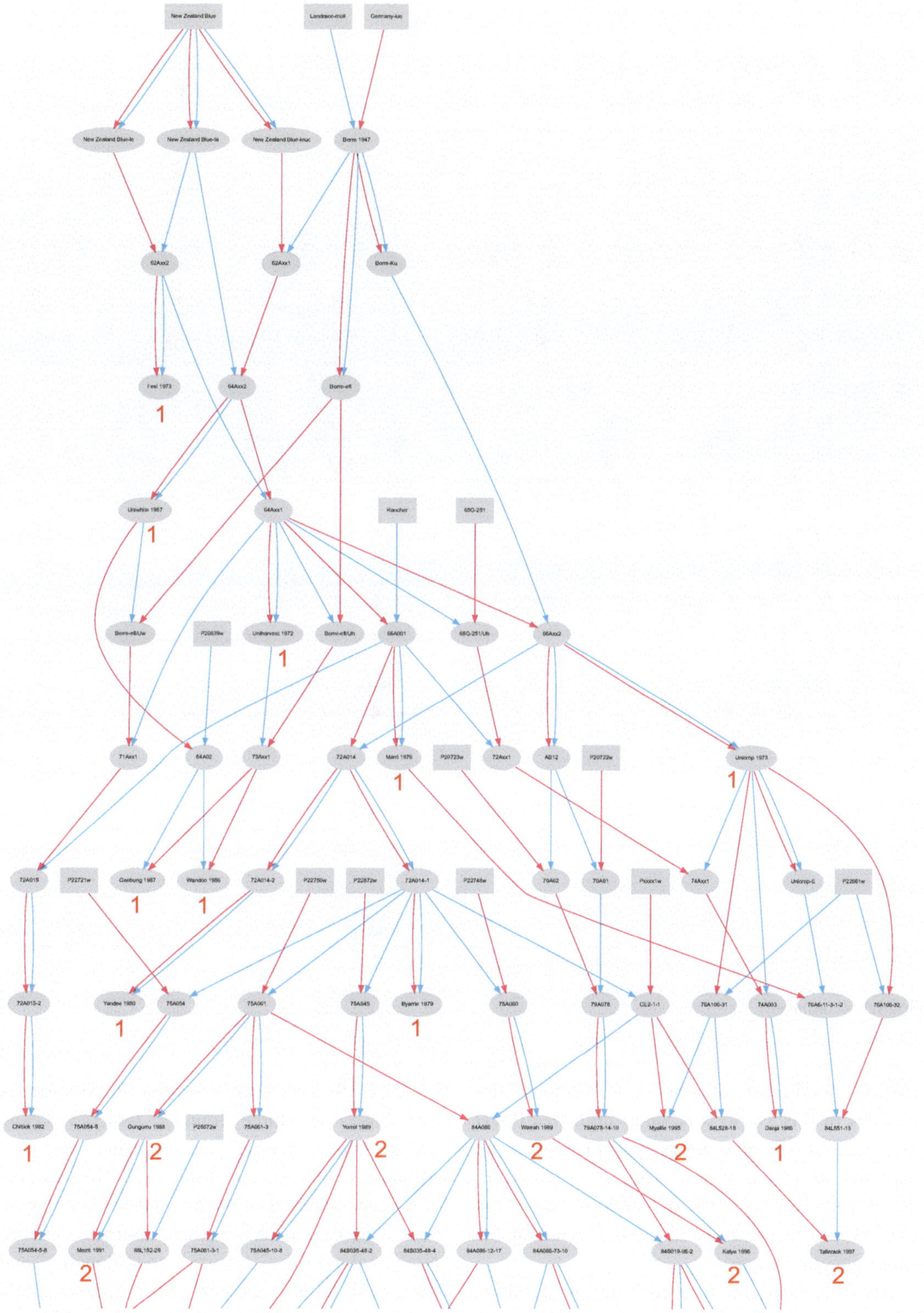

Fig. 1.1 Pedigree diagram of Australian narrow-leafed lupin varieties over four Phases of variety release, indicated by numerals 1, 2, 3, and 4 below the variety name. Female parents are indicated by red lines and male parents as blue lines. Selfing is indicated by both red and blue lines connecting from the parent

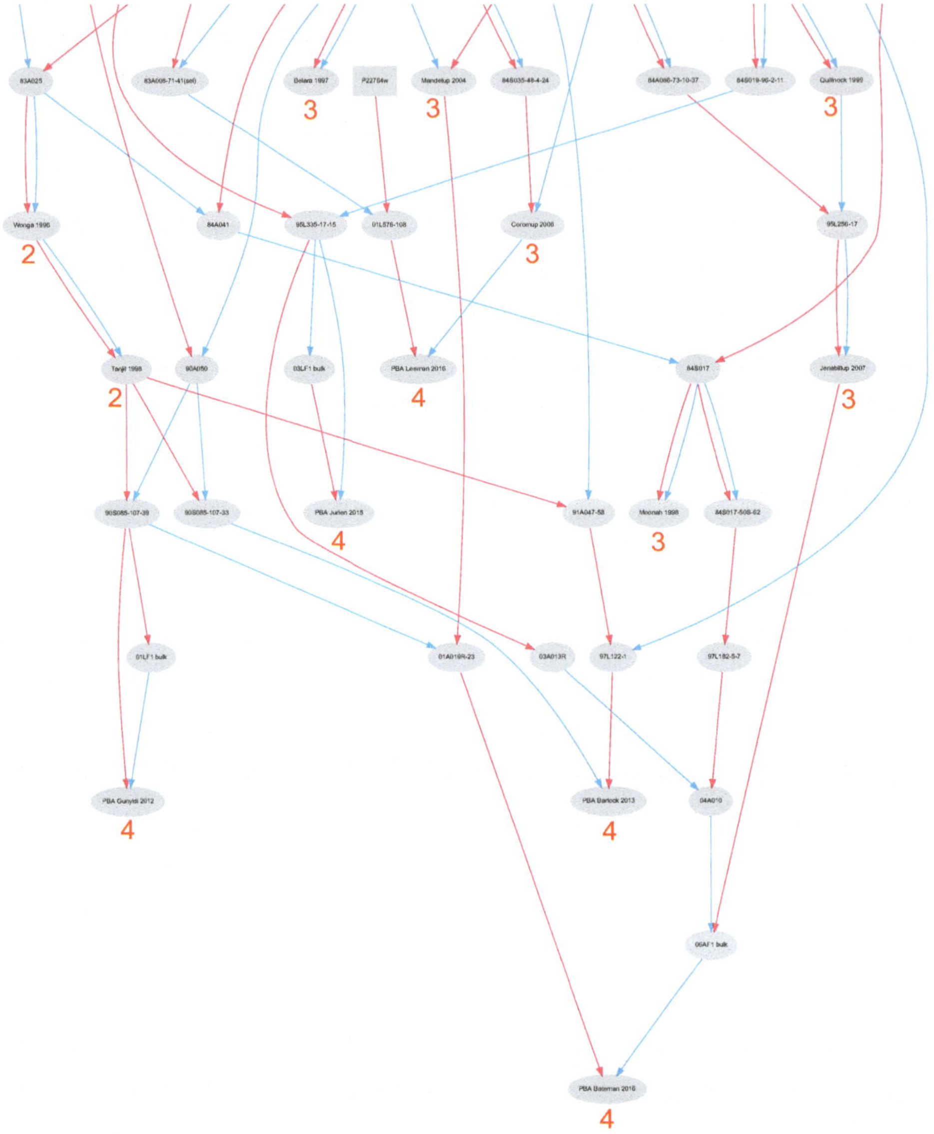

Fig. 1.1 (continued)

average coefficient of coancestry among the varieties in this Phase was again high ($f = 0.39$) (Fig. 1.2). On average, progeny of crosses in the Exploitation Phase were fixed for a common ancestor's allele at 39% of random loci. This level of kinship severely limits the potential for future genetic progress.

Another round of crossing with wild lupins helped to reduce population coancestry in varieties released during the Second Diversification Phase ($f = 0.24$). Second Diversification Phase varieties were based on three founder varieties "New Zealand Blue," "Borre," and "Rancher" and nine wild lupins (Fig. 1.2 and Table 1.2). The pedigree of varieties released in the Second Diversification Phase was dominated by three founder varieties "New Zealand Blue," "Borre," and "Rancher" and three wild lupin ecotypes (those used initially in formation of "Gungurru," "Yorrel," and "Myallie") (Fig. 1.2 and Table 1.2).

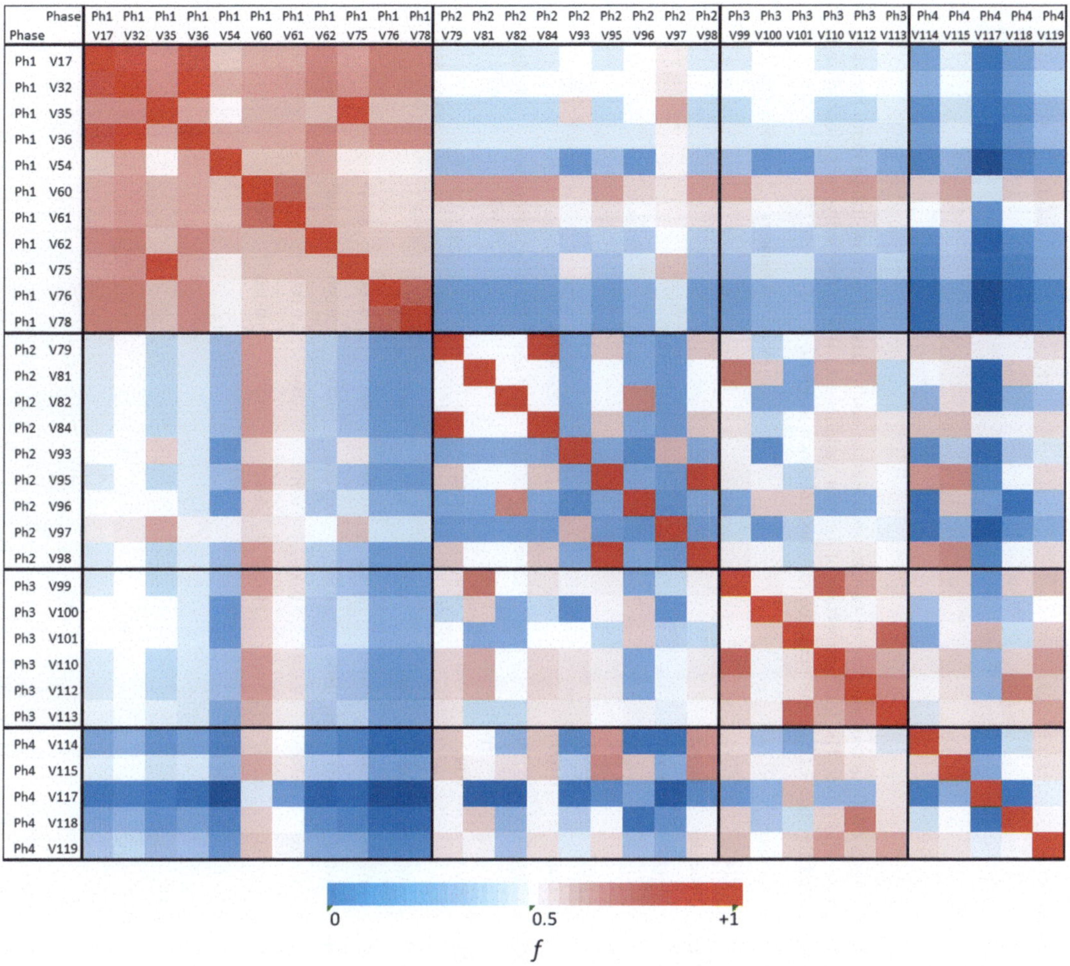

Fig. 1.2 Heat map of coefficients of coancestry (f) among 31 varieties released from the Australian lupin breeding program allocated during four Phases (Ph1 to Ph4) of variety release. Variety names associated with "V" numbers are listed in Table 1.1

1.2.3 Rate of Population Inbreeding

The rate of population inbreeding (ΔF) per cycle from the First Diversification Phase to the Exploitation Phase was calculated as follows (Falconer and Mackay 1996):

$$\Delta F = \frac{F_t - F_{t-1}}{1 - F_{t-1}}$$

From the data presented above for f (= F) in these two phases, and assuming two breeding cycles between these phases, then the average ΔF per cycle was 0.09, equivalent to 9% fixation of alleles per cycle due to random drift.

The effective population size (N_e) in the Exploitation Phase was estimated in an idealized population as follows (Falconer and Mackay 1996):

$$\Delta F = \frac{1}{2N_e}$$

With average ΔF per cycle equal to 0.09, N_e in the Exploitation Phase was estimated to be 5.4.

Table 1.2 The average proportion of ancestral founder varieties and wild types (WT) in the pedigree of varieties released in each Phase of narrow-leafed lupin breeding in Australia is based on average coefficient of coancestry from 4 founders and 11 wild types. The total proportion of founder varieties and WT in the pedigree is close to unity in Phases 1, 2, and 3, indicating that most of the ancestral parents are accounted for, but 0.12 of the pedigree remains unknown in Phase 4

Phase	No. varieties	Variety name:	New Zealand Blue	Borre	Rancher	65G-251	Proportion founders	Number founders
		Variety number	VI	V4	V14	V23		
		Variety status	Founder	Founder	Founder	Founder		
1	11		0.579	0.250	0.117	0.006	0.952	4
2	9		0.225	0.181	0.106	0.000	0.512	3
3	6		0.219	0.182	0.097	0.000	0.498	3
4	5		0.171	0.133	0.086	0.000	0.391	3

Phase	No. varieties	Variety name	P20722w	P20723w	Pxxxx1w	P20639w	P22661w	P22750w	P22872w	P22748w	P22721w	P26672w	P22764w	Proportion WT	Number WT
		Variety number	V18	V19	V28	V29	V30	V39	V40	V41	V42	V71	V72		
		Variety status	WT	WT	WT	WT	WT	WT	WT	WT	WT	WT	WT		
1	11		0.000	0.000	0.000	0.045	0.000	0.000	0.000	0.000	0.000	0.000	0.000	0.045	1
2	9		0.014	0.014	0.042	0.000	0.056	0.167	0.056	0.083	0.056	0.000	0.000	0.486	8
3	6		0.052	0.052	0.135	0.000	0.000	0.146	0.104	0.000	0.010	0.000	0.000	0.500	6
4	5		0.021	0.021	0.063	0.000	0.000	0.178	0.052	0.012	0.045	0.041	0.050	0.483	9

The published values of N_e in Thoroughbred horses was close to 100 (Corbin et al. 2010) and 95 in the Meatlinc sheep breeding program in the UK (Avendaño et al. 2003). The median effective population size in zoo animals was 22.6, which was considered low (Boakes et al. 2007). The effective population size in Australian narrow-leafed lupin breeding seems very low in comparison to animal breeding programs. In genetic conservation of endangered species, an effective population size of less than 50 is considered detrimental to long-term persistence; such populations are in danger of extinction (Rutledge et al. 2017). While self-pollinating crops are not in danger of extinction (inbreeding depression is not an issue), the rate of loss in genetic diversity of narrow-leafed lupin breeding is very high and detrimental to long-term genetic gain. The efficiency of molecular genetic techniques such as marker-assisted selection, gene mapping, genome-wide diversity analysis and genomic selection will decay rapidly as allelic diversity is removed from the population by random drift.

1.2.4 Founders and Migrants Contributing to the Pedigree

Another approach to estimating effective population size is to consider the number of founders or migrants in the pedigree of crop varieties. In varieties released during the Second Diversification Phase of Australian narrow-leafed lupin breeding (2007–2016), 68% of the pedigree was based on three founders and three migrant wild ecotypes (Table 1.2). This is similar to the situation reported in soybean in the USA, where five introductions accounted for 55% of the pedigree of public cultivars in the 1990s (Gizlice et al. 1994), and canola breeding in Australia, where four founder parents in 1970 made up more than 50% of the pedigrees of varieties released in 2000, and 11 founders made up 98.7% of the pedigree (Cowling 2007). The effective population size in these self-pollinating crops is much smaller than in animal breeding examples cited above.

1.3 Characteristics of Varieties Released in the Australian Narrow-Leafed Lupin Breeding Program 1967–2016

Narrow-leafed lupins are unusual because they were domesticated in the mid-twentieth century, and because pedigrees are mostly in the public record for 31 varieties from the first domesticated variety "Uniwhite" in 1967 to "PBA Bateman" in 2016 (Table 1.1 and Fig. 1.1). The history of breeding in the Australian narrow-leafed lupin breeding program is well documented (Gladstones 1970, 1975; Cowling 1999).

1.3.1 Domestication (Pre-1967)

International exchange of lupin germplasm in the 1950s and 1960s was very important for the domestication of *L. angustifolius* and its successful development as a competitive grain legume crop (Gladstones 1970). Scientists in eastern and western Europe, USA, and Australia exchanged valuable germplasm and research results, including domestication and disease resistance genes, despite major geographical and political barriers. Such international germplasm exchange remains vitally important for increasing the genetic diversity and future breeding prospects of international lupin breeding programs.

German researchers discovered a low-alkaloid natural mutant *iuc* in the 1930s. This gene was combined with a permeable seed natural mutant *moll* in the Swedish sweet forage variety "Borre" (1947) (Gladstones 1975). *Ku*, an early flowering natural mutant discovered in Australia in "Borre", was important for adaptation of narrow-leafed lupins to the climate of southern Australia (Gladstones 1970). A similar early-flowering gene *Julius* (*Jul*) was identified by Polish breeders in the Russian variety Krasnolistny (Mikołajczyk 1966), and this permitted spring sowing of narrow-leafed lupins in northern Europe. Recently, *Jul* and *Ku* were shown to have unique deletions at the same *LanFTc1* locus (Taylor et al. 2019). Other interesting deletion alleles were also discovered recently at this locus in wild ecotypes of narrow-leafed lupin (Taylor et al. 2019).

Gladstones (1970) also discovered natural mutants for non-shattering pods (*le*, *ta*) and white flowers and seeds (*leuc*) in the bitter fodder variety "New Zealand Blue". In effect, two foundation varieties "New Zealand Blue" and "Borre" contributed all the genes in the first fully domesticated sweet narrow-leafed lupin varieties "Uniwhite" (1967) (*iuc*, *mollis*, *ta*, *leuc*), "Uniharvest" (1971) (*iuc*, *mollis*, *ta*, *le*, *leuc*), and Unicrop (1973) (*iuc*, *mollis*, *ta*, *le*, *leuc*, *Ku*) (Table 1.1 and Fig. 1.1). Later varieties in the Foundation Phase incorporated disease resistance from USA variety "Rancher" (Gladstones 1975).

Global collaboration in lupin breeding allowed the domestication of sweet narrow-leafed lupins in Australia and in several other countries in the 1960s and 1970s. This set the stage for the Foundation Phase of narrow-leafed lupin breeding in Australia.

1.3.2 Phase 1: Foundation (1967–1987)

The Foundation Phase includes varieties released from 1967 to 1987. This Phase includes the first fully domesticated varieties "Uniwhite" (1967), "Uniharvest" (1971), and "Unicrop" (1973) (Fig. 1.1). Improved varieties "Marri" (1976), "Illyarrie" (1979), "Yandee" (1980), and "Danja" (1986) included contributions of disease resistance and yield from "Rancher" and frost tolerant germplasm "65G-251" from the USA. "Rancher" has survived in the pedigree of recent varieties (Figs. 1.1 and 1.2). A chemical mutant for mid-season flowering time (*efl*) was selected from "Borre" and used to breed mid-season flowering "Chittick" (1982) and "Wandoo" (1986) (Cowling 1999); the latter variety was withdrawn immediately due to susceptibility to cucumber mosaic virus (Cowling 1999). Finally, late-flowering "Geebung" (1987) was released in eastern Australia to replace "Uniharvest". Several successful varieties from the Foundation Phase, such as "Marri," "Chittick," and "Danja" did not contribute as parents to the First Diversification Phase (Fig. 1.1).

1.3.3 Phase 2: First Diversification (1987–1998)

The varieties of the First Diversification Phase were mostly the progeny of 1975 crosses of foundation variety 72A14-10 (the "Illyarrie" progenitor) with wild lupin ecotypes (Table 1.1). Domestication genes were reselected in the selfed progeny of each cross, which required substantial commitment inside the breeding program to select for soft seeds, low alkaloids, white flowers and seeds, and non-shattering pods. The first variety to be released was "Gungurru" in 1988 followed by "Yorrel" (1989), "Warrah" (1988), and "Merrit" (1991) (Cowling and Gladstones 2000). At least two ecotypes contributed to "Wonga" (1996), "Tanjil" (1998), "Myallie" (1995), "Kalya" (1996), and "Tallerack" (1997). A late flowering variety "Jindalee" was also registered for release in this Phase, but has an unknown pedigree (IP Australia 2019). It was not used in further cross-breeding and is not included in this analysis.

Selection for disease resistance was a feature of the First Diversification Phase. "Gungurru" (tested as breeding code 75A61-3) was moderately resistant to Phomopsis (Cowling and Wood 1989), and stubble derived from "Gungurru" displayed reduced lupinosis toxicity to sheep (Cowling et al. 1988). Phomopsis resistance removed a major impediment to the adoption of lupins by farmers—prior to "Gungurru," lupinosis was a serious mortality risk to sheep grazing lupin stubbles (Cowling and Gladstones 2000).

"Yorrel" and "Warrah" were progeny of 72A14-10 crossed to a different wild ecotype; each showed unique attributes derived from the wild parent, and each had improved resistance to Phomopsis. "Merrit" (1991) was a reselection from "Gungurru".

Strong selection occurred in the First Diversification Phase for anthracnose resistance, "Wonga" and its single plant selection "Tanjil" were resistant, "Kalya" was moderately resistant,

"Gungurru" and "Merrit" were moderately susceptible, and "Myallie" was susceptible (Garlinge 2005).

Resistance to brown spot and Pleiochaeta root rot were also selected during this Phase (Cowling et al. 1997), and moderate resistance was found in "Myallie," "Kalya," and "Tallerack" (Garlinge 2005).

Moderate resistance to seed transmission of cucumber mosaic virus was found in "Danja," whereas "Yorrel" and "Gungurru" were moderately susceptible and "Wandoo" was susceptible (Jones and Cowling 1995).

Aphid susceptibility was found in some varieties, and "Tallerack" was not promoted due to its susceptibility.

Disease resistance and higher yield contributed to the expansion of narrow-leafed lupin production in Australia in the 1990s (Cowling 1999). In 2004, approximately one million tonnes of lupins were produced in Western Australia, and 97% of the lupin area was sown to First Diversification Phase varieties (Garlinge 2005).

1.3.4 Phase 3: Exploitation (1998–2007)

Varieties released in the Exploitation Phase such as "Belara" (1997), "Moonah" (1998), "Quilinock" (1999), "Mandelup" (2004), "Coromup" (2006), and "Jenabillup" (2007) were derived from intermatings of high-yielding and disease-resistant parents from the First Diversification Phase. Every variety in the Exploitation Phase included ancestry of at least two wild lupin ecotypes. However, several successful varieties in the First Diversification Phase did not contribute as parents to the Exploitation Phase (Stefanova and Buirchell 2010) (Fig. 1.1).

Varieties released in this Phase showed substantial genetic improvements in grain yield, resistance to anthracnose (Stefanova and Buirchell 2010), and moderate tolerance of the herbicide metribuzin was selected following mutation breeding (Si et al. 2011). The exception was "Quilinock" which was very susceptible to anthracnose (Garlinge 2005).

1.3.5 Phase 4: Second Diversification (2007–2016)

Varieties released after 2007 such as "PBA Gunyidi" (2012), "PBA Barlock" (2013), "PBA Jurien" (2015), "PBA Leeman" (2016), and "PBA Bateman" (2016) include new wild ecotypes in the pedigree and complex pedigrees. Some of these varieties were selected for specific traits such as moderate metribuzin herbicide tolerance and disease resistance. There is some missing pedigree information due to complex pedigrees (Table 1.1 and Fig. 1.2). The missing information is due to complex crossing within the current pedigree, so the level of inbreeding is most likely underestimated in this part of the analysis.

1.3.6 Move to the Private Sector

Recently, the Australian narrow-leafed lupin breeding program was transferred from the public sector to the private sector, and no information exists on the genetic diversity retained for crossing in the private breeding program. This follows a trend in Europe of privatization of lupin breeding into companies in Germany and Poland. Genomic analysis could be used in future to estimate coancestry and effective population size, if relevant germplasm is made available to researchers.

1.4 Genetic Progress in Lupin Breeding

1.4.1 Australia

Substantial genetic progress has been achieved for yield and disease resistance in narrow-leafed lupin breeding in Australia during the 50 years of this study. Improvements are evident in grain yield, and resistance to Phomopsis, brown spot, CMV, and anthracnose (Cowling and Gladstones 2000). There was an 81% increase in grain yield in 34 years from "Unicrop" (1973) to "Mandelup" (2004) in trials of historical lupin varieties

(Stefanova and Buirchell 2010), although the yield improvement of new cultivars was only evident at high seeding rates (Cowling et al. 1998b). Tolerance to the herbicide metribuzin was included in the Second Diversification Phase (Si et al. 2011; Stefanova and Buirchell 2010). It appears that small-effective population size has not limited genetic progress in lupin breeding. This begs the question, then, why should we be concerned about low-effective population size in lupin breeding?

The number of breeding cycles during 50 years of this study was between 8 and 10 (range 4–8 years per cycle). There was a high rate of population inbreeding between Phases 2 and 3 (2 cycles). The rapid rate of population inbreeding ($\Delta F = 0.09$) in this period suggests an effective population size of approximately five, and a rapid approach to a genetic improvement plateau. It is also noted that a total of 15 "founder" or "migrant" genotypes contributed to the pedigree of released varieties after 50 years of lupin breeding, and 7 of these genotypes contributed to 73% of the pedigree in Phase 4 (Table 1.2). Compared with typical animal breeding programs (Goddard and Hayes 2009), very few individuals contribute to the lupin pedigree.

The proportion of founder varieties in the pedigree decreases over time as the proportion of wild types increases. However, even in Phase 4, the average coefficient of coancestry with founder varieties is 0.391, or 39.1% of the alleles are identical by descent to founder alleles (Table 1.2).

The major reason for concern about the small-effective population size in lupin breeding is the limit this imposes on long-term genetic progress. In a stochastic model of pea breeding for several traits in an economic index (Cowling et al. 2017), the model with high selection pressure and truncation selection appeared to have equal genetic gain as the model with optimal contributions selection for the first 10 cycles. After 10 cycles, equivalent to 40 or 50 years of lupin breeding, the high selection pressure scenario based on truncation selection reached a premature plateau in genetic improvement. In the pea model with high selection pressure there were 40 founder parents (Cowling et al. 2017), but only 4 founder parents and 11 wild types contributed to released varieties in the Australian lupin breeding program during 50 years (Table 1.2). Lupin breeding is therefore facing the risk of a premature plateau in genetic improvement, and new genetic diversity should be incorporated into the program as soon as possible.

1.4.2 Global Genetic Advances

A few narrow-leafed lupin breeding programs exist globally, normally alongside other lupin or pulse breeding programs.

There is renewed interest in lupin production and breeding in Europe due to the need for locally-grown plant protein and sustainable cropping (Lucas et al. 2015). The focus of breeding in northern Europe has been on early maturing *L. angustifolius* (Murphy-Bokern et al. 2017). Active breeding and release of varieties occurs in Germany at Saatzucht Steinach, where the most recent variety is "Mirabor" (www.saatzucht.de/english/grossleguminosen/leguminosengross.html). In France, the company Jouffray-Drillaud (https://www.jouffray-drillaud.com/accueil-en.html) lists two white lupin (*L. albus*) varieties. An EU research project "LIBBIO" is focused on selection in Andean lupin, *L. mutabilis* (https://www.libbio.net/).

In Poland, Poznan Plant Breeders (https://phr.pl/en/) and HR Smolice (https://www.hrsmolice.pl/pl/) have several varieties of narrow-leafed and yellow lupin in their portfolios. Out of the total grain legume acreage in Poland of 200,000 ha, yellow lupin occupies 15% and narrow-leafed lupin 30% (Prof Wojciech Swiecicki, *pers. comm.*).

In Russia, narrow-leafed lupin breeding is carried out mainly by State institutions. The State Register 2018 lists 25 varieties of *L. angustifolius*, 10 varieties of *L. albus*, and 10 varieties of *L. luteus* (Dr. Galina Gataulina, *pers. comm.*). Breeding institutes include the Former

All-Russian Research Institute of Lupin (Briansk region), The Russian State Agrarian University—Moscow Timiryazev Academy, All-Russia Research Institute of Grain Legumes and Groat crops, and Leningrad Research Institute of Agriculture. The sown area of lupin in the Russian Federation in 2017 amounted to 120,000 hectares, 5 times more than in 2012 (Dr. Galina Gataulina, *pers. comm.*).

In Chile, three breeding programs exist at Campex Baer (*L. angustifolius*, *L. albus*, and *L. mutabilis*), INIA Carillanca (*L. albus*), and Centro de Genómica Nutricional Agroacuícola (CGNA, *L. luteus*). Minor breeding programs on *L. mutabilis* occur in Peru, Ecuador, and Bolivia. The salmon industry generates a strong demand for lupin in Chile (Dr. Erik von Baer, *pers. comm.*).

1.5 Technologies to Improve Long-Term Genetic Gain

Four factors are important for long-term genetic gain for several commercially-important traits—genetic diversity, accurate prediction of breeding values, moderate selection pressure based on an economic index, and optimal mating designs (Cowling et al. 2017). Accuracy of prediction of breeding values has increased with the use of best linear unbiased prediction (BLUP) and GBLUP (genomic BLUP). However, while BLUP selection accelerated genetic gain compared with pre-BLUP breeding, it also increased the rate of population inbreeding (Avendaño et al. 2003). Independent culling on phenotype, the most common method of selection in crop breeding, is not conducive to long-term genetic gain (Cowling and Li 2018). The best outcome in the long run will be achieved with some form of optimal contributions selection for an economic index composed of weighted BLUP values for all traits (Cowling et al. 2017).

The first challenge in lupin breeding is to increase genetic diversity without reducing performance in the elite breeding program. Non-adapted germplasm may contain potentially valuable alleles, but how can these alleles be accessed without reducing commercial value in the breeding program through linkage drag? The answer is to migrate a small portion at a time of the genome of non-adapted lines into the elite breeding program. A BC_2-based migration scheme was proposed for introducing small portions of wild lupin genome into elite germplasm (Cowling et al. 2009). This also permitted selection of the key domestication traits during the BC_2-procedure. The scheme was introduced into the Australian lupin breeding program where it helped to increase protein content in near-elite BC_2 lines from crosses with high protein wild lupins (Buirchell 2008; Berger et al. 2013).

A major recommendation of this Chapter is that small isolated breeding programs, such as the Australian narrow-leafed lupin breeding program, while achieving good genetic progress after the domestication bottleneck, should increase their effective population size through migration from the large and diverse wild and landrace germplasm pool (Berger et al. 2012) or from contemporary breeding programs in other countries. Without much larger allelic diversity in the program, new technologies such as genomic selection will be economically inefficient. New genetic diversity should be incorporated into the elite program in small proportions (Cowling et al. 2009) to avoid reducing genetic gain in the breeding program. This process will be greatly assisted by optimal contributions selection to optimize mating designs for long-term genetic gain (Cowling et al. 2017; Cowling and Li 2018).

1.6 Conclusions and Recommendations

This Chapter describes population coancestry during 50 years of breeding in the Australian narrow-leafed lupin breeding program after the domestication bottleneck in the mid-twentieth century. In the Foundation Phase, immediately after the domestication bottleneck (1967–1987), the effective population size was very low (2 or 3 founder varieties in the pedigree). Several wild ecotypes were crossed with Foundation Phase varieties, and this reduced population coancestry in the First Diversification Phase (1987–1998). Population coancestry increased in the Exploitation Phase (1998–2007), and more wild lupins were introduced in the Second Diversification Phase (2007–2016). The genetic diversity available in the Australian narrow-leafed lupin breeding program in 2016 is relatively low compared with animal breeding programs, but similar to other crop breeding programs such as canola breeding in Australia. This raises a major challenge for lupin breeders—how to increase genetic diversity and improve long-term genetic gain without reducing the performance of the elite breeding program? The most promising approach is to use low rates of migration based on BC_2-derived progeny of crosses with wild lupins or with breeding lines exchanged between international breeding programs, and optimized mating designs based on optimal contributions selection.

Collaboration and germplasm exchange among lupin breeders and geneticists in the mid-twentieth century promoted the domestication of narrow-leafed lupins. The three founder varieties of Australian narrow-leafed lupin breeding were from Europe and USA. Since the domestication bottleneck, there is no evidence of migration of germplasm from international breeding programs. However, active narrow-leafed lupin breeding continues in Germany, Poland, Russia, and South America. It is highly likely that all international lupin breeding programs have low-effective population size, but most have unique genetics—hence, it would be mutually beneficial for all programs to undertake regular germplasm exchange. Even low levels of international germplasm exchange would improve long-term prospects for lupin genetic improvement in the twenty-first century. International exchange helped narrow-leafed lupins through the domestication bottleneck, and international exchange is the key to the recovery of effective population size after the domestication bottleneck.

Acknowledgements I received substantial help from Australian and international colleagues who provided material for this chapter. Dr. Bevan Buirchell kindly filled in some gaps in the pedigree record. Mr Paul McGowan, Senior Technical Officer (Bioinformatics) Agri-Science Queensland, kindly developed the pedigree diagram shown in Fig. 1.1. I thank international colleagues who sent substantial amounts of information, including Dr. Fred Stoddard (University of Helsinki, Finland), Dr. Galina Gataulina (The Russian State Agrarian University—Moscow Timiryazev Academy), Dr. Erik von Baer (Semillas Baer, Chile), and Prof. dr. hab. Wojciech Święcicki (Institute of Plant Genetics, Polish Academy of Sciences).

References

Avendaño S, Villanueva B, Woolliams JA (2003) Expected increases in genetic merit from using optimized contributions in two livestock populations of beef cattle and sheep. J Anim Sci 81:2964–2975

Berger JD, Buirchell BJ, Luckett DJ, Nelson MN (2012) Domestication bottlenecks limit genetic diversity and constrain adaptation in narrow-leafed lupin (*Lupinus angustifolius* L.). Theor Appl Genet 124:637–652

Berger JD, Clements JC, Nelson MN, Kamphuis LG, Singh KB, Buirchell BJ (2013) The essential role of genetic resources in narrow-leafed lupin improvement. Crop Pasture Sci 64:361–373

Boakes EH, Wang J, Amos W (2007) An investigation of inbreeding depression and purging in captive pedigreed populations. Heredity 98:172–182

Buirchell BJ (2008) Narrow-leafed lupin breeding in Australia -where to from here? In: Palta JA, Berger JB (eds) Lupins for health and wealth. Proceedings of the 12th international lupin conference, Fremantle, Western Australia, 14–18 Sept 2008, pp 226–230. https://www.cabdirect.org/cabdirect/abstract/20103116770. Accessed 6 May 2019

Clements JC, Buirchell BJ, Yang H, Smith PMC, Sweetingham MW, Smith CG (2005) Lupin, Chapter 9. In: Singh RJ, Jauhar PP (eds) Genetic resources, chromosome engineering, and crop improvement: grain legumes, vol I. CRC Press, pp 291–396

Corbin LJ, Blott SC, Swinburne JE, Vaudin M, Bishop SC, Woolliams JA (2010) Linkage disequilibrium and historical effective population size in the Thoroughbred horse. Animal Genetics 41:8–15

Cowling WA (1999) Pedigrees and characteristics of narrow-leafed lupin cultivars released in Australia from 1967 to 1998. Bulletin 4365, Agriculture Western Australia

Cowling WA (2007) Genetic diversity in Australian canola and implications for crop breeding for changing future environments. Field Crops Res 104:103–111

Cowling WA, Allen JG, Wood PMcR (1988) Resistance to Phomopsis stem blight reduces the lupinosis toxicity of narrow-leafed lupin stems. Aust J Exp Agric 28:195–202

Cowling WA, Buirchell BJ, Falk DE (2009) A model for incorporating novel alleles from the primary gene pool into elite crop breeding programs while reselecting major genes for domestication or adaptation. Crop Pasture Sci 60:1009–1015

Cowling WA, Buirchell BJ, Tapia ME (1998a) Lupin. Lupinus spp. Promoting the conservation and use of underutilized and neglected crops. 23. Institute of Plant Genetics and Crop Plant Research, Gatersleben/ International Plant Genetic Resources Institute, Rome, Italy, 105 pp

Cowling WA, Huyghe C, Swiecicki W (1998b) Lupin breeding, Chapter 4. In: Gladstones JS, Atkins CA, Hamblin J (eds) Lupins as crop plants—biology, production and utilization. CAB International UK, pp 93–120

Cowling WA, Gladstones JS (2000) Lupin breeding in Australia. In: Knight R (ed) Linking research and marketing opportunities for pulses in the 21st century. Kluwer Academic Publishers, Netherlands, pp 541–547

Cowling WA, Li L (2018) Turning the heat up on independent culling in crop breeding. In: Hermesch S, Dominik S (eds) Breeding focus 2018—reducing heat stress. Animal Genetics and Breeding Unit, University of New England, Armidale, Australia, pp 119–134

Cowling WA, Li L, Siddique KHM, Henryon M, Berg P, Banks RG, Kinghorn BP (2017) Evolving gene banks: improving diverse populations of crop and exotic germplasm with optimal contribution selection. J Exp Bot 68:1927–1939

Cowling WA, Sweetingham MW, Diepeveen D, Cullis BR (1997) Heritability of resistance to brown spot and root rot of narrow-leafed lupins caused by *Pleiochaeta setosa* (Kirchn.) Hughes in field experiments. Plant Breed 116:341–345

Cowling WA, Wood PMcR (1989) Resistance to Phomopsis stem and pod blight of narrow-leafed lupin in a range of environments and its association with reduced Phomopsis seed infection. Aust J Exp Agric 29:43–50

Falconer DS, Mackay TFC (1996) Introduction to quantitative genetics, 4th edn. Longman, Harlow, UK

Garlinge J (2005) 2005 crop variety sowing guide for Western Australia. Department of Agriculture and Food, Western Australia, Perth. Bulletin 4655. https://researchlibrary.agric.wa.gov.au/cgi/viewcontent.cgi?referer=&httpsredir=1&article=1156&context=bulletins. Accessed 6 May 2019

Gizlice Z, Carter TE Jr, Burton JW (1994) Genetic base for North American public soybean cultivars released between 1947 and 1988. Crop Sci 34:1143–1151

Gladstones JS (1970) Lupins as crop plants. Field Crop Abstr 23:123–148

Gladstones JS (1975) Lupin breeding in Western Australia: the narrow-leaf lupin (*Lupinus angustifolius*). J Dep Agric West Aust, Series 4, 16:44–49. https://researchlibrary.agric.wa.gov.au/journal_agriculture4/vol16/iss2/4. Accessed 6 May 2019

Gladstones JS (1998) Distribution, origin, taxonomy, history and importance, Chapter 1. In: Gladstones JS, Atkins CA, Hamblin J (eds) Lupins as crop plants—biology, production and utilization. CAB International UK, pp 1–39

Goddard ME, Hayes BJ (2009) Mapping genes for complex traits in domestic animals and their use in breeding programmes. Nat Rev Genet 10:381–391

IP Australia (2019) Plant breeders rights—search plant breeders rights. https://www.ipaustralia.gov.au/plant-breeders-rights. Accessed 6 May 2019

Jones RAC, Cowling WA (1995) Resistance to seed transmission of cucumber mosaic virus in narrow-leafed lupins (*Lupinus angustifolius*). Aust J Agric Res 46:1339–1352

Mikołajczyk J (1966) Genetic studies in *Lupinus angustifolius*. Part III. Inheritance of the alkaloid content, seed hardness and length of the growing season in blue lupin. Genet Pol 7:181–196

Mousavi-Derazmahalleh M, Bayer PE, Nevado B, Hurgobin B, Filatov D, Kilian A, Kamphuis LG, Singh KB, Berger JD, Hane JK, Edwards D, Erskine W, Nelson MN (2018) Exploring the genetic and adaptive diversity of a pan-Mediterranean crop wild relative: narrow-leafed lupin. Theor Appl Genet 131:887–901

Murphy-Bokern D, Stoddard FL, Watson CL (2017) Legumes in cropping systems. CAB International UK

Rutledge LY, Desy G, Fryxell JM, Middel K, White BN, Patterson BR (2017) Patchy distribution and low effective population size raise concern for an at-risk top predator. Divers Distrib 23:79–89

Si P, Pan G, Sweetingham M (2011) Semi-dominant genes confer additive tolerance to metribuzin in narrow-leafed lupin (*Lupinus angustifolius* L.) mutants. Euphytica 177:411–418

Stefanova KT, Buirchell B (2010) Multiplicative mixed models for genetic gain assessment in lupin breeding. Crop Sci 50:880–891

Święcicki W, Kroc M, Kamel KA (2015) Lupins, Chapter 6. In: de Ron AM (ed) 'Grain Legumes', Handbook of Plant Breeding, vol 10. Springer NY, pp 179–218

Taylor CM, Kamphuis LG, Zhang W, Garg G, Berger JD, Mousavi-Derazmahalleh M, Bayer PE, Edwards D, Singh KB, Cowling WA, Nelson MN (2019) INDEL variation in the regulatory region of the major flowering time gene *LanFTc1* is associated with vernalisation response and flowering time in narrow-leafed lupin (*Lupinus angustifolius* L.). Plant Cell Environ 42:174–187

Ecophysiology and Phenology: Genetic Resources for Genetic/Genomic Improvement of Narrow-Leafed Lupin

2

Candy M. Taylor, Lars G. Kamphuis, Wallace A. Cowling, Matthew N. Nelson and Jens D. Berger

Abstract

The narrow-leafed lupin (*Lupinus angustifolius* L.) is endemic to coarse-textured neutral to acid sands across the Mediterranean basin, distributed over temperature and rainfall gradients leading to increasing N–S terminal drought. *L. angustifolius* has a conservative reproductive strategy compared to other Old World species, with relatively early phenology, and a high proportion of physical dormancy in the seed. Nevertheless, appropriate phenology is the key adaptation to the terminal drought stress gradient across the species' distribution, with flow on effects for water use, stress onset and productivity. Lupins have evolved early phenology in low-rainfall environments of the Mediterranean region, facilitating escape from the annual summer drought, where the opportunity cost of reduced fitness (yield potential) is partially mitigated by higher reproductive investment (harvest index). In high-rainfall environments, ecotypes are later, producing more biomass, leading to greater yield potential, associated with higher water use and earlier stress onset under water deficit. Domesticated lupins were selected for low or no vernalisation requirement in northern Europe and southern Australia, which mimicked the early, drought escape reproductive strategy of low-rainfall ecotypes, but may limit their yield potential in higher rainfall environments of Mediterranean-type environments. Breeding options for later flowering cultivars for long-season environments are constrained by the vernalization response: all late flowering cultivars (*lanFTc1*) are vernalization-sensitive types and are very late flowering in environments where the vernalisation requirement is not readily satisfied.

C. M. Taylor · W. A. Cowling
UWA School of Agriculture and Environment, The University of Western Australia, Perth, WA 6009, Australia
e-mail: candy.taylor@research.uwa.edu.au

W. A. Cowling
e-mail: wallace.cowling@uwa.edu.au

L. G. Kamphuis · M. N. Nelson · J. D. Berger (✉)
Commonwealth Scientific and Industrial Research Organisation, Agriculture & Food, Floreat, WA 6014, Australia
e-mail: jens.berger@csiro.au

L. G. Kamphuis
e-mail: lars.kamphuis@csiro.au

M. N. Nelson
e-mail: matthew.nelson@csiro.au

L. G. Kamphuis
Centre for Crop and Disease Management, Curtin University, Bentley, WA 6102, Australia

L. G. Kamphuis · W. A. Cowling · M. N. Nelson
The UWA Institute of Agriculture, The University of Western Australia, Perth, WA 6000, Australia

M. N. Nelson
Natural Capital and Plant Health, Kew Royal Botanic Gardens, Wakehurst Place, Ardingly RH17 6TN, UK

K. B. Singh et al. (eds.), *The Lupin Genome*, Compendium of Plant Genomes,
https://doi.org/10.1007/978-3-030-21270-4_2

The mutation *efl* provided a lower vernalization requirement, but produced unreliable yields. Potentially there are levers for delaying flowering time that may be revealed in phenological studies of wild *L. angustifolius*. We have discovered alternative alleles at the *LanFTc1* locus that varyies in vernalisation-responsiveness, implying that there are other, more subtle phenology regulators to be uncovered in the species.

2.1 Adaptation and Phenology

Narrow-leafed lupin (*Lupinus angustifolius* L.) presents a very strong case for appropriate phenology as the key adaptive trait of Mediterranean annuals, with evidence in both wild and domesticated populations. *L. angustifolius* has a native Mediterranean distribution along a clearly defined north–south terminal drought stress gradient from northern Iberia and southern France through to northern Africa and the south-eastern Mediterranean (Fig. 2.1 and Table 2.1). Vegetative phase temperatures rise along this stress gradient, alongside decreasing (but increasingly variable rainfall) particularly in the reproductive phase (Table 2.1).

Wild *L. angustifolius* responds to this gradient by flowering and maturing increasingly early as rainfall decreases (Berger et al. 2017; Clements and Cowling 1994; Gladstones and Crosbie 1979; Lema et al. 2005). Moreover, there is a trade-off between seed size, early vigour and phenology in the species. Low-rainfall habitats have selected for early flowering, large seed size, high early vigour and high harvest index (and vice versa for higher rainfall habitats) (Berger et al. 2017). These are all ruderal traits sensu Grime (1979) that facilitates drought escape through early establishment and completion of a short lifecycle, whereby the opportunity cost of limited time for biomass accumulation is defrayed by high reproductive investment. Phenology appears to play a key role in this trade-off.

Narrow-leafed lupins from high-rainfall habitats delay flowering to escape frost (Table 2.1) and to match the extended potential growing season provided by high-rainfall climates. An extended growing season allows the development of competitive traits *sensu* Grime (1979), such as extensive root systems and larger canopies, as discussed below. In the Mediterranean climate, late flowering even in high-rainfall habitats is likely to result in variable, often relatively short reproductive phases, given high, rising spring temperatures (Table 2.1) and sandy, low

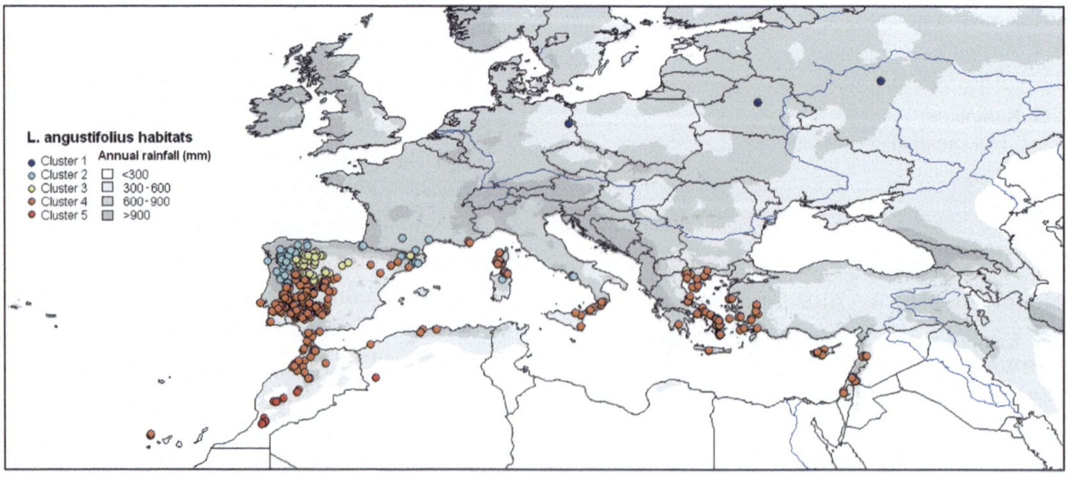

Fig. 2.1 Narrow-leafed lupin collection sites cluster strongly along a north–south terminal drought stress gradient around the Mediterranean basin. See also Table 2.1. *Source* Berger et al. (2008)

Table 2.1 *L. angustifolius* collection site habitats sampled in the Australian lupin collection cluster along a north–south terminal drought stress gradient, ranked by cluster number. Modified from: Berger et al. (2008)

Cluster	Description	Alt. (m)	Frost (days/season)	Temperature (°C)			Precipitation (mm)				Day length
r				Vegetative phase	Reproductive phase	Reproductive phase slope (°C/day)	Pre-season	Vegetative phase	Reproductive phase	Coefficient of variation (%)	Vegetative phase
1	Central Europe: long day, short season, no term drought	116	14	10.8	17.1	−0.02	283	110	212	52.2	14.2
2	Nth Med: med alt, cool frosty veg phase, low-term drought	457	60	8.9	16.2	0.09	61	536	309	72	10.1
3	Nth Med: high alt, cold, v. frosty veg phase, med-term drought	903	95	6.4	16.5	0.11	42	303	153	77.7	10.4
4	Central Med: av. climates, med-term drought	319	26	12.4	16.2	0.1	33	321	227	88.7	10.3
5	North Africa: high alt, variable rain, high-term drought	1,033	30	11.1	16	0.09	27	147	129	140.6	10.6

water-holding soils. As a result, high-rainfall ecotypes are subject to selection pressure for smaller seeds, which will fill more quickly in short, variable reproductive phases (Berger et al. 2017).

More detailed studies of lupin adaptation to the aforementioned Mediterranean stress gradient have elevated the role of phenology in specific adaptation. In both yellow and narrow-leafed lupin, the shorter vegetative phase of low-rainfall ecotypes produces less above and below ground biomass, and lower leaf area than high-rainfall ecotypes (Berger and Ludwig 2014; Berger et al. 2018). As a result, low-rainfall ecotypes tend to be less productive (lower seed yield and number) than high-rainfall ecotypes, albeit with a greater reproductive investment (higher harvest index). Lower biomass and leaf area has ramifications for water use and stress onset under water deficit. In both yellow and narrow-leafed lupin, low-rainfall ecotypes have lower transpiration rates and later stress onset compared to high-rainfall ecotypes (Berger and Ludwig 2014; Berger et al. 2018). In *L. luteus*, which has higher rates of water use and earlier stress onset than *L. angustifolius*, high-rainfall ecotypes have additional specific adaptation unrelated to phenology which allows them to maintain higher leaf hydration at low leaf water potential than low-rainfall ecotypes are able to (Berger and Ludwig 2014). Presumably this mitigates the profligate water use of high-rainfall ecotypes in yellow lupin, reducing the impact of self-induced transient drought between rainfall events. There is no evidence for similar specific adaptation in *L. angustifolius,* leading us to conclude that appropriate phenology is the key adaptive trait to rainfall gradients in this species (Berger et al. 2018).

These studies of wild germplasm reinforce the need for breeders to match phenology with target agricultural environments in order to minimize risk of terminal drought, and maximize yield potential in the crop. To date, breeders have taken the conservative, early phenology approach. Most modern yellow and narrow-leafed lupin cultivars resemble low-rainfall ecotypes in terms of phenology, biomass production and partitioning, water use and stress onset (Berger et al. 2017, 2018; Berger and Ludwig 2014).

Historically, there were good reasons for this. Early attempts at lupin domestication in central Europe failed because the indeterminate nature of the late flowering, vernalization responsive spring-sown crop delayed maturity, returning low and variable yields (Hondelmann 1984). With no terminal drought signal to end the summer growing season, European lupin cultivation was risky until the introduction of earlier maturing yellow lupins (Hondelmann 1984). Earliness and early vigour were quickly adopted as an enduring European breeding criteria, as evidenced by the important loci listed in Table 2.2, and in cultivar names, such as Pflugs Allerfrüheste (Plough's Earliest) (Gladstones 1970). Thus, the European breeding strategy was to select for early flowering to ensure timely maturity even in the absence of terminal drought in autumn, following spring and early summer sowing.

Early phenology was also a strong focus for Australian lupin improvement (Table 2.2). The Australian lupin industry is based in the south of the country in Mediterranean-type climates, more similar to the natural environment of the species in the Mediterranean region. Sowing occurs in autumn, with slow growth over a wet winter before the onset of a warm spring and terminal drought in late spring and early summer. Nevertheless, Australian lupin production climates differ from the Mediterranean region in one important respect: Australian winters tend to be warmer, particularly in Western Australia, where most of the country's lupin production occurs. As a consequence, the vernalisation signal that helps regulate flowering in many Mediterranean legumes, including wild lupin species, was an unreliable trigger in the Western Australian farming system (Perry and Poole 1975). Flowering time of vernalisation responsive cultivars may vary by up to 40 days between years in the same location, depending on winter timing and temperature (Berger et al. 2012b). The discovery

Table 2.2 An inventory of the known loci conferring phenological variation in the agriculturally significant old world lupin species, narrow-leafed lupin

Locus name	Phenotype	References
Kulikup (*Ku*)[a]	Early flowering and removal of vernalisation requirement	Gladstones and Hill (1969)
Julius (*Jul*)[a]	Early flowering and removal of vernalisation requirement	Mikołajczyk (1966)
efl	Intermediate flowering and reduction in vernalisation requirement	Gladstones (1977), Anonymous (1982)
retardans (*ret*)	Fast growth	Kepel (1952), Mikołajczyk (1963), Święcicki and Święcicki (1995)
properans (*prop*)[b]	Rapid growth, encourages branching on lower stem, and pleiotropic effects on leaf morphology and colouration	Kepel (1952), Mikołajczyk (1963), Święcicki and Święcicki (1995)
latifolius (*lat*)[b]	Rapid growth, encourages branching on lower stem, and pleiotropic effects on leaf morphology and colouration	Mikołajczyk (1963)
procerus (*proc*)	Fast growth	Mikołajczyk (1963), Święcicki and Święcicki (1995)
Det/*Rb1*[c,d]	Restricted branching growth habit	Gladstones (1994), Adhikari et al. (2001), Oram (2002)
Deter[c]	Restricted branching growth habit	Święcicki and Święcicki (1995), Adhikari et al. (2001), Oram (2002)
Rb2	Restricted branching growth habit	Adhikari et al. (2001)
mut-1/*rb3*[c,e]	Restricted branching growth habit	Adhikari et al. (2001), Oram (2002)
rb4	Restricted branching growth habit	Adhikari et al. (2001)
rb5	Restricted branching growth habit	Adhikari et al. (2001)
No upper branches (*nub*)[c]	Restricted branching growth habit	Oram (2002)
Primary and secondary inflorescences only (*paso*)[c]	Restricted branching growth habit	Oram (2002)

[a]The *Ku* and *Jul* loci are thought to be different alleles (or variants) of the same locus (Taylor et al. 2019)
[b]The *lat* locus is presumed to be the same as the *prop* locus
[c]The *Det*, *Deter*, *mut-1*, *nub* and *paso* loci are thought to be different alleles of the same locus (Adhikari et al. 2001; Oram 2002)
[d]*Det* was renamed as *Rb1* by Adhikari et al. (2001)
[e]*mut-1* was renamed as *rb3* by Adhikari et al. (2001)

and utilization of the vernalisation-silencing *Ku* allele was the single most significant event in Australian lupin breeding (see Sect. 2.2.2). Without *Ku*, Australian lupin production would be limited to the colder eastern regions, currently accounting for approximately 15% of Australian production (ABARES 2012), and southern Western Australia.

Unsurprisingly, after the identification of *Ku*, Australian lupin cultivars were almost exclusively vernalisation insensitive (Berger et al. 2012b). While lupin breeders did release later flowering varieties, such as Chittick (see subsequent discussion of *efl* in Sect. 2.2.2), Wandoo, Geebung and Jindalee in the 1980s through to 2000s; these were generally not competitive with early flowering *Ku* genotypes (Cowling 1999; Stefanova and Buirchell 2010). Subsequently, newer *Ku* cultivars became earlier, harvest index improved, and there was strong genetic gain in the warm, short-season northern Western Australian grainbelt (Berger et al. 2012a, 2012b). In so doing, breeders turned the narrow-leafed lupin into a vernalisation insensitive, highly temperature responsive crop. Indeed, modern Australian lupin cultivars such as Mandelup are as temperature responsive as southern Indian chickpea (Berger et al. 2011), a crop designed by ICRISAT breeders to flower in 4 weeks in their hot (mean temperature = 23–26 °C), terminal drought prone climate. Genetic gain in eastern Australia has been far less spectacular, and the difference between older, vernalisation responsive cultivars and modern *Ku* types is much smaller (Berger et al. 2012a). This is important because eastern Australia contains longer season environments with higher yield potential (Berger et al. 2012a). Similarly, within Western Australia, narrow-leafed lupin has made few inroads to the southern high-rainfall zone which offers greater productivity from a longer growing season, but which is prone to higher virus and fungal diseases plus nutrient deficiencies. In the Western Australian context, apart from the lower yielding vernalisation responsive *ku* types, breeders currently have no later flowering, long-season cultivars to offer to growers. Nevertheless, modelling suggests that these would be more productive in this region (Chen et al. 2016, 2017).

Lupin breeding would benefit from an improved understanding of phenology regulation, beyond the off/on vernalisation response provided by *Ku/ku*. Our studies with wild germplasm confirm that lupin species regulate phenology very tightly along drought stress gradients (Berger et al. 2017). To the best of our knowledge, these germplasm are all vernalisation responsive *ku* types, implying that there are other, more subtle phenology regulators to be uncovered in the species.

2.2 Germplasm Resources: Phenotypic and Genetic Variation for Flowering Time

Since the first reported collection of *Lupinus* germplasm by Nikolai Vavilov in the 1920s, approximately 33,000 accessions of various lupin species (not accounting for duplicates) have been accumulated by more than 20 substantially sized and independent gene banks across the globe (Berger et al. 2013). These collections serve as invaluable resources to breeding programmes as they seek to introduce new variation for traits, including phenology. Here we describe the phenotypic and genetic variation for phenology, particularly flowering time, present in current global holdings of wild and domestic narrow-leafed lupin germplasm, and its relationship to crop adaptation.

2.2.1 Phenological Variation Found in Wild Germplasm Collections

Considerable phenotypic variation for phenology and responsiveness to environmental signals promoting flowering exists among the genetically diverse wild populations of narrow-leafed lupin native to the Mediterranean region (Fig. 2.2). This variation has been consistently and significantly associated with different geographic and climatic variables, plus morphological traits such as seed size, in several germplasm evaluations (Clements and Cowling 1994; Gladstones and Crosbie 1979; Lema et al. 2005) and phenological and/or genetic studies (Berger et al. 2017; Mousavi-Derazmahalleh et al. 2018; Talhinhas et al. 2006).

There is a strong west–east Mediterranean gradient of genetic diversity and phenology across the natural Mediterranean distribution of *L. angustifolius* (Mousavi-Derazmahalleh et al. 2018). Portuguese and Spanish populations from

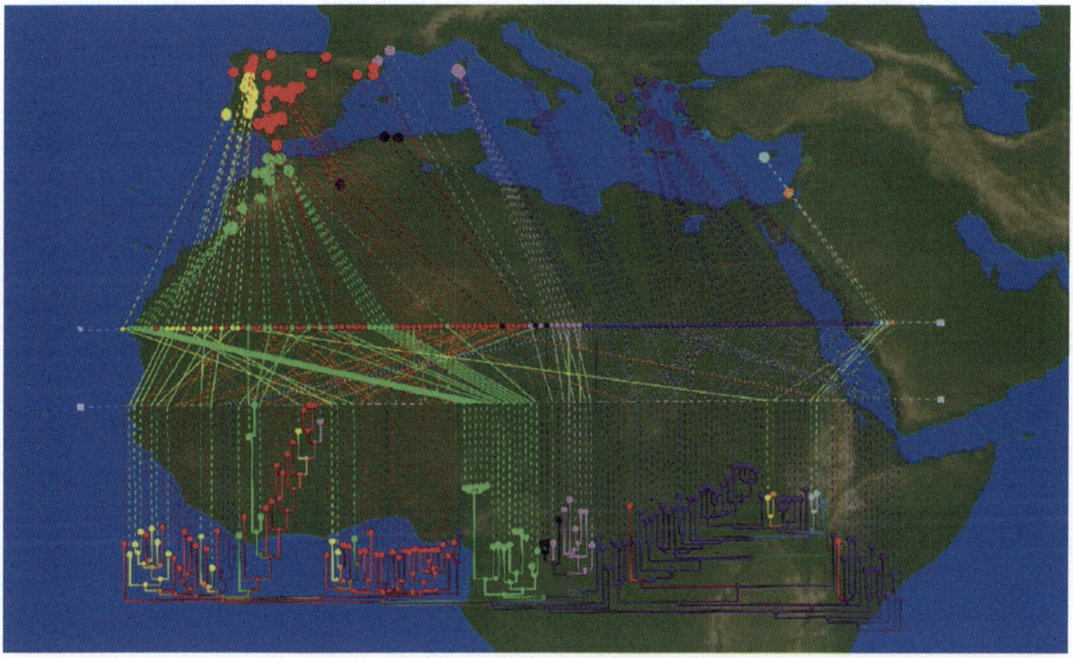

Fig. 2.2 Representation of the genetic diversity and phylogeny of wild narrow-leafed lupin accessions from native habitats, including within Portugal (yellow), Spain (red), France (pink), Italy (dark blue), Greece (purple), Turkey (sky blue), Cyprus (turquoise blue), Syria (dark green), Israel (orange), Algeria (black) and Morocco (light green). *Source* Mousavi-Derazmahalleh et al. (2018)

the Iberian Peninsula in the western Mediterranean have the latest phenology observed for narrow-leafed lupins, with late flowering and low early vigour (Berger et al. 2017). Additionally, germplasm from these origins has the strongest vernalisation requirement and/or inductive long day photoperiod response (Lema et al. 2005). These traits have evolved as mechanisms of adaptation to long, wet and cold winter seasons subject to severe frost events, particularly in mountainous regions at higher latitudes. Similarly, populations from the western and south-western coasts of Italy and the Islands of Sicily, Sardinia and Corsica are characterized by generally late flowering and has strong requirements for vernalisation; though types with more moderate flowering times can be found in coastal regions. To the east of the Mediterranean, where the species is believed to have spread in recent evolutionary times (Mousavi-Derazmahalleh et al. 2018), several populations have been identified with earlier flowering and lower vernalisation response than their western Mediterranean counterparts. In particular, germplasm from low-lying coastal areas in Israel, Greece and the Middle East have some of the earliest flowering times observed in wild narrow-leafed lupins to date (Clements and Cowling 1994; Gladstones and Crosbie 1979) and high early vigour. Evolution of this shift in phenology is likely to have been driven by the climates and geology within these geographic areas, particularly within the Balkan Peninsula and Aegean Islands (Clements and Cowling 1994).

A strong north–south phenological gradient across the species' natural distribution is also observed. In comparison to populations from the north-western Mediterranean, those originating from coastal areas in Morocco and Algeria in Northern Africa have been shown to have greater early vigour, earlier flowering and less demand for vernalisation (Berger et al. 2017). This reflects the strong selection for drought escape with decreasing northern latitude. Interestingly, germplasm from the Anti-Atlas mountainous region of southern Morocco has also been identified as early flowering (Gladstones and Crosbie 1979). This phenotype is relatively unique for mountainous populations of narrow-leafed lupins and is potentially explained by the unique short-seasoned and dry yet cold climate of the Anti-Atlas region (Gladstones and Crosbie 1979).

2.2.2 Phenotypic and Genetic Variation Found in Fully- and Semi-domesticated Germplasm Resources

As discussed previously, early phenology was a strong focus for both European and Australian breeders (Table 2.2), and their focus was on advanced lines in their breeding programmes. While this has given us a good understanding of flowering regulation in domesticated lupin, it has provided no insight into the wild variation required to further improve the crop.

In Australian breeding programmes of narrow-leafed lupin, the most important phenological regulator for crop adaptation and grain production was the *Ku* allele for early flowering time. *Ku* is a natural spontaneous mutation that was discovered by Dr John Gladstones in a single plant within a Borre variety crop on a property in Kulikup, Western Australia (Gladstones and Hill 1969). The allele facilitates early flowering by silencing the vernalisation requirement of narrow-leafed lupin (Nelson et al. 2017). The discovery of *Ku* has had enormous implications for breeding both within Australia and Northern Europe. Extensive trials by Gladstones and Hill (1969) demonstrated that *Ku* advanced flowering time by two to three weeks with autumn sowing within the Western Australian grain belt, enabling better adaption to the warm, short growing season of the northern grain belt where the vernalisation requirement had previously been a limitation for reliable grain production.

In addition to modifying the vernalisation response, the *Ku* allele may also influence other environmental signalling pathways affecting flowering time. From experience in both controlled growth cabinet and field-based environments, accessions carrying *Ku* are less responsive to photoperiod than recessive *ku* types, in which time to flowering is reduced by long day (>12 h) relative to short day (<12 h) photoperiod treatments (Berger et al. 2012b; Christiansen et al. 2008; Rahman and Gladstones 1972, 1974). Meanwhile, flowering time appears to be controlled by accumulated thermal time (degree-days) in *Ku* genotypes and, therefore, flowering in these genotypes occurs more rapidly under warmer temperatures (Berger et al. 2012b; Reader et al. 1995). Interestingly, not all *Ku*-bearing genotypes are equal: there are significant differences in temperature response among *Ku* types, with later releases tending to be earlier (Berger et al. 2012b).

Shortly after *Ku* was discovered in Australia in 1961, a second dominant allele for early flowering was reported in Przebędowo, Poland (Mikołajczyk 1966). The allele, named *Julius* (*Jul*), was present in Krasnolistny, a Russian-bred variety presumed to have been derived from Palestinian or other eastern Mediterranean populations, which at the time were a source of early phenology for Polish and other European breeders. Subsequent research by Rahman and Gladstones (1972) revealed that the original *Ku* mutant line and an introduction believed to be Krasnolistny responded to photoperiod in a near-identical manner and that neither accession required vernalisation, suggesting that *Ku* and *Jul* may, in fact, be variants of the same gene. Recent genetic analysis assessing nucleotide sequence conservation of the gene underlying *Ku*, named *LanFTc1*, supports this hypothesis and indicates that the *Ku* and *Jul* are indeed likely to be variants of the same gene (Taylor et al. 2019).

The third known locus for flowering time in domesticated germplasm is the recessive locus, *efl*. Unlike *Ku* and *Jul*, *efl* was derived by artificial mutation (Anonymous 1982; Gladstones 1977). Shortly after the discovery of *Ku*, research commenced in Australia to seek alternative sources of earliness due to concerns that *Ku* varieties would potentially flower excessively early with early autumn sowing and germination in long-season production zones (Gladstones and

Hill 1969). In the early 1960s, seed from cv. Borre was treated with 0.24% aqueous ethylene imine and the *efl* mutant was selected as the best of 15 early mutant lines that survived the mutagenesis treatment (Gladstones 1977). A desirable outcome was achieved as *efl* reduced but did not fully eliminate the vernalisation requirement and advanced flowering time by roughly 10–14 days relative to *ku* types, such as cv. Uniharvest. The first released variety, Chittick, received glowing initial reviews and was considered to have a flowering time close to the described optimum for much of the southern Western Australian production region, in addition to the absence of agronomic problems associated with early sowing of *Ku* varieties, including reduced crop height and increased vulnerability to abiotic and biotic stresses caused by excessively early flowering (Anonymous 1982). Nevertheless, *efl* has only been selected in two mid-season Australian varieties, Chittick and Wandoo, released in 1982 and 1986, respectively, (Cowling 1999; Stefanova and Buirchell 2010). Although Chittick (*efl*) has only half the vernalization requirement of *ku* types such as Geebung, it has a relatively low-temperature response (Berger et al. 2012b). Accordingly, while *efl* provides a moderate 10–15 day flowering time delay in cool environments (vegetative phase < 13 °C), in the warmer northern grainbelt (>13 °C) flowering time can be delayed by up to 40 days compared to temperature responsive *Ku* types (Berger et al. 2012b). A breeding line (WL176) derived from a cross between Australian cultivar Illyarrie and a wild Moroccan accession, P22855, has been suggested to contain a naturally occurring *efl* mutation or allelic variant, though this has yet to be validated (Landers 1995). Jindalee has a slightly lower vernalization requirement than *ku* types such as Geebung (Berger et al. 2012b), and contains an unknown allele for later flowering (Taylor et al., unpublished).

Several other loci resulting in early phenology in narrow-leafed lupin, but which do not directly impact upon floral signalling pathways, have also been reported in the literature. One group of these loci target growth rates. The *retardans* (*ret*) locus promotes faster general growth in domestic accessions than wild Mediterranean accessions (Kepel 1952; Mikołajczyk 1963; Święcicki and Święcicki 1995). Although it does not appear to be actively selected, at least within the Australian breeding programme, it appears that a number of cultivars of European and Australian origin may have inherited the character. Examples of varieties carrying *ret* include: Münchenberger Blaue Süßlupine, from Germany; Danko, from Belarus; Obornicki, Emir, Polonez, Saturn, Mirela, all from Poland; and Unicrop, Uniharvest, Yandee, Gunguru, Chittick and Yorrel, all from Australia (Święcicki and Święcicki 1995). This gene has not been confirmed independently to occur in Australian varieties.

The second locus associated with growth rate, named *properans* (*prop*) by Kepel (1952) and Mikołajczyk (1963), facilitates rapid growth in young plants and has pleiotropic effects, including broadening and lightening the colour of leaves, and encouraging branching on the lower section of the main stem (Święcicki and Święcicki 1995). However, due to the rapid rate of growth in young plants and the branching character, genotypes with *prop*, such as Puławski Różowy Wczesny (Poland), are susceptible to lodging. For this reason, *prop* has rarely been selected in European breeding programmes, which have instead favoured other early phenological variation. The *prop* locus is thought to be identical to the *latifolius* (*lat*) locus (Święcicki and Święcicki 1995).

The final locus associated with growth rate in narrow-leafed lupin is *procerus* (*proc*). It is the oldest of the three growth rate-related loci having been first described in the mid-1920s (Święcicki and Święcicki 1995). Later research by Mikołajczyk (1963) revealed that unlike *prop*, *proc* results in a reduction of lateral branching, particularly on the upper main stem, and that the two loci are epistatic. The *proc* allele has rarely been selected in breeding programmes, though it has been bred in a limited number of older varieties from Germany (e.g. Gülzower Süße Blaue Lupine and Münchenberger Frohwüchsige Blaublühende), Russia (e.g. Nemchynovskii 560 and Skorospely) and Poland (e.g. Puławski Wysoki), plus was additionally reported to be in Australian variety Danja (Święcicki and Święcicki 1995), but this has not been confirmed independently.

The second group of loci that indirectly affect phenology are those that affect plant architecture, particularly growth habit and branching. Traditionally, lupin varieties are indeterminate like their wild relatives. That is, after a raceme develops on the main stem, vegetative growth will continue in the form of branching at leaf nodes (axils) on the main stem. These branches are known as first-order branches and will also terminate with a floral meristem. For indeterminate lupin varieties, the apical first-order branches that develop on the first three to four leaf nodes beneath the main stem inflorescence are generally the most important for yield (Dracup and Kirby 1996). Provided resources are adequate and the growing season is long enough, it is possible for several orders of branching to develop in an indeterminate plant (Dracup and Kirby 1993).

During the 1970s, European breeders developed an interest in producing determinate varieties with restricted branching (Święcicki and Święcicki 1995). The motive for this was to find alternative sources of early maturity that would ensure completion of grain filling before favourable environmental conditions cease each season. Induced mutations that caused single flowers to develop in the place of lateral shoot apices at leaf axils on the upper main stem were observed in two different instances, one occurring in Przebędowo (Poland) and the other in Zodino (Belarus). The Polish instance had been achieved in mutated seed of the Turkus variety, and the recessive mutation was named *mut-1* (Święcicki and Święcicki 1995). In the Belarussian instance, a rare dominant mutation arose, named *Deter*, in the mutant line later known as Lanedeks (Święcicki and Święcicki 1995). Naturally occurring mutations for restricted branching have also been reported to occur frequently in the literature. Perhaps most well-known is the incompletely dominant mutation, *Det*, which was observed in a breeding line of Dr Ian Forbes from the United States of America, and was documented by Dr John Gladstones (Gladstones 1994).

In trying to consolidate knowledge of genetic inheritance for restricted branching, Adhikari et al. (2001) conducted several crossing experiments between accessions carrying the above-described mutations in addition to several new spontaneous mutants found in the Australian Illyarrie variety. Their results led them to conclude that *mut-1* (which the authors renamed as *rb3*), *Det* (which the authors renamed as *Rb1*) and *Deter* were distinct alleles of the same locus. Additionally, three new-independent loci were reported in spontaneous mutants of Australian cultivar, Illyarrie: *Rb2*, *rb4* and *rb5*.

A further study by Oram (2002) confirmed through artificial mutagenesis that the locus containing the *mut-1/rb3*, *Det/Rb1* and *Deter* alleles is a common target for reduced branching in narrow-leafed lupin. Additionally, two new mutants, named *nub* (*no upper branches*) and *paso* (*primary and secondary inflorescences only*), were shown to have recessive alleles relative to the wild type. In terms of their relationship to the other mutant alleles, *nub* was considered dominant over *paso*, and in some cases both alleles were dominant over *Rb1*, although this dominance was not always observed. Clearly, the genetics underpinning restricted branching is complex and yet to be fully understood in narrow-leafed lupin.

In summary, a number of loci conferring early phenology have been identified in fully- and semi-domesticated narrow-leafed lupin germplasm (Table 2.2). The most significant of these in terms of breeding within Australian and Europe are those which facilitate early flowering by reducing vernalisation requirement, including *Ku*, *Jul* and *efl*. However, several other loci that result in early phenology by modifying growth rate and plant architecture, namely, branching habit, have also been selected in breeding programmes to a limited extent. Although these loci achieve early phenology, and thus crop adaptation, through a variety of means, one commonality is noted: that all of these loci have been known for some time, and have failed to find traction in lupin breeding programmes by contributing to ongoing cultivar development (Table 2.2). This is likely to be a consequence of the centre of lupin production and breeding shifting from Central Europe to Australia, and a relatively insular approach in both regions which has left potential candidate parents behind. For crop adaptation to continue to improve, it is important for breeders and researchers alike to turn to our international germplasm collections to identify new phenology regulating loci in order to appropriately match cultivar phenology to target environment to maximize yield potential.

2.3 Conclusions

Phenology arguably represents the most influential trait for the adaptation of wild narrow-leafed lupin populations to varying habitats within the species' natural distribution around the Mediterranean and of domestic cultivars to agricultural environments across the globe. Flowering time is an incredibly important aspect of phenology, and its regulation in response to environmental and endogenous signals ensures that plants flower at an appropriate time where abiotic resources and conditions are conducive to flower and grain production. Collections of wild and domestic narrow-leafed lupins contain a substantial amount of naturally evolved and, to a lesser extent, artificially induced phenotypic and genetic variation for traits directly or indirectly affecting phenology, such as initial growth rate, growth habit, flowering time, and receptiveness to environmental signals inductive of flowering. The genomic resources described elsewhere in this book, for example, the draft reference genome (Sect. 3.3) and ongoing efforts to develop a pan-genome for narrow-leafed lupin (Sect. 3.6.2) are likely to be helpful in providing insight into the underlying gene(s) that are contributing to these important phenology related traits (outlined in more detail in Chap. 9). Moreover, conserving these valuable germplasm collections is fundamental to the future successful breeding of adapted narrow-leafed lupin crops, as it is a continued research to underpin the regulation of more subtle phenological traits in order to broaden the genetic selection for crop adaptation beyond the current key early flowering time genes.

References

ABARES (2012) Australian commodity statistics 2012. Australian Bureau of Agricultural and Resource Economics and Sciences, Canberra

Adhikari KN, Galwey NW, Dracup M (2001) The genetic control of highly restricted branching in narrow-leafed lupin (*Lupinus angustifolius* L.). Euphytica 117:261–274

Anonymous (1982) Register of lupin cultivars (b) *Lupinus angustifolius* L. (narrow-leafed lupin) cv. Chittick e (Reg. No. b-10). J Aust I Agric Sci 48:122–124

Berger JD, Ludwig C (2014) Contrasting adaptive strategies to terminal drought stress gradients in Mediterranean legumes: phenology, productivity and water relations in wild and domesticated *Lupinus luteus* L. J Exp Bot 65:6219–6229

Berger JD, Ludwig C, Buirchell BJ (2008) Ecogeography of the old world lupins: characterising the habitat range. In: Palta JABJD (ed) Lupins for health and wealth. Proceedings of the 12th international lupin conference, Fremantle, Western Australia, 14–18 September 2008. International Lupin Association, Canterbury, New Zealand, pp 355–361

Berger JD, Milroy SP, Turner NC, Siddique KHM, Imtiaz M, Malhotra R (2011) Chickpea evolution has selected for contrasting phenological mechanisms among different habitats. Euphytica 180:1–15

Berger JD, Buirchell B, Luckett DJ, Nelson MN (2012a) Domestication bottlenecks limit genetic diversity and constrain adaptation in narrow-leafed lupin (*Lupinus angustifolius* L.). Theoret Appl Genet 124:637–652

Berger JD, Buirchell B, Luckett DJ, Palta JA, Ludwig C, Liu D (2012b) How has narrow-leafed lupin changed in its 1st 40 years as an industrial, broad-acre crop? A GxE-based characterization of yield-related traits in Australian cultivars. Field Crop Res 126:152–164

Berger JD, Clements JC, Nelson MN, Kamphuis LG, Singh KB, Buirchell B (2013) The essential role of genetic resources in narrow-leafed lupin improvement. Crop Pasture Sci 64:361–373

Berger J, Shrestha D, Ludwig C (2017) Reproductive strategies in Mediterranean legumes: trade-offs between phenology, seed size and vigor within and between wild and domesticated *Lupinus* species collected along aridity gradients. Front Plant Sci 8:548

Berger JD, Ludwig C, Whisson K (2018) Contrasting adaptive strategies to Mediterranean rainfall gradients: phenology, productivity, water–use and stress onset trade-offs in wild and domesticated *Lupinus angustifolius* L. J Exp Bot 65:6219–6229

Chen C, Berger J, Fletcher A, Lawes R, Robertson M (2016) Genotype × environment interactions for phenological adaptation in narrow-leafed lupin: a simulation study with a parameter optimized model. Field Crop Res 197:28–38

Chen C, Fletcher A, Lawes R, Berger J, Robertson M (2017) Modelling phenological and agronomic adaptation options for narrow-leafed lupins in the southern grainbelt of Western Australia. Eur J Agron 89:140–147

Christiansen JL, Jørnsgård B, Jørgensen ST, Bibby BM (2008) Photoperiod sensitivity in narrow-leafed lupin (*lupinus angustifolius* L.). Acta Agric Scand B-S P 58:365–371

Clements J, Cowling W (1994) Patterns of morphological diversity in relation to geographical origins of wild *Lupinus angustifolius* from the Aegean region. Genet Resour Crop Evol 41:109–122

Cowling WA (1999) Pedigrees and characteristics of narrow-leafed lupin cultivars released in Australia from 1967 to 1998. Western Australia Department of Agriculture, Bulletin 4365, South Perth, Australia

Dracup M, Kirby EJM (1993) Patterns of growth and development of leaves and internodes of narrow-leafed lupin. Field Crop Res 34:209–225

Dracup M, Kirby EJM (1996) Lupin development guide. University of Western Australia Press, Nedlands, Western Australia

Gladstones JS (1970) Lupins as crop plants. Field Crop Abstr 23:123–148

Gladstones JS (1977) The narrow-leafed lupin in Western Australia. Western Australia Department of Agriculture, Bulletin 3990, South Perth, Australia

Gladstones JS (1994) An historical review of lupins in Australia. In: Paper presented at the proceedings of the 1st Australian lupin technical symposium, Perth, Australia

Gladstones JS, Hill GD (1969) Selection for economic characters in *Lupinus angustifolius* and *L. digitatus*. 2. Time of flowering. Aust J Exp Agric Anim Husb 9:213–220

Gladstones JS, Crosbie GB (1979) Lupin wild types introduced into Western Australia to 1973. Department of Agriculture, Western Australia, Technical Bulletin No. 43, South Perth, Australia

Grime JP (1979) Plant strategies and vegetation processes. Wiley, New York

Hondelmann W (1984) The Lupin-ancient and modern crop plant. Theoret Appl Genet 68:1–9

Kepel W (1952) The morphological features of the narrow-leafed lupin (*Lupinus angustifolius*). The Poznań Society of Friends of Science, Department of Mathematical and Natural Sciences Publication of the Section of Agriculture and Sylviculture 1:182–186

Landers KF (1995) Vernalization responses in narrow-leafed lupin (*Lupinus angustifolius*) genotypes. Aust J Agric Res 46:1011–1025

Lema M, Santalla M, Rodiñno AP, De Ron AM (2005) Field performance of natural narrow-leafed lupin from the northwestern Spain. Euphytica 144:341–351

Mikołajczyk J (1963) Inheritance of the growth pattern in *Lupinus angustifolius* L. and species hybrids *L. angustifolius* x *L. linifolius*. Genet Pol 4:139–169

Mikołajczyk J (1966) Genetic studies in *Lupinus angustifolius*. Part. III. Inheritance of the alkaloid content, seed hardness and length of the growing season in blue lupin. Genet Pol 7:181–196

Mousavi-Derazmahalleh M et al (2018) Exploring the genetic and adaptive diversity of a pan-Mediterranean crop wild relative: narrow-leafed lupin. Theoret Appl Genet 131:887–901

Oram RN (2002) Two reduced branching mutants in *Lupinus angustifolius* L. SABRAO J Breed Genet 34:27–33

Perry MW, Poole ML (1975) Field environment studies on lupins.1. Developmental patterns in *Lupinus angustifolius* L., the effects of cultivar, site and planting time. Aust J Agric Res 26:81–91

Rahman MS, Gladstones JS (1972) Control of lupin initiation by vernalization, photoperiod and temperature under controlled envrionment. Aust J Exp Agric Anim Husb 12:638–645

Rahman MS, Gladstones JS (1974) Effects of temperature and photoperiod on flowering and yield components of lupin genotypes in the field. Aust J Exp Agric Anim Husb 14:205–213

Reader MA, Dracup M, Kirby EJM (1995) Time to flowering in narrow-leafed lupin. Aust J Agric Res 46:1063–1077

Stefanova KT, Buirchell B (2010) Multiplicative mixed models for genetic gain assessment in lupin breeding. Crop Sci 50:880–891

Święcicki W, Święcicki WK (1995) Domestication and breeding improvement of narrow-leafed lupin (*L. angustifolius* L.). J Applied Genet 36:155–167

Talhinhas P, Leitão J, Neves-Martins J (2006) Collection of *Lupinus angustifolius* L. germplasm and characterisation of morphological and molecular diversity. Genet Resour Crop Ev 53:563–578

Taylor CM et al (2019) INDEL variation in the regulatory region of the major flowering time gene *LanFTc1* is associated with vernalisaiton response and flowering time in narrow-leafed lupin (*Lupinus angustifolius* L.). Plant Cell Environ 42:174–187

Overview of Genomic Resources Available for Lupins with a Focus on Narrow-Leafed Lupin (*Lupinus angustifolius*)

3

Karam B. Singh, Rhonda C. Foley, Gagan Garg and Lars G. Kamphuis

Abstract

Narrow-leafed lupin (*Lupinus angustifolius*) is an important grain legume crop for some parts of the world like Australia and parts of Northern Europe where it can form an important part of sustainable farming systems, reducing the need for nitrogenous fertiliser, providing valuable disease breaks and boosting cereal yields. Through the genome revolution and in particular rapid advances in next-generation sequencing technologies, there are new powerful tools available to help with gene discovery and to rapidly accelerate pre-breeding and breeding programmes for narrow-leafed lupin and other lupin crops. This chapter provides an overview of the genomic resources available for narrow-leafed lupin with a focus on the current reference genome which underpins many of the other resources. The cultivar Tanjil was chosen as the reference accession for narrow-leafed lupin and a short-read sequencing approach coupled with BAC-end sequence data was used to assemble the first comprehensive reference genome for the species. This genome assembly captured ~610 Mb of the estimated 921 Mb genome of narrow-leafed lupin with an annotated gene set of 33,076 genes. The narrow-leafed lupin reference genome has provided valuable insight into narrow-leafed lupin evolution and important information on some of its key plant-microbe interactions. The chapter also touches on some of the genomic resources that are in the pipeline in lupins and describes the lupin genome portal, a web-based resource that houses genomic and related information for narrow-leafed lupin.

K. B. Singh (✉) · R. C. Foley · G. Garg · L. G. Kamphuis
Commonwealth Scientific and Industrial Research Organisation, Agriculture & Food, Floreat, WA 6014, Australia
e-mail: Karam.singh@csiro.au

R. C. Foley
e-mail: Rhonda.foley@csiro.au

G. Garg
e-mail: Gagan.garg@csiro.au

L. G. Kamphuis
e-mail: Lars.kamphuis@csiro.au

K. B. Singh · L. G. Kamphuis
Centre for Crop and Disease Management, Curtin University, Bentley, WA 6102, Australia

K. B. Singh · L. G. Kamphuis
The UWA Institute of Agriculture, The University of Western Australia, Perth, WA 6000, Australia

3.1 Introduction

Until the last decade or so, genetic and related molecular resources in plants were developed primarily for a few 'model' systems such as Arabidopsis and *Medicago truncatula* and for major crops such as maize, rice and soybean. The advent of the genomic revolution and in

K. B. Singh et al. (eds.), *The Lupin Genome*, Compendium of Plant Genomes,
https://doi.org/10.1007/978-3-030-21270-4_3

particular a suite of next-generation sequencing technologies that are both high-throughput and low cost has allowed plant scientists to do research at a genomic scale for numerous plant species. This has dramatically changed the plant research and breeding landscapes. Many plants previously considered 'orphan crops' from a molecular and genomic points of view, have become amenable to the development of genomic resources at a scale that greatly facilitates research and breeding efforts for these crops. Underpinning these developments was the generation of a reference genome for each plant species. This was needed to facilitate a range of research and molecular breeding activities. This includes being able to greatly expedite the functional analysis of genes, for example, through the application of approaches such as exome capture and gene editing to help with gene discovery and gene regulation. Comparative genomic approaches have also helped to give valuable insight into plant evolution including the surveying of sequence diversity in a crop species through large-scale sequencing of a range of accessions including wild relatives.

On the breeding side, the development of a high-quality reference genome for a crop species is essential to be able to capitalise effectively on modern breeding technologies that involve marker or even genotype selection coupled with rapid progress in high-throughput phenotyping methods. Thus a good reference genome underpins the development of molecular marker and genomic selection approaches involving genotyping arrays or direct sequencing of mapping populations, both capable of allowing the screening of large numbers of plants at an early stage of plant growth (Crossa et al. 2017; Varshney et al. 2013).

A group of crop plants where the genomic revolution has been quite impactful are legumes. Many legume species are pasture or grain crops and important for sustainable agriculture primarily through their ability to fix nitrogen (and thereby reduce the reliance on fertilisers) as well as their ability to provide valuable disease breaks when used in crop rotations. Lupins belong to the genus *Lupinus* in the genistoids clade of legumes, which diverged early in papillionoid legume evolution (Lavin et al. 2005). Lupins are widely distributed geographically, primarily in the Mediterranean region and North and South America, and inhabit a wide range of habitats (Drummond et al. 2012). Only a few lupin species have been domesticated and the two major lupin crops are white lupin (*L. albus*) and narrow-leafed lupin. While the grain of narrow-leafed lupin is still mainly used for animal feed, it has recently received strong interest for its potential as a human food and food additive due to the composition of the grain which has very little starch but substantial amounts of protein and fibre (Arnoldi et al. 2015; Lee et al. 2006). A range of genomic resources has been produced for lupins in the last few years, primarily for narrow-leafed lupin and many of these are described in this chapter and in some of the following chapters.

3.2 BAC Libraries Provided Early Insight Into The Narrow-Leafed Lupin Genome

Among the first genomic resources produced in lupins were two bacterial artificial chromosome (BAC) libraries for narrow-leafed lupin. BAC libraries, which typically have an insert size in the range of 100 kb or more have been very useful for a wide range of genetic and genomic studies in many model and crop species. Of particular importance has been their deployment in BAC-end sequencing projects which has allowed a range of molecular markers such as SSRs to be identified that have helped with the generation of genetic and physical maps (Varshney et al. 2010).

The first BAC library in narrow-leafed lupin was constructed for the cultivar Sonet using the *Hind*III restriction enzyme and had 6x coverage of the genome with an average insert size of 100 Kb (Kasprzak et al. 2006). As described in detail in Chap. 4 by Susek and Naganowska, BAC clones from this library were used in fluorescent in situ hybridisation (BAC-FISH) experiments for chromosome identification in narrow-leafed lupin (Lesniewska et al. 2011) and as tools for comparative chromosome mapping among lupin species. The second BAC library was constructed for cultivar Tanjil using the *Bam*HI restriction enzyme and had 12x coverage of the genome, with an average insert size of 99.7 Kb (Gao et al. 2011). 9,600 randomly selected BAC clones from this library were sequenced at 5′ and 3′ ends to give 13,985 bp of BAC-end sequences with usable sequence data and an average size of 683 bp which made up approximately 1% of the narrow-leafed lupin genome (Gao et al. 2011). These BAC-end sequences gave valuable, early insight into the composition of the narrow-leafed lupin genome in terms of G:C content, repetitive sequences (9,966 simple sequence repeat motifs were identified) and helped to guide the approach undertaken for sequencing the narrow-leafed lupin genome of Tanjil, using NGS technologies.

3.3 Narrow-Leafed Lupin Genome Sequences Including the Current Reference Genome

Yang and colleagues (2013b) published a preliminary draft narrow-leafed genomic assembly for cultivar Tanjil using whole genome sequencing approaches with short-read sequencing technology. This approach resulted in 26.9X coverage of the genome. Typically a genome is assembled where possible into contigs that represented contiguous genome sequences and scaffolds which consisted of contigs and gaps. A widely used criterion to judge the quality of a genome assembly is the scaffold N50 value. This value represents the median contig size of an assembly such that half the genome assembly is in contigs greater or equal to the N50 value. The preliminary genome assembly reported by Yang and colleagues (2013b) was quite fragmented with a scaffold N50 of 7,319 scaffolds. The draft genome assembly predicted 57,807 genes which was considerably higher than subsequent lupin assemblies. The draft assembly was useful in generating 8,224 sequence-defined molecular markers that were used to improve the narrow-leafed genetic map.

A more comprehensive draft genome assembly of narrow-leafed cultivar Tanjil, using a range of different insert size libraries and BAC-end sequencing data was generated by Hane and colleagues (Hane et al. 2017) with a 162.8X coverage of the genome. This assembly captured >98% of the gene content as demonstrated by CEGMA analyses (Parra et al. 2007). In addition 98.5–99.0% of reads from four transcriptome datasets from different tissue types were mapped back to the gene assembly. This assembly was initially generated using Paired-End Illumina data and then improved via scaffolding with additional sequence data obtained from larger Paired-End reads as well as data from Mate-Pairs and the BAC-end sequencing data described earlier. This gave an assembly totalling 810 Mb which was close to the estimated genome size. However, to minimise fragmentation and scaffolding errors, scaffolds less than 200 bp were removed, resulting in an assembly of 609 Mb and a scaffold N50 of 232 and scaffold N50 length of 703 Kb. These figures for the narrow-leafed genome assembly compared favourably with other draft genome sequences available at the time for legume crops (Hane et al. 2017), and the assembly has provided a useful reference genome for narrow-leafed lupin and to a lesser degree other lupin species.

Further analysis of the genome assembly revealed that the genome is highly repetitive with 57% of the genome comprised of repetitive elements. A combination of approaches and datasets were used to analyse the protein-coding genes and a total of 33,076 gene/proteins were annotated. 77% of the 610 Mb was assigned to a linkage group using a dense reference genetic map that was generated for the wild (P27255) × domestic (83A:476) recombinant inbred line (RIL) population (n = 150) (Hane et al. 2017). This reference genetic map consisted of 9,965 markers which were a combination of gene-based molecular markers (Kamphuis et al. 2015) supplemented with DArTSeq markers (Hane et al. 2017). The combination of this dense genetic map coupled with the Tanjil genome made it possible to significantly narrow down the

Table 3.1 Overview of the location of disease resistance genes and key domestication genes in the narrow-leafed lupin (*L. angustifolius*) genome

Trait	Locus name	LG	Region (Kb)	Markers provided to the breeding programme
Anthracnose resistance	*Lanr1*	NLL-11	388	yes
Phomopsis resistance[a]	*PhtjR*	NLL-05	127.6	yes
Bitterness locus	*Iucundus*	NLL-07	746	yes
Flowering time	*Ku*	NLL-10	413	yes
Pod shattering	*Tardus*	NLL-01	517.6	yes
Pod shattering	*Lentus*	NLL-08	387.1	yes
Soft seededness	*Mollis*	NLL-17	119.5	yes
Flower colour	*Leucospermus*	NLL-03	907.1	yes

[a]Molecular markers from (Yang et al. 2013a)

location of key domestication genes in narrow-leafed lupin as well as the two loci controlling resistance to anthracnose and phomopsis stem blight (see Table 3.1), as described in more detail later in this chapter. Flanking and co-segregating markers linked to these traits have been provided to narrow-leafed lupin breeders.

In addition to the reference genome, two other narrow-leafed lupin accessions were sequenced at 50X coverage (Hane et al. 2017). One was a wild accession (P27255) and the other the early Australian cultivar Unicrop. These datasets provided a large number of molecular makers in the form of indels and SNPs, and demonstrated that the wild accession was significantly divergent across all regions of the genome from Tanjil, while the two domesticated accessions had much lower levels of diversity. The reference genome and associated datasets provided valuable insights into narrow-leafed lupin evolution, as discussed in Chap. 7 by Cannon. Yang and colleagues (Yang et al. 2015) also performed whole genome re-sequencing at lower coverage for nine Australian lupin cultivars that were released from 1973 to 2007. A large number of SNPs were identified through the various pair-wise comparisons and 207,887 molecular markers were anchored on the narrow-leafed lupin genetic map. Further details of the results of this study are presented in Chap. 6 by Yang and Książkiewicz.

To assist with the annotation of the Tanjil reference genome, several transcriptome datasets were generated for major tissue types (leaf, stem, root, flower, seed) for Tanjil, Unicrop (an early Australian cultivar) and P27255 (Foley et al. 2015; Kamphuis et al. 2015) as well as a smallRNA dataset (DeBoer et al. 2019), as described in detail in Chap. 5 by Kamphuis et al.

3.4 Insight from the Reference Genome into Plant-Microbe Interactions for Narrow-Leafed Lupin

Legumes, like other plant species, have complex interactions with other organisms including hostile interactions with pathogens and pests, and beneficial interactions with symbiotic organisms including arbuscular mycorrhizal (AM) fungi and the unique relationship legumes have with nitrogen-fixing rhizobium bacteria. The association between plants and mycorrhizal fungi, which helps the plant with nutrient acquisition, is quite ancient and involves signalling interplay between the fungi and plant host to initiate the symbiosis and then complex signalling pathways in both the host and fungi to establish the symbiosis. Excellent progress on the gene networks used by the plant for a successful mycorrhizal symbiosis has been achieved in a number of plant species, including the model legumes *M. truncatula* and *Lotus japonicus* (MacLean et al. 2017). Interestingly in legumes, where the symbiotic relationship with rhizobium help fix nitrogen evolved much later than with mycorrhizal fungi, there is strong evidence that many of the early elements of the mycorrhizal fungi signalling pathway were recruited for the rhizobium symbiosis resulting in what has been termed the common symbiosis signalling pathway (Oldroyd 2013).

In contrast to most other legumes, lupins are unable to form symbiosis with mycorrhizal fungi. Using phylogenomic approaches, genes that are associated with and essential for mycorrhizal symbiosis were identified (Bravo et al. 2016). The narrow-leafed lupin genome was screened for the presence of mycorrhizal symbiosis genes, and shown to possess 20 out of 33 characterised mycorrhizal-association genes (Hane et al. 2017). These included genes involved in rhizobial-legume symbiosis, including genes also involved in calcium spiking such as CCaMK and cyclops, or biosynthesis, regulation or transport of plant hormones. However, narrow-leafed lupin lacked key genes required specifically for AM symbiosis but not nodulation, including *SbtM1*, *SbtM3*, *HA1*, *EXO70I*, *RAM2*, *PT4*, *STR1*, *STR2*, *RAM1*, *ERF1*, *RAD1* and *DIP1* (Hane et al. 2017). These included a number of grass transcription factors such as *Ram1* and *Ram2* which are required for arbuscular development, genes important for deposition of the periarbuscular membrane such as *EXO71* and genes like *PT4* important for nutrient transport between the symbiont and the host plant (see MacLean et al. (2017) and references within). Lupin species manage to do without the mycorrhizal symbiosis by employing a carboxylate-releasing strategy, supplemented in some cases, such as *L. albus*, with a P-mobilising cluster-root strategy, particularly during periods of low nutrient availability (Lambers et al. 2013, 2018).

The genome sequence for narrow-leafed lupin has also shed new light on the interaction with rhizobium. During nodulation, lupins become infected by rhizobia via intercellular penetration that occurs at the junction between the root hair base and adjacent epidermal cell rather than through intracellular infection threads through the root hair, as is the case with most other legumes (González-Sama et al. 2004). Moreover, lupin nodules are indeterminate but have characteristics that differ from other indeterminate nodules, for example, the outer cortex is important for their initiation and there are other morphological differences (González-Sama et al. 2004). Short infection-thread like structures have been observed in cortical cells but their importance is not clear (González-Sama et al. 2004; James et al. 1997; Tang et al. 1992). Despite this, all genes known to be required for rhizobial infection were present in narrow-leafed lupin, suggesting fundamentally conserved mechanisms underlying different infection modes (Hane et al. 2017). However, an interesting feature of lupin that distinguishes it from the model legumes *M. truncatula and L. japonicus,* is the division of infected cells which entails assortment of symbiosomes between daughter cells (González-Sama et al. 2004). This presumably involves elements of the cytoskeleton and their associated proteins for the distribution and segregation of symbiosomes. Hane and colleagues (Hane et al. 2017) speculated that the observed expansion of microtubule binding protein families could play a role in this phenomenon.

Legumes, like other plants, are under attack from a wide range of pathogens and pests, and deploy two major forms of defence responses to deter these attacks (Ramirez-Prado et al. 2018). These broadly speaking are a basal defence pathway that has components that are constitutively expressed as well as those induced by pathogen associated molecular patterns and an induced defence pathway, typically known as effector-triggered immunity (ETI) and relies on plant resistance (*R*) genes (Jones and Dangl 2006; Qi et al. 2011; Thomma et al. 2011). There was a significantly lower number of classical plant *R* genes of the nucleotide-binding site (NBS) leucine rich repeats (LRR) type in narrow-leafed lupin relative to other well studied legumes such as *Glycine max* and *M. truncatula* (Hane et al. 2017). To investigate this further, consensus sequences for coiled-coil, Toll and mammalian Interleukin1 receptor-NBS-LRR proteins, as well as NBS-LRR genes previously identified in *M. truncatula* were used to identify NBS-LRRs present in the narrow-leafed lupin genome (Hane et al. 2017). This predicted a total of 68 genes and this number was considerably less than those observed in other closely related legumes such as *M. truncatula* (764), soybean (506), pigeonpea (406) and chickpea (187) (Ameline-Torregrosa et al. 2008; Varshney et al. 2013).

The narrow-leafed lupin NBS domains identified above were compared with 32 NBS domain sequences that were derived from PCR amplification of the narrow-leafed lupin genome using degenerate primers derived from *M. truncatula* conserved NBS sequences (Foley and Singh, unpublished observations), but this did not identify additional resistance gene homologs. This supports the completeness of the narrow-leafed lupin genome assembly and that further NBS-LRR homologs were unlikely to have been missed. The distribution of the NBS-LRR genes across the 20 linkage groups of narrow-leafed lupin was also examined as shown in Table 3.2 and found to be well-spread across the narrow-leafed lupin genome with little evidence of clustering (Hane et al. 2017). Thus, of the 68 homologs, 56 were anchored in pseudo-chromosomes and were distributed across all linkage groups but linkage group 4. Clusters of three or more NBS-LRR homologs on a single scaffold were identified on linkage groups 2, 6, 8 and 13.

A major disease pressure on lupins, including narrow-leafed lupin is anthracnose, caused by *Colletotrichum lupini* (Talhinhas et al. 2016). The cultivar Tanjil is resistant to anthracnose and a single dominant resistance gene (*Lanr1*) was mapped to linkage group 11 (Kamphuis et al. 2015; Yang et al. 2013b). Using the improved genetic map (Hane et al. 2017) it was possible to

Table 3.2 Distribution of the 68 NBS-LRR genes across the 20 linkage groups of narrow-leafed lupin

Linkage group	Number of resistance gene homologs	Clusters on a single scaffold
NLL-01	1	
NLL-02	11	Two clusters of 6 and 3 genes
NLL-03	3	
NLL-04	3	
NLL-05	3	
NLL-06	8	Two clusters of 4 and 2 genes
NLL-07	1	
NLL-08	3	A cluster with 3 genes
NLL-09	5	
NLL-10	1	
NLL-11	3	
NLL-12	1	
NLL-13	6	A cluster with 3 genes
NLL-14	–	
NLL-15	–	
NLL-16	–	
NLL-17	1	
NLL-18	–	
NLL-19	4	
NLL-20	2	
Total	56	
Unassigned	12	
Total NBS-LRR genes	68	

define the location of *Lanr1* to a single scaffold (Scaffold_133) on the reference genome that spanned 388 Kb and contained 41 predicted genes (*Lup005013.1-Lup005054.1*) including an NBS-LRR resistance gene (*Lup005042.1*). Alignment of Lup005042.1 amino acid sequence from the four parents of the two RIL populations for narrow-leafed lupin showed conserved sequence for anthracnose resistant lines Tanjil and its close relative 83A:476 compared to the susceptible lines Unicrop and P27255, thus making Lup005042.1 an excellent candidate for *Lanr1* (Hane et al. 2017). Tanjil also has resistance to phomopsis stem blight caused by *Diaporthe toxica*, which is controlled by the dominant resistance gene *PhtjR* (Yang et al. 2013a). This locus was fine-mapped using marker sequences from Yang et al. (2013a) and found to contain 12 predicted gene annotations. However, unlike the *Lanr1* region, it did not harbour any NBS-LRR genes (Kamphuis and Singh, unpublished observations).

3.5 The Lupin Genome Portal

Many of the narrow-leafed genomic datasets described in the preceding sections are housed at the lupin genome portal (https://www.lupinexpress.org) (Hane et al. 2017). The lupin genome portal provides some useful background information about lupins and narrow-leafed lupin in particular. It also hosts a range of narrow-leafed lupin genomic datasets, which can also be queried by BLAST homology searching as illustrated in Fig. 3.1. The narrow-leafed lupin datasets include the reference genome including the genome annotation, the genome scaffold assembly and annotated gene sets and various transcriptome assemblies for narrow-leafed lupin as described in Chap. 5. These datasets can be queried using a built-in and interactive BLAST, GBrowse and CMap interfaces (Donlin 2009; Priyam et al. 2015; Youens-Clark et al. 2009) as illustrated in Fig. 3.2. The website also provides links to a range of other legume genomic resources to facilitate comparative genomic analysis in areas such as syntenic relationships and legume evolution. For example, syntenic maps to other legume species are available via the legume information system website (https://legumeinfo.org/) with direct links available on the lupin genome portal. A publication page on the website presents a collection of recent lupin publications under areas such as lupin genomic resources, genetic mapping and synteny, diversity and adaptation and molecular markers. Collectively, the resources hosted in the website can help with genomics-based breeding approaches.

3.6 What is in the Pipeline?

3.6.1 Generating a Gold Standard Reference Genome

The current narrow-leafed lupin Tanjil assembly was generated using a range of short-read sequencing data coupled with some BAC-end sequences, which represented best practice at that time (Hane et al. 2017). The Singh group at CSIRO is in the process of improving this draft assembly by supplementing the short-read sequencing data with >91 Gb of long-read sequence data generated using PacBio Sequel technology. This will be augmented using High-C technology to provide near contiguous chromosome sequences.

3.6.2 Generation of a Narrow-Leafed Lupin Pan-Genome

The generation of a pan-genome can be an important resource for a crop species where it can help with both research and pre-breeding/breeding activities. In crop plants, a pan-genome represents the entire gene set for a representative set of domesticated and/or wild accessions. It includes the core genome containing genes present in all the accessions sampled as well as an accessory genome that contains genes that are present only in some accessions among the sampled set. To generate a pan-genome resource for narrow-leafed lupin, the Singh group at CSIRO is in the process of re-sequencing over 50 narrow-leafed lupin accessions, representing genetically diverse lines

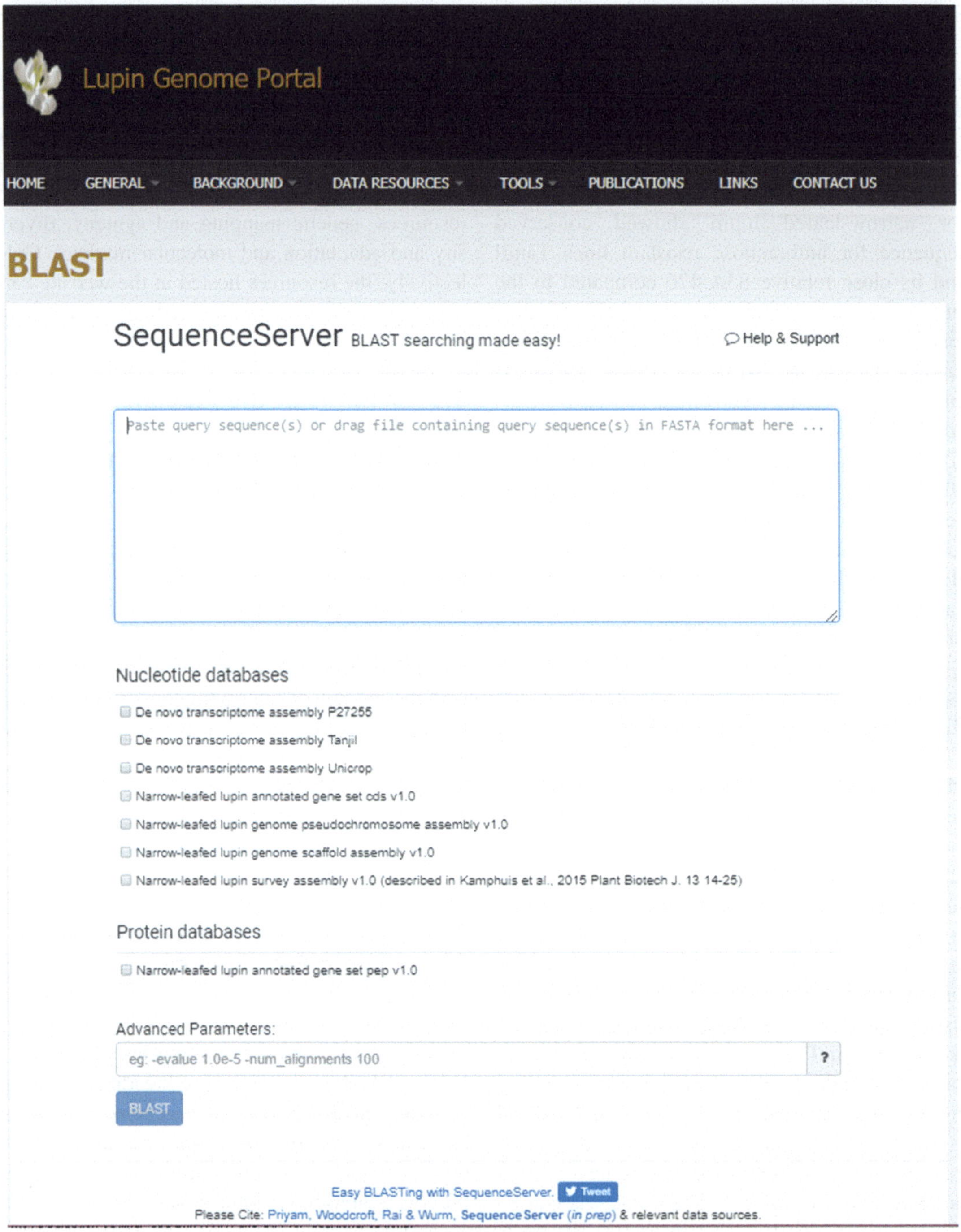

Fig. 3.1 A screen shot from the lupin genome portal showing some of the datasets present on this website

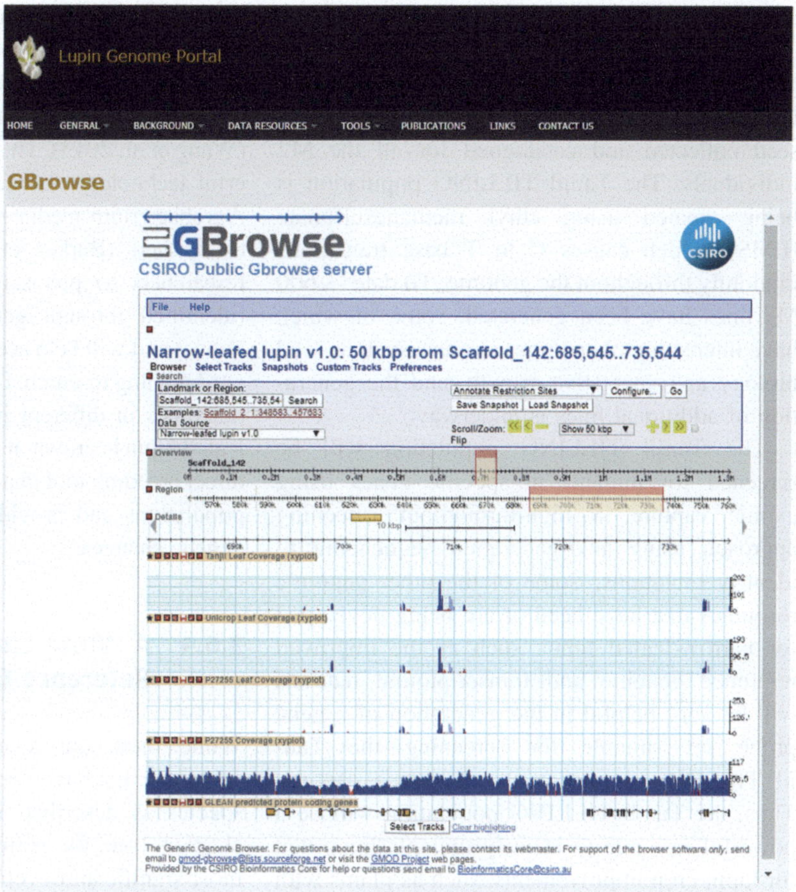

Fig. 3.2 A screen shot from the lupin genome portal showing the use of the GBrowse to navigate the narrow-leafed lupin genome

of wild and domesticated narrow-leafed lupin cultivars from Australian and European collections. While this project is still underway a preliminary pan-genome has been assembled ($\sim$885 Mb) with an annotated gene set of 40,117 genes. The majority of these genes were conserved across all narrow-leafed lupin accessions examined to date in the draft pan-genome. The current focus is on the further functional annotation, variant calling and presence/absence analysis, and adding more sequence information from additional lines of importance for narrow-leafed lupin breeding. This pan-genome resource will lay the foundation for a medium- to high-throughput genotyping platform for pre-breeding research as well as genomic selection in lupin breeding programmes through the discovery of evenly distributed SNPs across the narrow-leafed lupin pan-genome.

3.6.3 Reverse Genetic Resources to Exploit Genomic Advances in Narrow-Leafed Lupin

Reverse genetics has become a critical part of biological research activities and is a way of determining function for genes whose function are unknown. Reverse genetic approaches are varied but seek to generate a mutation(s) in a gene(s) and the subsequent analysis of possible mutant phenotypes. To facilitate reverse genetic approaches in narrow-leafed lupin, efforts are underway in the Singh group at CSIRO to develop a lupin targeting induced local lesions in genomes (TILLING) population in the cultivar Tanjil background. TILLING combines a mutagenesis approach to generate random mutations in genes, with DNA screening approaches of increasing sophistication and throughput to identify mutations in these genes. A TILLING population consisting typically of thousands of M2 individuals can be created with the DNA and seed collected and catalogued for all the M2 individuals. The Tanjil TILLING population is being created using ethyl methanesulfonate (EMS), which causes C to T base transitions randomly throughout the genome. To date ~600 M_2 lines have been generated, some of which have interesting phenotypes in areas such as seed biology and vegetative growth, and the generation of additional lines is underway.

The Tanjil TILLING population will be screened for mutants in specific genes using exome capture, a high-throughput screening approach using next-generation sequencing to identify mutations. Some of the other genomic resources that have been or are being developed for narrow-leafed lupin, such as the improved reference genome and transcriptomic datasets will be very helpful in the generation of a gene probe set suitable for screening the Tanjil TILLING population using exome capture. Thus, the Tanjil TILLING population will be a powerful resource to identify mutants important for lupin crop improvements, such as plants with altered alkaloid content, reduced allergenicity, improved seed composition, enhanced oil content as well as other traits important to the lupin breeding programs.

Another exciting development in the reverse genetic space is gene editing. Gene editing is a powerful technique that has been established in many plant species and has great potential to accelerate narrow-leafed lupin pre-breeding research by providing rapid proof-of-concept information for genes linked to important traits that can then be targeted by other approaches such as TILLING. Gene editing involves editing nucleases ('molecular scissors') that are used to insert, replace or remove DNA in a specific region (Bak et al. 2018). The nuclease creates a double-stranded break at a specific location in the genome (e.g. a specific target gene) and then capitalises on the plant cells endogenous machinery to repair in the induced break through either homologous recombination or non-homologous end-joining. There are currently at least four different classes of gene editing technology with the CRISPR technology leading the way (Wang et al. 2013). The establishment of this powerful technology for narrow-leafed lupins, where Agrobacterium-mediated transformation has been established (Barker et al. 2016), would allow researchers to precisely target genes of interest (identified through genomic analyses described above) and will help accelerate narrow-leafed lupin pre-breeding research. Depending on the regulatory outcomes in different jurisdictions where narrow-leafed lupin is grown, it may eventually be used as a direct breeding tool that can complement TILLING populations and provide a wider range of useful genetic changes.

3.6.4 A White Lupin (*L. albus*) Reference Genome

While there are a number of transcriptomic resources that have been developed in other lupin species as described in some of the following chapters (see, for example, Chaps. 6 and 12), the most significant development we are aware of at this stage in the lupin genome space is the development of a *L. albus* genome assembly. *L. albus* known also as white lupin is another important domesticated lupin species that is grown primarily in Europe. A striking feature of *L. albus* is its ability to produce cluster roots and it can therefore serve as an interesting model to study root development (Gallardo et al. 2019; Shane and Lambers 2005).

An initiative to sequence the *L. albus* genome that is being led by the group of Benjamin Péret at CNRS, France is well advanced as reported at the 12th Congress of the International Plant Molecular Biology in Montpellier in 2018 and International Legume Society meeting in 2019. The *L. albus* genomic team are using a combination of short- and long-read sequencing and the assembly is being further improved through the use of high resolution approaches, with optical and genetic maps. This has resulted in a final assembly of 451 Mb with nearly all of the assembly successfully mapped to the 25 pseudo-chromosomes of *L. albus*. A total of 38,258 coding genes were identified and annotated as well as 3,129 ncRNAs. The completeness of the assembly was further supported by BUSCO analysis against the Plantae BUSCO dataset which identified 94.6% of the genes in this dataset being present in the *L. albus* genome assembly. Like its close relative, narrow-leafed lupin, the *L. albus* genome contained significant numbers of repetitive elements as well as evidence for a substantial degree of gene duplication. The Péret group also reported on transcriptome datasets from ten organs of *L. albus*. Once completed and published the *L. albus* genome assembly and associated datasets will be valuable resources for lupin and legume biology in many areas.

3.7 Conclusions

The recent domestication of narrow-leafed lupin in major growing areas such as in Australia has been successful, but the genetic base has been shown to be narrow (Berger et al. 2013). Difficulties to broaden the genetic base through crosses with wild accessions and the need to retain the recessive domestication genes has held back the crop and there are still many agronomic limitations remaining (Berger et al. 2013). The revolution in sequencing technologies and high-throughput NGS genotyping in recent years poses exciting opportunities to accelerate yield gains for this important crop species and allows researchers and breeders to tackle some key adaptive, domestication and disease traits for different growing regions. As shown in Fig. 3.3, in the last few years there have been several valuable genomic resources developed for narrow-leafed lupin, and additional resources such as the pan-genome and TILLING population are under development. These genomic resources will accelerate efforts by researchers and breeders to improve narrow-leafed lupin and other lupin crops to help provide more of the valuable protein increasingly in demand globally and facilitate the establishment of more sustainable cropping systems where lupins crops are/can be grown.

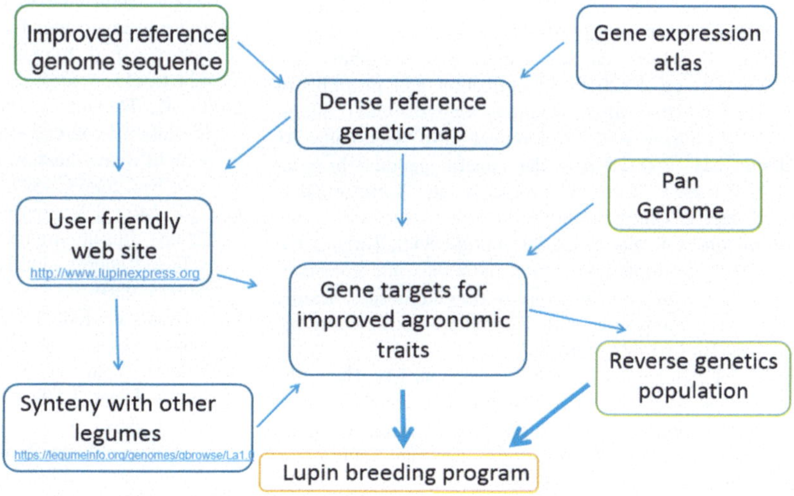

Fig. 3.3 Schematic representation showing some of the genomic resources already developed for narrow-leafed lupin, circled in blue and some under development, circled in green, and the arrows depicting how these resources are interconnected to help advance research and breeding activities

Acknowledgements Narrow-leafed lupin genomic work in the authors group has been supported by Grains Research and Development Corporation (GRDC), CSIRO, Curtin University and the University of Western Australia.

References

Ameline-Torregrosa C, Wang B-B, O'Bleness MS, Deshpande S, Zhu H et al (2008) Identification and characterization of nucleotide-binding site-leucine-rich repeat genes in the model plant *Medicago truncatula*. Plant Physiol 146:5–21

Arnoldi A, Boschin G, Zanoni C, Lammi C (2015) The health benefits of sweet lupin seed flours and isolated proteins. J Funct Foods 18:550–563

Bak RO, Gomez-Ospina N, Porteus MH (2018) Gene editing on center stage. Trends Genet 34:600–611

Barker SJ, Si P, Hodgson L, Ferguson-Hunt M, Khentry Y et al (2016) Regeneration selection improves transformation efficiency in narrow-leaf lupin. Plant Cell Tissue Organ Cult 126:219–228

Berger JD, Clements JC, Nelson MN, Kamphuis LG, Singh KB et al (2013) The essential role of genetic resources in narrow-leafed lupin improvement. Crop Pasture Sci 64:361–373

Bravo A, York T, Pumplin N, Mueller LA, Harrison MJ (2016) Genes conserved for arbuscular mycorrhizal symbiosis identified through phylogenomics. Nat Plants 2

Crossa J, Perez-Rodriguez P, Cuevas J, Montesinos-Lopez O, Jarquin D et al (2017) Genomic selection in plant breeding: methods, models, and perspectives. Trends Plant Sci 22:961–975

DeBoer K, Melser S, Sperschneider J, Kamphuis LG, Garg G et al (2019) Identification and profiling of narrow-leafed lupin (*Lupinus angustifolius*) micro-RNAs during seed development. BMC Genomics 20

Donlin MJ (2009) Using the generic genome browser (GBrowse). Current protocols in bioinformatics (Chap. 9: Unit 9.9-Unit 9.9)

Drummond CS, Eastwood RJ, Miotto STS, Hughes CE (2012) Multiple continental radiations and correlates of diversification in *Lupinus* (Leguminosae): testing for key innovation with incomplete taxon sampling. Syst Biol 61:443–460

Foley RC, Jimenez-Lopez JC, Kamphuis LG, Hane JK, Melser S et al (2015) Analysis of conglutin seed storage proteins across lupin species using transcriptomic, protein and comparative genomic approaches. BMC Plant Biol 15:106

Gallardo C, Hufnagel B, Casset C, Alcon C, Garcia F et al (2019) Anatomical and hormonal description of rootlet primordium development along white lupin cluster root. Physiol Plant 165:4–16

Gao L-L, Hane J, Kamphuis LG, Foley R, Shi B-J et al (2011) Development of genomic resources for the narrow-leafed lupin (*Lupinus angustifolius*): construction of a Bacterial Artificial Chromosome (BAC) library and BAC-end sequencing. BMC Genomics 12:521

González-Sama A, Lucas MM, De Felipe MR, Pueyo JJ (2004) An unusual infection mechanism and nodule morphogenesis in white lupin (*Lupinus albus*). New Phytol 163:371–380

Hane JK, Ming Y, Kamphuis LG, Nelson MN, Garg G et al (2017) A comprehensive draft genome sequence for lupin (*Lupinus angustifolius*), an emerging health food: insights into plant-microbe interactions and legume evolution. Plant Biotechnol J 15:318–330

James E, Minchin F, Iannetta P, Sprent J (1997) Temporal relationships between nitrogenase and intercellular glycoprotein in developing white lupin nodules. Ann Bot 79:493–503

Jones JDG, Dangl JL (2006) The plant immune system. Nature 444:323–329

Kamphuis LG, Hane JK, Nelson MN, Gao L, Atkins CA et al (2015) Transcriptome sequencing of different narrow-leafed lupin tissue types provides a comprehensive uni-gene assembly and extensive gene-based molecular markers. Plant Biotechnol J 13:14–25

Kasprzak A, Safar J, Janda J, Dolezel J, Wolko B et al (2006) The bacterial artificial chromosome (BAC) library of the narrow-leafed lupin (*Lupinus angustifolius* L.). Cell Mol Biol Lett 11:396–407

Lambers H, Clements JC, Nelson MN (2013) How a phosphorus-acquisition strategy based on carboxylate exudation powers the success and agronomic potential of lupines (*Lupinus*, Fabaceae). Am J Bot 100:263–288

Lambers H, Albornoz F, Kotula L, Laliberte E, Ranathunge K et al (2018) How belowground interactions contribute to the coexistence of mycorrhizal and non-mycorrhizal species in severely phosphorus-impoverished hyperdiverse ecosystems. Plant Soil 424:11–33

Lavin M, Herendeen PS, Wojciechowski MF (2005) Evolutionary rates analysis of Leguminosae implicates a rapid diversification of lineages during the tertiary. Syst Biol 54:575–594

Lee YP, Mori TA, Sipsas S, Barden A, Puddey IB et al (2006) Lupin-enriched bread increases satiety and reduces energy intake acutely. Am J Clin Nutr 84:975–980

Lesniewska K, Ksiazkiewicz M, Nelson MN, Mahe F, Ainouche A et al (2011) Assignment of 3 genetic linkage groups to 3 chromosomes of narrow-leafed Lupin. J Hered 102:228–236

MacLean AM, Bravo A, Harrison MJ (2017) Plant signaling and metabolic pathways enabling arbuscular mycorrhizal symbiosis. Plant Cell 29:2319–2335

Oldroyd GED (2013) Speak, friend, and enter: signalling systems that promote beneficial symbiotic associations in plants. Nat Rev Microbiol 11:252–263

Parra G, Bradnam K, Korf I (2007) CEGMA: a pipeline to accurately annotate core genes in eukaryotic genomes. Bioinformatics 23:1061–1067

Priyam A, Woodcroft BJ, Rai V, Munagala A, Moghul I et al (2015) Sequenceserver: a modern graphical user interface for custom BLAST databases. https://doi.org/10.1101/033142.

Qi YP, Tsuda K, Glazebrook J, Katagiri F (2011) Physical association of pattern-triggered immunity (PTI) and effector-triggered immunity (ETI) immune receptors in Arabidopsis. Mol Plant Pathol 12:702–708

Ramirez-Prado JS, Abulfaraj AA, Rayapuram N, Benhamed M, Hirt H (2018) Plant immunity: from signaling to epigenetic control of defense. Trends Plant Sci 23:833–844

Shane MW, Lambers H (2005) Cluster roots: a curiosity in context. Plant Soil 274:101–125

Talhinhas P, Baroncelli R, Le Floch G (2016) Antrhacnose of lupins caused by *Colletotrichum lupini*: a recent disease and a succssful worldwide pathogen. J Plant Pathol 98:5–14

Tang C, Robson A, Dilworth M, Kuo J (1992) Microscopic evidence on how iron deficiency limits nodule initiation in *Lupinus angustifolius* L. New Phytol 457–467

Thomma BPHJ, Nurnberger T, Joosten MHAJ (2011) Of PAMPs and effectors: the blurred PTI-ETI dichotomy. Plant Cell 23:4–15

Varshney RK, Glaszmann JC, Leung H, Ribaut JM (2010) More genomic resources for less-studied crops. Trends Biotechnol 28:452–460

Varshney RK, Song C, Saxena RK, Azam S, Yu S et al (2013) Draft genome sequence of chickpea (*Cicer arietinum*) provides a resource for trait improvement. Nat Biotechnol 31:240–246

Wang HY, Yang H, Shivalila CS, Dawlaty MM, Cheng AW et al (2013) One-step generation of mice carrying mutations in multiple genes by CRISPR/Cas-mediated genome engineering. Cell 153:910–918

Yang HA, Tao Y, Zheng ZQ, Shao D, Li ZZ et al (2013a) Rapid development of molecular markers by next-generation sequencing linked to a gene conferring phomopsis stem blight disease resistance for marker-assisted selection in lupin (*Lupinus angustifolius* L.) breeding. Theoret Appl Genet 126:511–522

Yang HA, Tao Y, Zheng ZQ, Zhang QS, Zhou GF et al (2013b) Draft genome sequence, and a sequence-defined genetic linkage map of the legume crop species *Lupinus angustifolius* L. PLoS One 8

Yang HA, Jian JB, Li X, Renshaw D, Clements J et al (2015) Application of whole genome re-sequencing data in the development of diagnostic DNA markers tightly linked to a disease-resistance locus for marker-assisted selection in lupin (*Lupinus angustifolius*). BMC Genomics 16:660.

Youens-Clark K, Faga B, Yap IV, Stein L, Ware D (2009) CMap 1.01: a comparative mapping application for the internet. Bioinformatics 25:3040–3042

4 Cytomolecular Insight Into *Lupinus* Genomes

Karolina Susek and Barbara Naganowska

Abstract
Lupins are a group within Genisteae interesting from the point of view of their complex evolutionary history, which is the result of their paleoploid origin (Atkins et al in Lupins as crop plants: biology, production, and utilization. CAB International, pp 67–92, 1998; Gladstones in Lupins as crop plants: biology, production, and utilization. CAB International, pp. 1–36, 1998). Contemporary species are characterised by a striking diversity of chromosome numbers. Their genomes vary in size and consist of many small and similar length chromosomes. The Old World lupins and New World lupins are different in their cytological traits such as the basic and somatic chromosome numbers. Here, we summarise the cytogenetic research on genome size estimation, chromosome number identification and integrative genetic and cytogenetic mapping. The importance of narrow-leafed lupin as a crop and the relative wealth of data concerning its cytogenetic and genetic characterisation have led to its role as a useful model species within the genus *Lupinus* and a reference for better understanding of legume genome evolution. Cytogenetic comparative mapping, using *L. angustifolius*-derived markers, has revealed the karyotype variation in lupins. Insight into chromosome rearrangements has led to a hypothetic model of lupin karyotype evolution. This research has established a starting point for the further analysis of the structure and diversity of lupin karyotypes in the facet of evolution.

K. Susek (✉) · B. Naganowska
Institute of Plant Genetics, Polish Academy of Sciences, Poznań, Poland
e-mail: ksus@igr.poznan.pl

B. Naganowska
e-mail: bnag@igr.poznan.pl

4.1 *Lupinus* spp. Genome Sizes and Chromosome Numbers

Until the year 2000, there were few reports concerning the nuclear DNA content of lupins. Data on genome size for a number of *Lupinus* spp., measured using cytophotometric methods and later flow cytometry, is available at a constantly updated database (https://data.kew.org/cvalues/CvalServlet?querytype=2). The nuclear genomes of lupins are small, only a few times larger than that of the model plant *Arabidopsis thaliana*. A complex analysis of the 2C DNA values for Old World lupins (OWL) and New World lupins (NWL) as well as an analysis of the relationships among the taxa within the genus showed a similar variation in the 2C DNA values in both groups ranged from 0.97 pg to 2.44 pg and from 1.08 pg to 2.68 pg, respectively (Naganowska et al. 2003, 2006). Furthermore, for the OWL species and botanical forms, there was no significant correlation between the DNA content and chromosome

K. B. Singh et al. (eds.), *The Lupin Genome*, Compendium of Plant Genomes,
https://doi.org/10.1007/978-3-030-21270-4_4

Table 4.1 Basic cytogenetic features of old world lupins based on Naganowska and Zielińska (2002), Hajdera et al. (2003), Naganowska and Kaczmarek (2005) and Susek et al. (2016)

Group	Species	Chromosome number (2n)	Genome size (pg/2C DNA)	Loci rDNA	
				45S	5S
Malacospermae (smooth-seeded)	*L. angustifolius*	40	1.89	1	1
	L. cryptanthus	40	1.86	2	2
	L. linifolius	40	1.88	–	–
	L. albus	50	1.16	1	1
	L. graecus	50	1.13	–	–
	L. termis	50	1.14	–	–
	L. vavilovi	50	1.14	–	–
	L. luteus	52	2.44	1–2	1
	L. hispanicus	52	2.15	1–2	1
	L. micranthus	52	0.98	1	1
	L. princei	38	0.97	1–2	1
Scabrispermae (rough-seeded)	*L. atlanticus*	38	1.42	2	1
	L. digitatus	36	1.37	2	1
	L. cosentinii	32	1.61	3	1 or 3
	L. pilosus	42	1.36	1–2	1
	L. palaestinus	42	1.39	2	1

number. The 'Homogeneous groups' that were determined by a statistical analysis coincided with the OWL taxonomic sections (Naganowska et al. 2003; Table 4.1). An interesting exception was found for the rough-seeded *L. princei*, which had the lowest 2C DNA value that was closer to the smooth-seeded species. Generally, the results confirmed the heterogeneous character of the smooth-seeded group of OWL and were interpreted as being a trace of several independent evolutionary lineages from the ancient rough-seeded stock. As to the NWL, an analysis of the 2C DNA content for 38 species and accessions (Naganowska et al. 2006), annuals and perennials (from different 'complexes'—Dunn 1984) revealed a relationship between genome size and the type of life cycle. However, the limited sampling of the NWL as well as their insufficiently defined taxonomy did not allow further conclusions on interspecific relationships in that group to be made. To the best of our knowledge, further analyses of the lupin DNA content on large samples have not been undertaken.

Cytological analyses of the karyotypes in the OWL group, which comprises just over a dozen herbaceous annuals, revealed seven different somatic chromosome numbers: 2n = 32, 36, 38, 40, 42, 50 and 52 as well as different basic chromosome numbers x = 5–9 and x = 13. In contrast to the OWL, the NWL are a large group of several hundred species (including multifoliolate or unifoliolate species that are considered to be more primitive), herbaceous or shrubby and annual, biennial or perennial species. This group has a low diversity of somatic chromosome numbers and their basic chromosome number is x = 6. Most species from western North America and the Andes have 2n = 48 (96) and the group of lupins from the eastern parts of North and South America mainly have 2n = 36. Lupins from the Florida region are exceptional with 2n = 52 that is typical for OWL (see review in Wolko et al. 2011).

4.2 Chromosome Markers for Cytogenetic Mapping of *Lupinus* Species

The development of molecular cytogenetics has enabled a deeper insight into the plant genome. One widely applied method that is used to study plant chromosomes is fluorescence in situ hybridisation (FISH). In lupins, FISH has been adapted in order to provide additional information about their genome organisation and chromosome identification. The first chromosome markers were established using double-target FISH, which permitted the chromosomes that carry the rDNA loci (clusters of rRNA genes), to be identified. Cytogenetic mapping revealed variation in rDNA loci among OWL and NWL (Hajdera et al. 2003; Kaczmarek et al. 2009; Naganowska and Zielinska 2002; Susek et al. 2016). The 45S rDNA markers have shown diverse loci numbers that range from one locus in *L. albus* (2n = 50), *L. angustifolius* (2n = 40), *L. micranthus* (2n = 52) and *L. mutabilis* (2n = 48), one to two loci in *L. hispanicus* (2n = 52), *L. luteus* (2n = 52), *L. princei* (2n = 38) and *L. pilosus* (2n = 42), two loci in *L. atlanticus* (2n = 38), *L. cryptanthus* (2n = 40), *L. digitatus* (2n = 36) and *L. palaestinus* (2n = 42) and up to three loci exclusively in *L. cosentinii* (2n = 32). Conversely, 5S rDNA has only been observed at one locus in 12 lupins that have been analysed (Naganowska and Kaczmarek 2005), Hajdera et al. (2003) and two loci in *L. cryptanthus* (Susek et al. 2016). No correlation between the numbers of rDNA loci and chromosome numbers was observed. Moreover, the signal of the 18–25S rDNA probe in lupin chromosomes was extremely large and covered most of the chromosome, including the nucleolar organiser region (Naganowska and Zielinska 2004). These two observations, together with overall low rDNA loci numbers, were interpreted as possibly being connected with the ancient origin of polyploidy in *Lupinus* spp., which caused different types of multiple chromosome rearrangements. The relatively low number of the polymorphic loci of both rDNA markers allowed the identification of a few chromosomes. A maximum of six pairs of 32 chromosomes in *L. cosentinii* (Hajdera et al. 2003; Naganowska and Zielińska 2002) and two pairs of 52 chromosomes, e.g. in *L. micranthus*, were identified (Naganowska and Zielińska 2002). Thus, additional markers were required to continue lupin molecular karyotyping in order to determine a chromosome-based phylogeny and evolution.

The progress of the genomic era has provided new approaches for cytogenetic mapping studies. For lupins, the first bacterial artificial chromosome (BAC) library of the *L. angustifolius* nuclear genome (Kasprzak et al. 2006) was an important tool at this stage. BAC clones from this library were used for FISH (BAC-FISH) experiments which allowed them to be mapped on the chromosomes of *L. angustifolius*. It was shown by the nature of the hybridisation signals obtained that the BACs represented either single locus or multiple loci on the chromosomes. However, while both BAC types contributed to a general characterisation of the *L. angustifolius* genome, only the single-locus BACs were informative for chromosome identification. These chromosome-specific cytogenetic BAC markers established a foundation for assigning the genetic linkage groups (LGs) to the chromosomal maps of *L. angustifolius* (Lesniewska et al. 2011). Among the twelve clones that produced single-locus signals, eight were localised on three chromosomes. Based on BAC-end sequences, the genetic markers were generated and these chromosomes (Lang06, Lang08 and Lang17) were assigned to three linkage groups (NLL-06, NLL-08 and NLL-17, respectively) of the genetic map that was published by Nelson et al. (2010). Importantly, complementary genetic and cytogenetic mapping illustrated the powerful approach to hook the genomic regions carrying important genes, such as the *SymRK* gene (Lesniewska et al. 2011). Combined mapping of narrow-leafed lupin genome resulted in the generation of 18 BAC-derived cytogenetic chromosome markers that identified a further nine chromosomes (Ksiazkiewicz et al. 2015). Later, more chromosome markers were generated as part of research whose goal was to characterise the important processes or metabolic pathways in

L. angustifolius, and these markers were introduced into the genome map. The BACs analysed were carrying sequences connected with Phomopsis stem blight resistance (Ksiazkiewicz et al. 2013), sequences of genes of flowering induction (Ksiazkiewicz et al. 2016; Nelson et al. 2017) and genes of chalcone isomerase (Przysiecka et al. 2015).

4.3 Cytogenomic Map of *L. angustifolius* as a Basic Tool for Comparative Studies

Cytogenetic mapping of *L. angustifolius* genome provided a comprehensive, integrated genome map (Fig. 4.1) with all chromosome pairs identified with specific markers and assigned to narrow-leafed lupin linkage groups (Wyrwa et al. 2016). Specific chromosomes were identified by BAC-FISH through cross-hybridisation with one to twelve BACs. In addition, the comprehensive map provided information on the linkage mapping of the 5S and 45S rDNA loci in NLL-02 and NLL-16, respectively. Thus, a nucleolar organiser was localised as well (Wyrwa et al. 2016). Finally, the map of the narrow-leafed lupin genome (Wyrwa et al. 2016) was precisely re-integrated based on assignment of chromosomes to pseudomolecules using whole BAC sequences (WBS). The newly obtained WBS (PacBio) along with other BAC sequences (NGS platform, Illumina) available for lupins in the NCBI database have expanded the cytogenomic resources. This established an efficient set of tools for integrating chromosomal maps and pseudomolecules. Seven chromosomes were assigned to other pseudomolecules or were saturated by additional BAC markers in comparison to Wyrwa et al. (2016). Integrative mapping also enabled the mapping of clones that had not yet been assigned to linkage groups, e.g. the inclusion of a previously unassigned cluster group (Cluster-2) containing markers of agronomic traits, i.e. the flowering time to pseudomolecule NLL-20 (Susek et al. 2019).

4.4 Karyotype Variation and Rearrangements Among Lupins

Comparative genome mapping using FISH has aimed to identify various chromosomal rearrangements within the species of the genus *Lupinus*, using the *L. angustifolius* karyotype as a reference. The first comparative BAC-FISH mapping was performed for four wild OWL, namely *L. cryptanthus* (2n = 40), *L. micranthus* (2n = 52), *L. pilosus* (2n = 42) and *L. cosentinii* (2n = 32) and one wild NWL, *L. multiflorus* (2n = 36) (Susek et al. 2016). This approach revealed different patterns of BAC-FISH signals on the chromosomes as a single locus and/or a single hybridisation signal with additional signals obtained with BAC clones containing repetitive sequences. It is noteworthy that some clones were not able to yield detectable signals, e.g. in

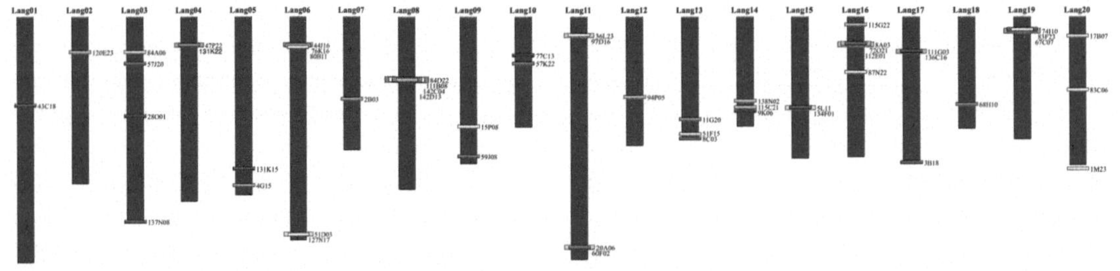

Fig. 4.1 Ideogram of *L. angustifolius* chromosomes (Lang01 to Lang20). The karyotype was created based on reciprocal assignment to the whole genome of *L. angustifolius*. The BAC clones are indicated by rectangles with a unique colour, according to their position in the genome. The overlapping rectangles are shown overlapping BAC positions in *L. angustifolius*. All of the chromosomes are drawn to scale, whereby the Mb units refer to the *L. angustifolius* genome sequence (Susek et al. 2019)

L. cosentinii. These differences in the arrangement of the BACs in the chromosomes of the wild species exposed various sequence reshufflings in the lupin genomes. For example, the highest level of similarity of the BAC pattern was found between *L. angustifolius* and *L. cryptanthus*. This result reflects the high genome homology between these two species as *L. cryptanthus* is considered to be a wild botanical form of *L. angustifolius* (Naganowska et al. 2003). In other species, the chromosome rearrangements were more variable, thus suggesting that multiple structural reshufflings may have occurred in their karyotypes during their evolution. These rearrangements along with BAC sequence analyses shed light on a complex evolutionary history, which possibly involved chromosomal changes such as fusions/fissions and repetitive sequence amplifications. To determine chromosome variation within OWL species, and propose the model of their karyotype evolution, twelve species and/or wild botanic forms were analysed. These were: *L. angustifolius* (2n = 40) and *L. cryptanthus* (2n = 40), *L. luteus* (2n = 52) and *L. hispanicus* (2n = 52) and *L. albus* (2n = 50) and *L. graecus* (2n = 52), as well as other wild species such as *L. micranthus* (2n = 52), *L. atlanticus* (2n = 38), *L. digitatus* (2n = 36), *L. cosentinii* (2n = 32), *L. pilosus* (2n = 42) and *L. palaestinus* (2n = 42) were analysed. Of the 52 *L. angustifolius* markers, the ones between 19 and 41 were developed for a particular species (Susek et al. 2019). Development of chromosomal markers for these species enabled tracking chromosomal rearrangements across OWL. Based on the different localisation of BACs in the studied lupins, in comparison to *L. angustifolius*, four types of chromosome changes were identified. For example, one *L. angustifolius* chromosome (Lang06), carrying three clones in the arm A and two clones in the arm B, was represented by two chromosomes in related lupins. The clones from the arm A hybridised to one chromosome of *L. luteus*, *L. albus* and *L. graecus*. By contrast, one clone from the arm B was observed in other chromosomes of these three species, respectively. Additionally, several BAC markers allowed cultivated species to be distinguished from their wild forms (Susek et al. 2019).

Comparative analyses of the OWL at the chromosome level revealed an evolutionary diversification of their genomes, but the direction of these changes can only be deduced from their phylogeny. The hypothesis on the course of the evolution of the lupin karyotype was discussed (Susek et al. 2016, 2019). It was assumed from the phylogenetic trees (Drummond et al. 2012) that the karyotype with 52 chromosomes, which corresponded to the *L. micranthus* one, may have evolved by numerous fusions (~7.5 MYA) that led to the intermediate karyotype with 40 chromosomes of the *L. angustifolius* type (~4.5 MYA), which was followed by a further reduction to 32 chromosomes (*L. cosentinii*, ~4.0 MYA). However, fission events cannot be ruled out, for example, the karyotypes with 32 chromosomes (*L. cosentinii*) could have arisen from the karyotypes with 42 chromosomes (*L. pilosus*, ~3.0 MYA) assuming that one of the *L. cosentinii* chromosomes would correspond to two chromosomes of *L. pilosus*. These hypotheses were also supported by the genetic mapping by Kroc et al. (2014), who concluded that *L. angustifolius* underwent a duplication and/or triplication of the genome and by the reference genome assembly (Hane et al. 2017) as discussed in Chap. 7.

A model of chromosomal rearrangements was also proposed that illustrated the changes in the number and/or structure of the chromosomes. It was assumed that these differences occurred as a result of the loss or insertion of DNA sequences through numerous deletion, insertion and duplication mutations. Chromosome reduction from 2n = 52 to 2n = 36 might shape the lupin genomes. Species with 2n = 52 chromosomes could have evolved from an 'ancestral' species with a chromosome number of 2n = 54. Species with a lower number of chromosomes have probably undergone three rounds of Whole Genome Duplication (WGD), as a result of which, species with a chromosome number of 2n = 48, 2n = 42 and 2n = 36 evolved, and the accompanying events of aneuploidy led to the formation of karyotypes with the number of chromosomes of

2n = 52, 50, 40, 38, 32 such as *L. luteus* (2n = 52), *L. albus* (2n = 50), *L. angustifolius* (2n = 40), *L. atlanticus* (2n = 38) and *L. cosentinii* (2n = 32). The hypothesis also assumes that the basic number of chromosomes in the Old World lupins is x = 6 (Susek et al. 2019).

Comparative cytogenetic observations were also implemented in the analyses of the relationships between *L. angustifolius* and two other crops, *L. albus* and *L. luteus*. FISH using centromeric BAC probes showed that the centromeric regions are more similar between *L. angustifolius* and *L. luteus* than between *L. angustifolius* and *L. albus*. However, the centromeric sequence composition (short AGG and GATAC repeats) can be different in *L. angustifolius* and *L. luteus* since the organisation of a centromere-specific region may have diverged in the *Lupinus* genus (Wyrwa et al. 2016).

4.5 Variation in Chromatin Modifications Within Lupins

Studies of epigenetic variation at the level of chromosomes are of particular interest for the species that are recognised as polyploids (Wendel et al. 2016). The variation of epigenetic changes in lupins was divulged by characterisation of the methylation profiles of the genomic DNA of the crop and wild lupins, which indirectly illustrates the genomic changes that have resulted from evolution and/or adaptation. The chromosomal pattern of DNA methylation demonstrated a variable distribution in all the lupins studied, which suggests that there were various changes in DNA methylation and genome reorganisation that affected the 5mC signal in lupins (Susek et al. 2017). In addition, an examination of the chemical modifications of DNA and the histone chromatin composition revealed differences among the lupin genomes at chromosome level. The distribution of these modifications was cytogenetically compared between *L. angustifolius* and both of its crop relatives, i.e. *L. albus* and *L. luteus* as well as with the model legume *M. truncatula*. The euchromatin-specific marker H3K4me2 was identified in all of the metaphase chromosomes of the analysed species. Its distribution pattern was diverse, especially in the terminal parts of the chromosome arms. However, the distribution of H3K4me2 was restricted to the terminal regions of both chromosome arms in *L. angustifolius* and *M. truncatula*, while in *L. albus* and *L. luteus*, this histone modification was mostly observed within one chromosome arm. Furthermore, immunodetection of the heterochromatin specific marker H3K9me2 showed a similar pattern of distribution along the distal and proximal regions of both chromosome arms with weaker immunofluorescence signals around the centromeres in *L. angustifolius* and *M. truncatula* but only in the distal parts of the chromosome arms in *L. albus*. Surprisingly, no histone modification of H3K9me2 was observed in the *L. luteus* chromosomes. It could be hypothesised that this species established a unique complex chromatin modification system during evolution. (Susek et al. 2017).

4.6 Conclusion

The significant development of *Lupinus* spp. cytogenetics has mainly been due to the use of *L. angustifolius* chromosome-specific BAC markers (Fig. 4.2a). This enabled the identification of chromosome regions harbouring genes of interest and assigned linkage groups/pseudomolecules to their corresponding chromosomes. The integration of whole genome sequence and chromosome maps was an effective and comprehensive research strategy in determining the correctness of the *L. angustifolius* genome sequence. The cytogenomic map of *L. angustifolius* and chromosomal markers are also tools in studies of the evolutionary processes in lupins and other Genisteae. Lupin karyotypes underwent various and numerous chromosomal rearrangements during evolution. The variation in the number and structure of chromosomes might be a result of whole genome duplication/triplication events along with aneuploidy events, where a reduction in chromosome number seems to be the basic trend of chromosomal changes in the OWL. Further comparative

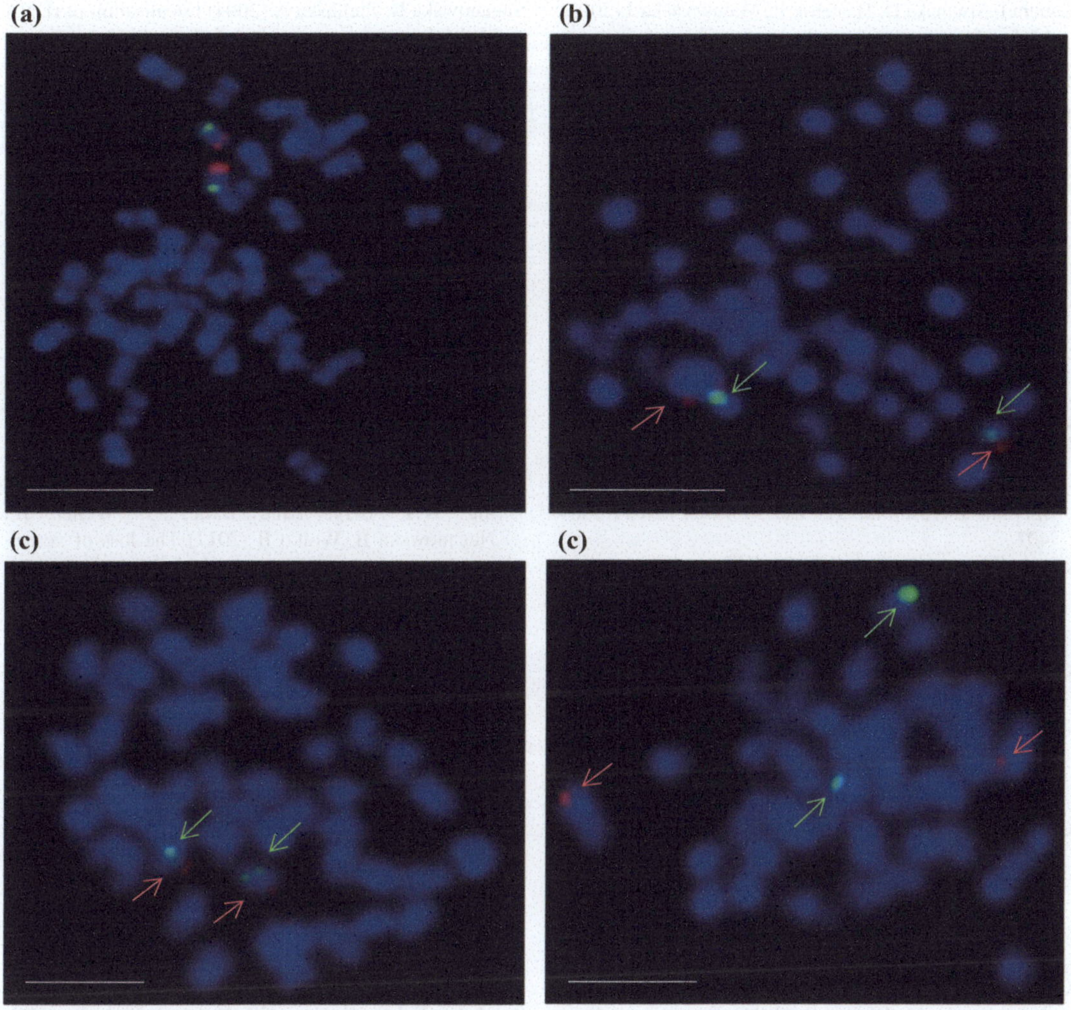

Fig. 4.2 Comparative FISH using BAC clones from the *L. angustifolius* chromosome Lang17 (**a**) that hybridises to the chromosomes of *L. albus* (**b**), *L. luteus* (**c**) and *L. digitatus* (**d**). BAC1 and BAC2 localisation in three crop and wild lupin chromosomes correspond to either the green or red dots, respectively, and are indicated by arrows. Chromosomes visualised in blue, scale bar: 5 μm

cytogenetic analyses may facilitate an understanding of the mechanisms that were involved in shaping legume chromosomes.

References

Atkins CA, Smith PMC, Gupta S, Jones MGK, Caligari PDS (1998) Genetics, cytology and biotechnology. In: Gladstones JS, Atkins CA, Hamblin J (eds) Lupins as crop plants: biology, production, and utilization. CAB International, pp 67–92

Drummond CS, Eastwood RJ, Miotto ST, Hughes CE (2012) Multiple continental radiations and correlates of diversification in *Lupinus* (Leguminosae): testing for key innovation with incomplete taxon sampling. Syst Biol 61:443–460

Dunn DB (1984) Cytotaxonomy and distribution of new world lupin species. In: Proceedings of the 3rd international lupin conference, 4–8 June1984, La Rochelle, France, pp 67–85

Gladstones JS (1998) Distribution, origin, taxonomy, history and importance. In: Gladstones JS, Atkins CA, Hamblin J (eds) Lupins as crop plants: biology, production, and utilization. CAB International, pp 1–36

Hajdera I, Siwinska D, Hasterok R, Maluszynska J (2003) Molecular cytogenetic analysis of genome structure in *Lupinus angustifolius* and *Lupinus cosentinii*. Theor Appl Genet 107:988–996

Hane JK, Ming Y, Kamphuis LG, Nelson MN, Garg G, Atkins CA, Bayer PE, Bravo A, Bringans S, Cannon S et al (2017) A comprehensive draft genome sequence for lupin (*Lupinus angustifolius*), an emerging health food: insights into plant–microbe interactions and legume evolution. Plant Biotechnol J 15:318–330. https://doi.org/10.1111/pbi.12615

Kaczmarek A, Naganowska B, Wolko B (2009) Karyotyping of the narrow-leafed lupin (Lupinus angustifolius L.) by using FISH, PRINS and computer measurements of chromosomes. J Appl Genet 50:77–82

Kasprzak A, Safar J, Janda J, Dolezel J, Wolko B, Naganowska B (2006) The bacterial artificial chromosome (BAC) library of the narrow-leafed lupin (*Lupinus angustifolius* L.). Cell Mol Biol Lett 11:396–407

Kroc M, Koczyk G, Swiecicki W, Kilian A, Nelson MN (2014) New evidence of ancestral polyploidy in the Genistoid legume *Lupinus angustifolius* L. (narrow-leafed lupin). Theor Appl Genet 127:1237–1249

Ksiazkiewicz M, Wyrwa K, Szczepaniak A, Rychel S, Majcherkiewicz K, Przysiecka L, Karlowski W, Wolko B, Naganowska B (2013) Comparative genomics of Lupinus angustifolius gene-rich regions: BAC library exploration, genetic mapping and cytogenetics. BMC Genomics 14:79

Ksiazkiewicz M, Zielezinski A, Wyrwa K, Szczepaniak A, Rychel S, Karlowski W, Wolko B, Naganowska B (2015) Remnants of the legume ancestral genome preserved in gene-rich regions: insights from Lupinus angustifolius physical, genetic, and comparative mapping. Plant Mol Biol Rep 33:84–101

Ksiazkiewicz M, Rychel S., Nelson MN, Wyrwa K, Naganowska B, Wolko B (2016) Expansion of the phosphatidylethanolamine binding protein family in legumes: a case study of Lupinus angustifolius L. FLOWERING LOCUS T homologs, LanFTc1 and LanFTc2. BMC Genomics 17:820

Lesniewska K, Ksiazkiewicz M, Nelson MN, Mahe F, Ainouche A, Wolko B, Naganowska B (2011) Assignment of 3 genetic linkage groups to 3 chromosomes of narrow-leafed lupin. J Hered 102:228–236

Naganowska B, Kaczmarek A (2005) Repetitive DNA sequences in cytogenetic studies of the Lupinus genome. In: Santen EV, Hill GD (eds) 11th international lupin conference, Guadalajara, Jalisco, Mexico. International Lupin Association, Canterbury, New Zealand, pp 27–29

Naganowska B, Zielinska A (2002) Physical mapping of 18S–25S rDNA and 5S rDNA in *Lupinus* via fluorescent in situ hybridization. Cell Mol Biol Lett 7:665–670

Naganowska B, Zielinska A (2004) Localisation of rDNA in the Lupinus genome during the cell cycle. J Appl Genet 45:189–193

Naganowska B, Wolko B, Sliwinska E, Kaczmarek Z (2003) Nuclear DNA content variation and species relationships in the genus *Lupinus* (Fabaceae). Ann Bot 92:349–355

Naganowska B, Wolko B, Śliwińska E, Kaczmarek Z, Schifino-Wittmann M (2006) 2C DNA variation and relationships among new world species of the genus *Lupinus* (Fabaceae). Plant Syst Evol 256:147–157

Nelson MN, Moolhuijzen PM, Boersma JG, Chudy M, Lesniewska K, Bellgard M, Oliver RP, Swiecicki W, Wolko B, Cowling WA, Ellwood SR (2010) Aligning a new reference genetic map of *Lupinus angustifolius* with the genome sequence of the model legume *Lotus japonicus*. DNA Res 17:73–83

Nelson MN, Ksiazkiewicz M, Rychel S, Besharat N, Taylor CM, Wyrwa K, Jost R, Erskine W, Cowling WA, Berger JD, Batley J, Weller JL, Naganowska B, Wolko B (2017) The loss of vernalization requirement in narrow-leafed lupin is associated with a deletion in the promoter and de-repressed expression of a Flowering Locus T (FT) homologue. New Phytol 213:220–232

Przysiecka L, Ksiazkiewicz M, Wolko B, Naganowska B (2015) Structure, expression profile and phylogenetic inference of chalcone isomerase-like genes from the narrow-leafed lupin (*Lupinus angustifolius* L.) genome. Front Plant Sci 6:268

Susek K, Bielski WK, Hasterok R, Naganowska B, Wolko B (2016) A first glimpse of wild lupin karyotype variation as revealed by comparative cytogenetic mapping. Front Plant Sci 7:1152

Susek K, Braszewska-Zalewska A, Bewick AJ, Hasterok R, Schmitz RJ, Naganowska B (2017) Epigenomic diversification within the genus *Lupinus*. PLoS ONE 12:e0179821

Susek K, Bielski W, Czyż KB, Hasterok R, Jackson SA, Wolko B, Naganowska B (2019) Impact of chromosomal rearrangements on the interpretation of lupin karyotype evolution. Genes 10:259. https://doi.org/10.3390/genes10040259

Wendel JF, Jackson SA, Meyers BC, Wing RA (2016) Evolution of plant genome architecture. Genome Biol 17:37. https://doi.org/10.1186/s13059-016-0908-1

Wolko B, Clements JC, Naganowska B, Nelson MN, Yang H (2011) *Lupinus*. In: Kole C (ed) Wild crop relatives: genomic and breeding resources. Springer, Berlin, pp 153–206

Wyrwa K, Ksiazkiewicz M, Szczepaniak A, Susek K, Podkowinski J, Naganowska B (2016) Integration of *Lupinus angustifolius* L. (narrow-leafed lupin) genome maps and comparative mapping within legumes. Chromosome Res

Transcriptome Resources Paving the Way for Lupin Crop Improvement

5

Lars G. Kamphuis, Rhonda C. Foley, Karen M. Frick, Gagan Garg and Karam B. Singh

Abstract

A range of transcriptomic resources have been generated for lupins from expressed sequenced tag (EST) libraries to the more recent next generation RNA sequencing libraries. This chapter will describe these resources and how they have been utilized to (a) generate gene-based molecular markers, (b) assist with the annotation of the reference genome for narrow-leafed lupin (*Lupinus angustifolius*), and (c) addressed specific research questions that assess global expression under different conditions and/or tissue types. For white lupins (*L. albus*) these include transcriptome studies using RNA sequencing libraries to investigate cluster root formation and the plants phosphate uptake status, for narrow-leafed lupins investigations into smallRNAs, seed storage protein and alkaloid content in the grain, and for yellow lupin (*L. luteus*) investigations into organ abscission. While transcriptomics in lupins is still in its infancies compared to larger pulse crops, lupin transcriptome resources will no doubt grow and lay strong foundations for lupin crop improvement.

5.1 Background

Lupins are a minor agricultural crop around the globe, with the majority of crop production occurring in Australia, where they are grown in rotation with wheat, barley, and canola. This provides the other crops a break from various diseases. The other rotational benefit of lupins is that they have the ability to fix atmospheric nitrogen through their beneficial interaction with rhizobia (Berger et al. 2012b). Four lupin species can be considered fully domesticated, which include pearl lupin (*Lupinus mutabilis*), yellow lupin (*L. luteus*), white lupin (*L. albus*), and narrow-leafed lupin (*L. angustifolius*) (see Chap. 8). Domestication of narrow-leafed lupins is relatively recent when compared with other broadacre crops, with breeding commencing in the 1900s and the first domesticated varieties

L. G. Kamphuis (✉) · R. C. Foley · G. Garg · K. B. Singh
CSIRO Agriculture and Food, Wembley, WA, Australia
e-mail: lars.kamphuis@csiro.au

R. C. Foley
e-mail: Rhonda.foley@csiro.au

G. Garg
e-mail: gagan.garg@csiro.au

K. B. Singh
e-mail: karam.singh@csiro.au

L. G. Kamphuis · K. B. Singh
Centre for Crop and Disease Management, Curtin University, Bentley, WA, Australia

K. M. Frick
Department of Plant and Environmental Science, University of Copenhagen, Copenhagen, Denmark
e-mail: kmf@plen.ku.dk

K. B. Singh et al. (eds.), *The Lupin Genome*, Compendium of Plant Genomes,
https://doi.org/10.1007/978-3-030-21270-4_5

being released in Australia in the 1960s. Narrow-leafed lupin is currently the most widely grown lupin species and current varieties have a very narrow genetic base where a strong focus has been on adaptation to short-season environments (Berger et al. 2012a). The majority of narrow-leafed lupin crop production is used for animal feed, however in recent years the crop has attracted attention for use in aquaculture and human consumption. This is because the narrow-leafed lupin grain is high in protein ($\sim$30%) and dietary fiber, low in starch and is gluten free (Foley et al. 2015). While the current genetic base of modern narrow-leafed lupin varieties is narrow, there is a wealth of genetic diversity present in lupin germplasm collections, thus offering tremendous potential to increase yield and resistance to both the biotic and abiotic stresses that this crop faces (Mousavi-Derazmahalleh et al. 2018).

The age of genomics has offered opportunities to decode crop genomes and investigate specific research questions in the crop directly, rather than through the use of model plant systems such as *Arabidopsis thaliana* and *Medicago truncatula*. DNA sequencing commenced in the 1970s with the Sanger chain termination method being the first approach to sequence DNA in a reliable and reproducible manner (Zimmermann et al. 1988). This led to the production of the so-called "first generation" sequencing instruments from 1987 onward (McGinn and Gut 2013). In 2005 the Genome Analyser instruments were released, significantly increasing the sequencing throughput as the short read massive parallel sequencing initiated the "next generation sequencing" (NGS) era in genomics (McGinn and Gut 2013). Since 2005, the data output of NGS instruments more than doubled each year providing vast datasets to broaden our understanding of structural and functional genomics of crop plants. NGS has rapidly led to extensive genome projects for important pulse crops such as soybean (Schmutz et al. 2010) and chickpea (Varshney et al. 2013) and while similar efforts have been made and are ongoing in lupins, the investment in these projects is reflected by the size of the crop when compared to other pulses. For narrow-leafed lupin survey genome sequences were generated in 2013 and 2015 (Kamphuis et al. 2015; Yang et al. 2013), followed by a more comprehensive draft genome assembly in 2017 (Hane et al. 2017) reviewed in Sect. 3.3 and a genome sequencing project for white lupin is currently underway (Benjamin Péret, pers comm; Sect. 3.6.4). Third generation single cell sequencing will likely offer opportunities to improve genome assemblies in pulse crops even further. NGS technologies have also proven to be effective for whole transcriptome sequencing also known as RNA sequencing or RNASeq. RNASeq has allowed a more precise and sensitive method to determine gene expression levels compared to older array-based techniques such as oligo- and microarrays. This book chapter will focus on and summarize the use of sequencing-based gene expression studies in lupin species and how these have been utilized to address research questions that aid lupin crop improvement.

5.2 The Use of Transcriptomics in Lupins

Research that involves looking at global gene expression in a species is often referred to as transcriptomics. Such studies in the early days of cDNA/gene sequencing involved Sanger sequencing of specific cDNA products, termed "expressed sequence tags", ESTs. Another approach, albeit not used in lupins, is oligo or microarrays, where picomolar amounts of specific gene sequences are placed on a slide as probes to which cDNA samples from a specific library are hybridized. The successful hybridization of expressed genes in libraries are visualized with fluorescent labels. More recently, sequencing-based methods (RNASeq) have become the norm to investigate expression patterns by sequencing RNA libraries using NGS technologies. RNA sequencing data can be used for a variety of applications, and in lupins they have been used to assist genome annotations, develop gene-based molecular markers as well as investigate specific research questions from a global gene expression level through the comparison of different accessions and or treatments. The latter is called differential gene expression analysis and such studies have been conducted in a few different lupin species (Table 5.1). These RNA sequencing datasets are commonly deposited in the NCBI GenBank as short read archives (SRAs) and studies to date and their associates SRAs are summarized in Table 5.1. The subsequent paragraphs will outline transcriptome studies in various lupin species.

Table 5.1 Overview of published next generation sequencing (NGS) derived transcriptome datasets in the various lupin species and their associated GenBank BioProject or Short Read Archive identifiers

Authors	Year	Species	Transcriptome dataset	BioProject ID	Short read archive
Parra-Gonzalez et al.	2012	*Lupinus luteus*	Young leaves, buds, flowers, and seeds	–	SRA055806
O'Rourke et al.	2013	*Lupinus albus*	Leaves, roots, and cluster roots	PRJNA144943	SRX087928-SRX087939
Wang et al.	2014	*Lupinus albus*	Root tips, juvenile cluster Roots, and mature cluster roots	–	–
Secco et al.	2014	*Lupinus albus*	Root tips, immature cluster Roots, and mature cluster roots	–	SRA145661
Cannon et al.	2015	*Lupinus angustifolius*	Leaves	PRJEB8056	ERX651085
Cannon et al.	2015	*Lupinus polyphyllus*	Leaves	PRJEB8056	ERX651026
Kamphuis et al.	2015	*Lupinus angustifolius*	Root, stem, leaf, flower, and seed	PRJNA248164	SRX547928-SRX547934
Foley et al.	2015	*Lupinus albus*	Seed	PRJNA271721	SRX832018-SRX832021
Foley et al.	2015	*Lupinus angustifolius*	Seed	PRJNA271721	SRX832022-SRX832027
Foley et al.	2015	*Lupinus cosentinii*	Seed	PRJNA271721	SRX832028, SRX832029
Foley et al.	2015	*Lupinus luteus*	Seed	PRJNA271721	SRX832030, SRX832031
Foley et al.	2015	*Lupinus mutabilis*	Seed	PRJNA271721	SRX832032, SRX832033
Fischer et al.	2015	*Lupinus angustifolius*	Infected leaf tissue with *Colletotrichum lupini*	–	–
Nevado et al.	2016	Several *Lupinus* spp.	Stem and leaves	–	SAMN04869551-SAMN04869616
Glazinska et al.	2017	*Lupinus luteus*	Flowers, pods, and flower pedicels	PRJNA285604	SRX1069734
Yang et al.	2017	*Lupinus angustifolius*	Small pods, large pods without seeds, seeds, flowers, pedicels, leaves, stems, and roots	PRJNA386115	SRX3415722
Książkiewicz et al.	2017	*Lupinus albus*	Young leaves, floral buds, and developing pods	PRJNA380248	SRX2663946, SRX2663947
Kroc et al.	2019a, b	*Lupinus angustifolius*	Leaves	PRJNA389154	SRR5723679-SRR5723682
Venuti et al.	2019	*Lupinus albus*	Roots under Fe and P deficiency	PRJNA445290	SRX3832719-SRX3832730

5.2.1 Development of Gene-Based Molecular Markers in Narrow-Leafed Lupin

Some of the early use of ESTs was to convert these expressed sequences into gene-based molecular markers. The rationale behind this approach is that molecular markers in genes could lead to the identification of perfect markers that control traits of interest. One approach is to obtain ESTs for two different lupin lines, identify polymorphisms between the two for expressed genes, design primer sequences that span the polymorphism, and genotype them by polymerase chain reaction (PCR), Sanger sequencing and/or restriction digests and gel electrophoresis. Such markers have been successfully employed for *L. albus* (Croxford et al. 2008), *L. luteus* (Parra-González et al. 2012), and *L. angustifolius* (Fischer et al. 2015; Nelson et al. 2006, 2010). An alternative approach is to identify length-based polymorphisms from ESTs which was used by Fischer et al. (2015) where ESTs were subjected to a simple sequence repeat finder and ESTs with such repeats converted into length-based polymorphic PCR markers. This approach yielded 17 expressed sequence tags-simple sequence repeat (EST-SSR) markers that were genotyped on an F_2 population segregating for anthracnose resistance (Fischer et al. 2015), where one such marker (LJM7_1) is tightly linked to resistance. The study by Nelson et al. (2010) generated 42 primer pairs which were designed on the basis of sequence information of *M. truncatula* and *Pisum sativum* EST sequences using an intron-targeted strategy to amplify legume-wide conserved single or low copy genes. While the efforts by Fischer and Nelson were based on ESTs from other legumes, Tian and associates (2009) generated a cDNA library for the white lupin cultivar "Lupro 2085" from roots obtained at 5, 10, 15, and 20 days after planting, where equal amounts of RNA for each time point were pooled to generate the cDNA library. A total of 8,000 clones were sequenced and after quality and vector sequence removal a total of 5,150 ESTs remained. These were assembled into 540 contigs and 1,915 singletons totaling 2,455 unigenes. Of these white lupin root ESTs 82 were converted to SSR markers for diversity assessments in white lupin (Tian et al. 2009).

5.2.2 Development of Gene-Based Molecular Markers in Yellow Lupin

The first published transcriptome data for yellow lupin was a set of cDNA libraries generated from young leaves, buds, flowers, and seeds of the accession "A25 Variation" (Parra-González et al. 2012). Two distinct tissue pools were generated for cDNA library construction with the first containing young leaves, flowers, and buds and the second seed from different developmental stages. Both cDNA libraries were sequenced on a Roche 454-sequencer, which yielded 205 and 530 Mb of sequence data for the respective libraries. The young leaves, flowers, and buds library produced 604,869 reads that were assembled into 26,975 contigs, whereas the seed library generated 1,345,892 reads assembled into 43,674 contigs. When the data from both libraries were pooled this resulted in the assembly of 71,655 contigs. Since the different libraries were likely to contain transcript and alternative splice variants the contigs were clustered into 55,309 isotigs, where 38,200 translated into proteins with 8,741 being full-length proteins. Subsequently, the authors determined which isotig sequences contained simple sequence repeats and identified 2,572 of these with di-, tri-repeats being the most frequent totaling 83.1% of all repeats identified in the EST sequences. To convert these sequences into molecular markers the list of ESTs was narrowed down further to 375 by selecting EST-SSRs with at least seven repeat units and sufficient flanking sequences for primer design. Of these 222 were polymorphic across a set of six genetically diverse *L. luteus* accessions, 130 were monomorphic and 23 failed to amplify. Of the 222 polymorphic markers a subset of 50 EST-SSRs were used to genotype a population of 64 yellow lupin accessions from Poland, Ukraine, Russia, Spain, Germany,

Morocco, Belarus, Portugal, Netherlands, Israel, Hungary, and Chile and assess their diversity. Phylogenetic analysis did not show any clear geographical patterns but at least six distinct clades were formed (Parra-González et al. 2012). The lack of clear geographical patterns is not surprising given the intermingled breeding history of material that was assessed and the fact that yellow lupin has a broad distribution across the Mediterranean region.

5.2.3 Using Molecular Markers to Generate Genetic Maps in Lupins

While the study by Parra-Gonzalez and colleagues focused on genetic diversity assessment, these EST-SSRs can also be utilized to generate genetic linkage maps and this has been achieved for various lupin species including narrow-leafed lupin (Nelson et al. 2006, 2010) and white lupin (Phan et al. 2007). Often times these EST-SSRs are also transferrable across lupin species and Parra-Gonzalez et al. (2012) observed that EST-SSRs from yellow lupin also amplified PCR fragments in two other lupin species in *L. hispanicus* and *L. mutabilis* with 33 and 14 of the 50 EST-SSR primers tested amplifying in the respective species. The genetic maps from white lupin (Phan et al. 2007) and narrow-leafed lupin (Nelson et al. 2006, 2010) have EST-derived markers that amplify in both species. The gene-based PCR primers used in these studies were generated to anneal to conserved exon sequences identified from *Lupinus* species and *M. truncatula* ESTs and span across intron regions predicted based on the *M. truncatula* genome sequence. A total of 2,492 *Lupinus* EST sequences were assessed and 280 were identified as having conserved exons between the species and spanning introns (Nelson et al. 2006). Of the 280 primer pairs, 230 amplified single amplicons in narrow-leafed lupin, while 28 had more than one amplicon. The single amplicons were sequenced on the parents of a recombinant inbred line (RIL) population (83A:476 × P27255), where 75 were successfully converted into single nucleotide polymorphism (SNP) markers and nine were length polymorphic markers. The remaining 2,192 *Lupinus* ESTs were assessed for the presence of simple sequence repeats and 28 were successfully converted of which 14 were genotyped on the RIL population (Nelson et al. 2006).

An updated genetic map published in 2010 added an additional set of 19 previously unpublished EST-derived Sequence-Tagged_Sites (STS) markers to the genetic map (Nelson et al. 2010) and an additional 54 gene-based PCR markers were added in 2014 (Kroc et al. 2014). In white lupin EST-derived markers have also been utilized to generate a genetic map for a RIL population derived from an Ethiopian landrace "P27174" and the "Kiev mutant" (Phan et al. 2007). A total of 626 intron spanning markers were tested including the 280 primers tested in the study by Nelson et al. (2006) on the parents of the white lupin RIL population. Of the 626 markers that were tested, 378 produced single amplicons of which 112 were determined as polymorphic between the parents and 105 were successfully placed on the genetic map (Phan et al. 2007). More recently, a total of 3,597 transcriptome derived markers were added to this genetic map, which allowed the genetic mapping of four key agronomic traits in resistance to anthracnose, resistance to phomopsis stem blight, seed alkaloid content, and vernalization requirement (Książkiewicz et al. 2017).

As the cost of NGS sequencing started to come down, a number of projects commenced generating RNA sequencing datasets to generate gene-based molecular markers. The first parallel sequencing-based approach was described earlier where 454 sequencing of cDNA libraries was employed by Parra-González et al. (2012) in yellow lupin. The first comprehensive NGS sequencing approach in narrow-leafed lupin generating Illumina RNA TruSeq libraries was employed by Kamphuis et al. (2015). The authors generated tissue-specific libraries for root, stem, leaf flower and different development seed stages for the reference cultivar "Tanjil", a closely related cultivar "Unicrop" and a wild accession "P27255". In addition to these libraries a lower density RNA TruSeq library from leaf tissue of the accession "83A:476", a sister line of cultivar "Wonga" was utilized in this study which was generated as part of the 1000 Plants initiative (https://www.onekp.com; Cannon et al. 2015). Together these four narrow-leafed lupin lines form the parents of two recombinant inbred line populations and the datasets generated were employed to identify gene-based polymorphisms for molecular marker design: A de novo transcriptome for reference cultivar "Tanjil" was assembled comprising 104,766 sequences with a total length of 51.6 Mb, where 63,271 discrete genes were predicted. Interestingly, the leaf and stem tissue showed highly similar expression patterns, whereas 52 tissue-specific transcripts were identified for the flower (28), seed (17) and roots (7). Reads of the RNA sequencing libraries from lines "Unicrop", "P27255", and "83A:476" were aligned to the "Tanjil" transcriptome assembly, which identified around two million SNPs and around 100,000 insertion–deletion (indel) polymorphisms across the four narrow-leafed lupin lines. Here the wild accession "P27255" had the most polymorphisms relative to "83A:476" which form the parents of a "wide cross" compared to the more closely related parents of the "narrow cross" cultivars "Tanjil" and "Unicrop". When aligning the identified polymorphic SNPs and indels for both RIL populations to the survey assembly the authors demonstrated that the in silico designed markers were evenly distributed across the genome (Kamphuis et al. 2015), which would help in their use in narrow-leafed lupin breeding efforts.

To generate high-quality gene-based molecular markers, very stringent primer design criteria were used which resulted in a total of 63,269 standard PCR-based in silico primers, of which 57,615 were SNPs and 5,654 were length polymorphisms (Kamphuis et al. 2015). This in silico list was reduced by identifying those markers that are polymorphic for both RIL populations. While 838 indel sites were polymorphic between both parent pairs (with 239 having a ≥ 2 bp indel), 11,149 SNPs were polymorphic. Since the breeding program (at the Department of Agriculture and Food of Western Australia) at the time was utilizing the Fluidigm SNP genotyping platform, efforts were made to convert the SNPs into Fluidigm assays, which resulted in 13,517 bi-allelic assays in silico. To validate these gene-based molecular markers a set of 96 Fluidigm SNP markers and a set of 239 indel markers were genotyped using the Multiplex Ready PCR approach (Hayden et al. 2008), optimized and described in Gao et al. (2011) for narrow-leafed lupin SSR markers. This resulted in 88 (92%) Fluidigm SNP markers and 170 (81%) indel markers being successfully genotyped on 152 individuals of the 83A:476 × P27255 reference genetic map (Table 5.2). An additional 381 RNA sequencing derived SNP markers were placed on the genetic map by Hane et al. (2017). The addition of the gene-based markers to the genetic map resolved the placement of two clusters that were not assigned to a linkage group. Furthermore, the gene-based markers were linked to traits of interest such as anthracnose resistance and alkaloid content (Kamphuis et al. 2015).

Fischer and colleagues generated in-depth RNA sequencing datasets of two different narrow-leafed lupin lines, which differ in their resistance to anthracnose, caused by the hemi-biotrophic pathogen *Colletotrichum lupini* (Fischer et al. 2015). These included the susceptible *L. angustifolius* cultivar "Arabella" and the resistant breeding line "Bo7212", where conidial spore suspensions of *C. lupini* were sprayed on the plants and leaf material was collected at 4, 8, 24, and 48 h post inoculation. These samples were pooled for RNA extraction and subsequent sequencing. Subsequently, SNP polymorphisms between transcripts of the two narrow-leafed lupin lines were identified and converted into molecular markers using high resolution melt (HRM) analysis for linkage association with anthracnose resistance. A total of 92 RNASeq derived HRM markers were successfully genotyped on a segregating F2 population derived from lines "Arabella" and "Bo7212". The authors identified that resistance to anthracnose in "Bo7212" was located on

Table 5.2 Overview of gene-based molecular markers across different lupin species derived from either expressed sequenced tags (ESTs) or RNA sequencing data

Author	Year	Species	Marker derived from	Number of new markers generated	Number of new markers placed on a genetic map
Nelson et al.	2006	*L. angustifolius*	ESTs	112	98
Phan et al.	2007	*L. albus*	ESTs	112	105
Tian et al.	2009	*L. albus*	ESTs	82	–
Nelson et al.	2010	*L. angustifolius*	ESTs	42	19
Parra-Gonzalez et al.	2012	*L. luteus*	ESTs	222	–
Parra-Gonzalez et al.	2012	*L. hispanicus*	ESTs	33	–
Parra-Gonzalez et al.	2012	*L. mutabilis*	ESTs	14	–
Kroc et al.	2014	*L. angustifolius*	EST	54	54
Fischer et al.	2015	*L. angustifolius*	ESTs	17	17
Fischer et al.	2015	*L. angustifolius*	RNASeq data	2	92
Kamphuis et al.	2015	*L. angustifolius*	RNASeq data	335	258
Hane et al.	2017	*L. angustifolius*	RNASeq data	384	381
Książkiewicz et al.	2017	*L. albus*	RNASeq data	4,830	3,597

linkage group NLL-11 and subsequently utilized three RNASeq indel markers from the Kamphuis et al. (2015) study for further fine-mapping and investigation to determine if the source of anthracnose resistance present in cultivar "Tanjil" conferred by *Lanr1* is the same locus as resistance conferred to anthracnose identified in the "Bo7212" genetic background. The authors determined that the resistance in Bo7212 is caused by a distinct locus from *Lanr1* and hence named the anthracnose resistance locus identified in the "Bo7212" background *LanrBo* (Fischer et al. 2015).

In summary, a number of EST and RNASeq derived molecular markers exist for various lupin species for genetic diversity assessment and generation of genetic maps for linkage association. For narrow-leafed lupin, a total of 427 RNASeq derived molecular markers exist and 171 EST-derived markers of which 258 and 117 have been placed on the reference genetic map, respectively (Kamphuis et al. 2015; Table 5.2). For white lupin a total of 194 EST-derived molecular markers exist of which 98 have been placed on a genetic map (Phan et al. 2007; Table 5.2), whereas 4,830 transcriptome-derived markers were generated or which 3,597 were placed on the genetic map (Książkiewicz et al. 2017). Some of the markers amplify successfully across lupin species as shown for markers across *L. luteus, L. hispanicus* and *L. mutabilis* (Parra-González et al. 2012) as well as markers for *L. angustifolius* and *L. albus* (Nelson et al. 2006; Phan et al. 2007). With NGS sequencing becoming more affordable RNASeq-derived molecular markers will complement other NGS sequencing strategies to generate dense genetic maps paving the way for markers associated with traits and subsequent gene identification to generate perfect molecular markers for breeding programs.

5.2.4 Use of Transcriptome Data for Annotation of the Narrow-Leafed Lupin Genome

RNA sequencing has emerged as a powerful technique for genome-wide detailed analysis of the transcriptionally active genes and as such has been utilized in narrow-leafed lupin for gene annotation of the most recent genome assembly (Hane et al. 2017). The first assembly released in

2013 had no RNA sequencing data to assist the annotation process and relied on gene identification through homology to the Arabidopsis and soybean (*Glycine max*) annotated gene sets (Yang et al. 2013). The latter resulted in the prediction of 57,807 genes with an average transcript length of 2,038 bp and coding sequence of 1,033 bp. This is a significantly higher number of predicted genes compared to other annotated legume genomes (Schmutz et al. 2010, 2014; Varshney et al. 2012, 2013; Young et al. 2011) and the current narrow-leafed lupin genome assembly that has a total of 33,076 protein coding genes. The current assembly utilized EST and RNAseq datasets, proteome data as well as in silico prediction using homology to proteins of Arabidopsis, chickpea, common bean, medicago, pigeonpea, and soybean (Hane et al. 2017). The RNA sequencing datasets provided support through the alignment of this expression data to the gene annotation where the transcriptomes of the different tissue types including root, stem, leaf, flower, and seed from cultivars "Tanjil" and "Unicrop" and wild accession "P27255" were used. The 33,076 genes have an average transcript length of 3,673 bp and an average coding sequence of 1,289 bp. Furthermore 80.4% (26,580 genes) have a functional annotation assigned and genes with tissue-specific expression were identified (Hane et al. 2017). The combination of using gene prediction tools and EST, RNASeq and proteome support thus has significantly improved the annotation of the narrow-leafed lupin genome, with efforts to improve the assembly and annotation further currently underway (Kamphuis, Foley and Singh, unpublished data).

5.2.5 Identification of Seed Storage Proteins in Lupins

One of the key attributes of the lupin grain is its high protein content, an important source of nutrition for animal feed, aquaculture, and more recently human consumption. Therefore, efforts have been undertaken to identify the key proteins in the lupin grain. The major seed proteins in pulses are storage proteins which accumulate during seed development which are rapidly utilized during seed germination as a source of nitrogen and carbon during early seedling growth (Duranti et al. 2008).

Foley and colleagues generated a cDNA library derived from seeds of the *L. angustifolius* cultivar "Tanjil", 20–26 days post anthesis, which resulted in a library of approximately one million clones with an average insert size of 1800 bp (Foley et al. 2011). A total of 2,395 of these clones yielded high-quality Sanger sequence data and these EST sequences were compared to seed storage protein sequences in the NCBI database. This resulted in the identification of three alpha, seven beta, two gamma, and four delta conglutin seed storage proteins, which included 11 novel conglutins. These 16 conglutins showed strong homology to known seed storage proteins from other legumes and validation by quantitative PCR showed that these conglutins were all expressed during seed development (Foley et al. 2011).

A more in-depth transcriptome study was initiated thereafter where RNA sequencing libraries were developed for maturing seed between 20 and 26 days post anthesis for five different lupin species to gain insight into the complexity of the conglutin family in lupins, with regard to their phylogeny, gene structure, expression during maturation, and protein abundance in fully mature lupin grain (Foley et al. 2015). The maturing seed transcriptomes for three narrow-leafed lupin accessions, "Tanjil", "Unicrop", and "P27255" were aligned to the survey genome and de novo transcriptome (Kamphuis et al. 2015), which confirmed the presence of the previously identified 16 conglutins in narrow-leafed lupin and their presence in all three accessions, albeit with some nucleotide sequence differences. No additional conglutin genes were identified in the narrow-leafed lupin transcriptome datasets. A consensus narrow-leafed lupin sequence for each of the 16 conglutins was generated and the reads of the other lupin species (two white lupin accessions, one yellow and pearl lupin accession, and one wild lupin species *L. cosentinii*) aligned to obtain sequence information and diversity of the conglutins in the other species. In all species homologous gene sequences were identified, which allowed subsequent phylogenetic comparisons. These showed that for the alpha conglutins there are three subgroups that have duplicated and diverged prior to speciation, which is also the case in other legumes such as medicago, pea, peanut, and soybean (Foley et al. 2011). The same was observed for gamma conglutins with the "Old World" lupins where two subgroups exist, whereas "New World" lupin *L. mutabilis* gamma are more closely related to one another. The phylogenetic relationship among both the beta and delta conglutins was not as clear cut when compared to the alpha and gamma conglutins. While some beta and delta conglutins have diverged prior to lupin speciation others appear to have diverged after speciation (Foley et al. 2015).

Subsequently, the expression levels of the conglutins across the different species were investigated and it was found that within a species similar RNA expression levels were observed for each of the conglutins across the accessions, whereas distinct expression levels were observed between lupin species. For example, over 60% of conglutin transcripts in *L. luteus* and *L. cosentinii* were delta conglutins, whereas gamma conglutin expression was highest in *L. albus* and *L. cosentinii.* In contrast beta conglutin expression was low in *L. cosentinii* compared to all other species. On an individual conglutin expression profile, differences were observed between species. These differences raised the question whether this translates to differences in protein abundance in the mature lupin grain, as protein abundance can be affected by a number of factors such as transcription, translation and protein turnover. Therefore, the protein concentration of mature seed was determined and total protein content across the different species ranged from 30 to 44%. Strong correlation between gene expression levels of maturing seed and protein content in mature grains were observed for the species *L. albus, L. angustifolius* and *L. luteus*, while this was not the case for *L. cosentinii* and *L. mutabilis. L. cosentinii* had the lowest percentage of protein content and the highest level of conglutin transcript abundance in maturing seed, while *L. mutabilis* was the opposite. The generation of antibodies to the different conglutins demonstrated that conglutin proteins are present only in the seed and that they accumulate during seed development with low levels detected between 4 and 20 days after anthesis, thereafter increasing in amount till the mature seed stage (Foley et al. 2015).

Further work is required to investigate the protein composition of the mature lupin grain and to determine whether translation regulation, post-translation modification, or other factors determine the final composition of a mature lupin grain. With a narrow-leafed lupin Targeting Induced Local Lesions in Genomes (TILLING) population under development, it would also be worth determining what effect knocking out individual conglutin genes will have on the overall protein content in a mature lupin grain. Future work could thus allow one to alter the protein composition in the lupin grain, removing those conglutins that cause allergenicity (such as the beta conglutins) and increase those that are optimal and beneficial for both animal and human nutrition (such as the gamma conglutins).

5.2.6 Cluster Root Formation and Phosphorus Acquisition in White Lupin

One of the special features of white lupin is the formation of cluster roots to increase the root surface area and more efficiently take up phosphorus. Phosphorus (or orthophosphate, the form which is taken up by plants) is often considered one of the most limiting nutrients restricting plant growth and development. Therefore cluster root biology in white lupin has become a model system to investigate orthophosphate uptake (Cheng et al. 2011). To date, four transcriptome sequencing studies have been conducted to investigate the regulatory network of cluster root development and the uptake of orthophosphate

(O'Rourke et al. 2013; Secco et al. 2014; Venuti et al. 2019; Wang et al. 2014).

The study by O'Rourke and associates (2013) investigated the genome-wide expression profiles in cluster roots, normal roots, and leaves under orthophosphate rich and in orthophosphate deficient scenarios. Since no genome sequence for white lupin was available the authors generated a de novo transcriptome assembly using the triplicate libraries of leaf and root samples under orthophosphate sufficient and deficient conditions, totaling ~368 million reads. The samples were collected 16 days after emergence, meaning that the phosphate deficient samples would have formed and developed functional cluster roots. Combined with 8,441 publicly available ESTs, the assembly generated 125,821 unique transcripts spanning 145 Mbp with an average length of 1,155 bp. This resource was the first gene index for white lupin increasing the available gene sequence for the species ~14-fold with 63% of transcripts assigned a putative functional annotation. Subsequently, the authors conducted a differential expression analysis where 2,128 transcripts were differentially expressed in response to orthophosphate deficiency, where 1,342 and 904 were differentially expressed in leaves and roots (e.g., normal roots vs cluster roots), respectively. In the leaves 987 transcripts were upregulated in orthophosphate sufficient conditions versus 355 in orthophosphate deficient conditions. Of the 904 transcripts differentially expressed between normal roots and cluster roots, 396 were upregulated in the normal roots in orthophosphate sufficient conditions, whereas 535 transcripts were upregulated in orthophosphate deficient cluster roots. Among the differentially regulated transcripts in deficient conditions were transcription factors (110) and high-affinity phosphate transporters (155), which exhibited distinct patterns between leaves and roots. A total of 23 transcription factor families were differentially regulated, with the bHLH (14) and AP2_EREB (17) transcription factor families the largest two responding to the orthophosphate deficiency in the roots, whereas the MYB transcription factor family with 33 transcripts was the largest in the leaves. Furthermore, transcripts associated with reactive oxygen species produced under orthophosphate deficient conditions were identified, including glutathione S-transferase, glutathione peroxidase, glyoxalase, ferritins, and NADPH-oxidase transcripts. Comparisons of the white lupin transcriptome to orthophosphate deficient libraries from Arabidopsis and potato identified 12 conserved transcripts that are differentially expressed across all three species, making them perfect biomarkers to monitor the phosphorus status of lupin plants.

While the initial study by O'Rourke et al. (2013) looked at differential responses in white lupins 16 days after emergence in cultivar "Ultra", Wang and colleagues (2014) generated samples from tissues of the cultivar "Feodora" 20 days after planting under orthophosphate deficient conditions. They harvested tissue from different developmental segments of the roots, which included pre-emergent root segments of the first-order laterals without root tips, juvenile cluster roots that had not reached their final length and mature cluster roots located basal to juvenile cluster roots. The three libraries yielded approximately 147 million reads and were aligned to the white lupin transcript database generated and described by O'Rourke et al. (2013). This resulted in 103,147 unique transcripts, which were subjected to differential expression analysis in the three different root segments. While no details were provided in regards to the number of differentially expressed transcripts between the different root samples, the authors showed that ethylene plays a key role in cluster root maturation, whereas transcripts related to abscisic and jasmonic acid were upregulated in mature cluster roots, while auxin and cytokinin receptors were most strongly expressed during cluster root initiation (Wang et al. 2014). This is in accordance with a previous study that found that transcripts involved in ethylene biosynthesis and auxin metabolism and sensing were upregulated in orthophosphate deficient cluster roots, and that cytokinin degradation is essential for cluster root development (O'Rourke et al. 2013). The importance of cytokinin degradation was demonstrated in two ways, firstly through exogenous application of synthetic cytokinin which inhibited cluster root formation and secondly through RNA interference of a cytokinin oxidase which was highly upregulated in orthophosphate deficient cluster roots, which altered cluster root formation patterns (O'Rourke et al. 2013). The importance of auxin has been demonstrated through the identification of highly expressed transcripts involved in auxin homeostasis in the RNA sequencing study by O'Rourke et al. (2013) and Secco et al. (2014). When auxin is applied exogenously to white lupins it promotes cluster root formation, whereas addition of auxin-influx or -efflux transport inhibitors to growth media suppresses cluster root formation (Meng et al. 2013).

In terms of metabolic changes during cluster root development upregulation of transcripts involved in orthophosphate recycling, such as transporters and enzymes involved in the synthesis of sulfolipids and down-regulation of orthophosphate dependent glycolytic pathways for carbohydrate catabolism were observed (Wang et al. 2014). Metabolic pathways involved in biosynthesis of organic acids were upregulated resulting in root exudates for phosphorus mobilization, while down-regulation of enzymes in the TCA cycle enzymes, suggesting citrate accumulation and exudation is important, the accumulation of citrate in mature clusters was confirmed by formazan staining. Another organic compound that is known to accumulate to facilitate phosphorous mobilization is malate. The transcript encoding malate dehydrogenase was highly upregulated in juvenile clusters and declined during cluster root maturation, whereas a phosphoenolpyruvate carboxylase was highly expressed in mature clusters supporting the exudative burst of malate to mobilize phosphorus (Wang et al. 2014).

A third white lupin RNA sequencing study was conducted by Secco and associates (2014) who generated datasets of root tips and mature roots of orthophosphate sufficient plants and root tips, immature cluster roots and mature cluster roots of orthophosphate deficient plants using the cultivar "Kiev mutant". This yielded ~133 million paired-end reads which were de novo assembled (46,383 transcripts with an average length of 896 bp) and subsequently merged with the white lupin gene index to generate an improved gene index for white lupin with a total of 65,097 transcripts. The improved white lupin gene index totals a length of ~105.8 Mbp and the transcripts have an average length of 1,625 bp and is named LAGI02 (Secco et al. 2014). The five different libraries generated were subsequently aligned to the new gene index and differential expression analysis conducted, where 28,915 transcripts were expressed in all five samples and a total of 835 were differentially expressed by eightfold or more in at least one pair-wise comparison, which were the focus of further analysis. The study showed that phosphate deficiency induced changes in gene expression were very similar when comparing root tip and mature roots of sufficient and deficient orthophosphate treatments, respectively. In contrast, tissue-specific differences related to different developmental stages of cluster root formation were observed. Like the previous two studies these included transcripts associated with the TCA cycle and glycolysis leading to the production of organic acids. Interestingly, their data suggest that the root tip is not producing and exporting malate and citrate. The study furthermore confirmed the role of auxin and certain transcription factors in the initiation and formation of cluster roots (Secco et al. 2014).

The fourth study by Venuti and colleagues (2019) identified features in common between iron and phosphorus acquisition mechanisms in white lupin. Here, white lupin plants were grown under phosphorous or iron-deficient conditions for 32 days and samples harvested for RNA sequencing from the apex of control and iron-deficient roots, as well as phosphorus-deficient and iron-deficient cluster roots. This identified 1,623–5,416 differentially expressed transcripts for the different treatments relative to the control apex root sample. Most enriched categories in the iron and phosphorous deficient samples were related to metabolic reprogramming such as glycolysis, the TCA cycle, amino acid synthesis, secondary metabolism, and cell wall synthesis. Further dissection of the differentially expressed genes showed that genes involved in the iron acquisition were upregulated in both iron and phosphorous deficient roots (Venuti et al. 2019).

In summary, the four white lupin RNA sequencing studies are all centric to cluster root research and the generated expression atlas resulting from these studies form the foundation to investigate the intricate network of signaling components involved in cluster root initiation, development, and maturation. It will allow the development of plants with improved phosphorus uptake, where white lupin cluster roots will continue to play an essential role as a model system to investigate this further. These will be aided by the development of additional white lupin genomic resources such as the generation of a reference genome assembly, which is currently underway (Benjamin Péret, pers. comm.; Sect. 3.6.4).

5.2.7 Alkaloid Biosynthesis in Narrow-Leafed Lupin

Quinolizidine alkaloids are secondary metabolites produced in all lupin species, which are detrimental to animal and human health but provide the plant protection from herbivore predation. In the 1930s the first low-alkaloid lupin varieties were produced and these "sweet" varieties led to the adoption of lupin in both Europe and Australia predominantly for animal feed (Frick et al. 2017). Modern-day "sweet" varieties generally have a low alkaloid content, however they are not alkaloid-free. As such an industry threshold of 0.02% alkaloid content in the lupin grain has been put in place in Australia and some European countries, and the grain must remain below this threshold in order for it to be used for higher-value food and feed purposes (Cowling and Tarr 2004). While in most growing seasons narrow-leafed lupin crops remain below this threshold, at times "sweet" varieties exceed the threshold. Which environmental factors cause this spike in alkaloids has been focus of recent research, where transcriptome investigations have been utilized (Frick et al. 2017).

Transcriptome studies in narrow-leafed lupin have aided the generation of gene-based molecular markers (Kamphuis et al. 2015) and assisted the annotation of the current genome assembly (Hane et al. 2017). In addition, the transcriptome resources described by Kamphuis et al. (2015) have been utilized to investigate the genetic control of alkaloid biosynthesis in narrow-leafed lupin (Frick et al. 2018; Yang et al. 2017) by comparing bitter to sweet transcriptome datasets. The study by Yang et al. (2017) generated cDNA libraries with RNA isolated from large pods, large seeds, flowers, pedicels, leaves and roots of the bitter variety "Oskar" to generate both Illumina short reads and PacBio long reads (Yang et al. 2017). A de novo assembly of the PacBio reads resulted in 10,661 transcripts after which the short reads were aligned to these transcripts and the unaligned reads de novo assembled into 181,904 contigs. The combined bitter transcriptome for cultivar "Oskar" thus has 192,565 transcript sequences with an average transcript length of 1,564 bp, totaling 175.9 Mb of sequence (Yang et al. 2017). The bitter transcriptome of "Oskar" was subsequently compared to the sweet transcriptome of cultivar Tanjil using the raw data from Kamphuis et al. (2015). Since genes involved in secondary metabolism are often co-expressed, co-expression analysis was utilized to compare the transcript profiles of known alkaloid biosynthesis genes such as *LaLDC*, a lysine decarboxylase gene, to other transcripts which resulted in the identification of 33 genes with similar expression profile, including one gene with homology to a copper amine oxidase. This copper amine oxidase was named *LaCAO* and qRT-PCR data confirmed co-expression of this gene with *LaLDC*. Subsequently, the sequence of *LaCAO* was aligned to other copper amine oxidases involved in secondary metabolism and this revealed strong conservation of the motifs and residues which interact with copper ions suggesting it is an active enzyme. Localization studies through transient expression of a fused green fluorescent protein to the LaCAO protein and the use of a peroxisomal fluorescent marker showed that LaCAO targets the peroxisome. Lysine decarboxylase (LDC) is the first enzyme in the alkaloid biosynthetic pathway (Bunsupa et al. 2012), which converts lysine into cadaverine. This thus raised the question whether the identified copper amine oxidase (LaCAO) could oxidize cadaverine into 1-piperideine efficiently and this was indeed demonstrated through GC-MS studies with purified LaCAO protein (Yang et al. 2017). The second enzyme in the narrow-leafed lupin alkaloid biosynthesis pathway has thus been identified and characterized. Of the other genes that were co-expressed with *LaLDC* and *LaCAO* were potential biosynthetic, regulatory, and transporter genes such as a geraniol 8-hydroxylase-like gene, a myb and bHLH transcription factor and several lipid transporters. An independent study by Frick et al. (2018) also identified *LaCAO* as co-expressed with *LaLDC* in the bitter accession "P27255". These authors utilized transcriptome datasets of bitter and sweet varieties to identify additional candidate genes involved in alkaloid biosynthesis (Table 5.3). This resulted in the identification of a berberine bridge-like enzyme (*LaBBE-like*) two major latex-like proteins (*LaMLP1* and *LaMLP2*), which were co-expressed with *LaLDC* (Frick et al. 2018). Two additional MLP-like genes located adjacent to *LaMLP2-like* in the narrow-leafed lupin genome also displayed similar expression patterns. The expression of these transcripts and two other known alkaloid biosynthesis genes (*LaHMT/HLT* and *LaAT*) (Bunsupa et al. 2011) were validated in sweet and bitter lines by qPCR. *LaLDC, LaCAO, LaAT,* and three of the *LaMLP-like* genes showed similar expression patterns in the bitter line compared to the sweet cultivars, with strongest expression observed in the bitter green tissue (leaf and stem) (Frick et al. 2018). *LaHMT/HLT*, *LaBBE-like,* and *LaMLP3-like* displayed separate expression patterns to the other alkaloid biosynthesis genes and for *LaBBE-like* and *LaMLP3*-like expression in leaf, stem, flower, and developing seed was observed. This strong expression of the identified transcripts of *LaLDC, LaCAO, LaAT,* and three of the *LaMLP-like* genes (Table 5.3) in the leaf and stem tissue of the bitter accession compared to sweet cultivars correlated with the total alkaloid content in the grain at harvest (Frick et al. 2018). The unchanged expression for the majority of these genes in the seed during seed development between bitter and sweet accessions furthermore suggests that in narrow-leafed lupin the majority of the alkaloids are transported to the grain rather than produced within the seed.

The expression patterns of the functionally characterized genes *LaLDC, LaCAO,* and *LaHMT/HLT* were subsequently investigated following applications of both abiotic stresses such as drought and temperature (Frick et al. 2018) as well as biotic stress from aphid predation (Frick et al. 2019). Both drought and increased temperature as well as their combination had a cultivar specific effect on alkaloid production when three different cultivars, "Danja", "Tanjil", and "Tallerack" were investigated. For cultivar "Danja" grain alkaloid content

Table 5.3 Overview of characterized and candidate genes involved in alkaloid biosynthesis in narrow-leafed lupin (*L. angustifolius*), their gene identifier, scaffold and chromosome location in the Tanjil genome (Hane et al. 2017)

Gene/Locus name	Lupin gene ID	Scaffold ID	Linkage group
LaLDC or *LaL/ODC*	*Lup009726.1*	Scaffold_190	NLL-15
LaCAO	*Lup000530.1*	Scaffold_10_8	NLL-15
LaHMT/HLT	*Lup002250.1/Lup002251.1*	Scaffold_423	NLL-04
LaAT	*Lup021586.1*	Scaffold_40_1	NLL-16
LaBBE-like	*Lup003016.1*	Scaffold_12_32	NLL-07
LaMLP1-like	*Lup019334.1*	Scaffold_336	NLL-10
LaMLP2-like	*Lup015922.1*	Scaffold_29_1	NLL-06
LaMLP3-like	*Lup015921.1*	Scaffold_29_1	NLL-06
LaMLP4-like	*Lup015923.1*	Scaffold_29_1	NLL-06

increased under increased temperature, whereas increased temperature affected grain alkaloid content for cultivar “Tanjil”. For the third cultivar “Tallerack” neither temperature, nor drought had an effect on grain alkaloid content. While *LaLDC* expression has been demonstrated as a good indicator for alkaloid content between bitter and sweet lines (Frick et al. 2018; Yang et al. 2017), its expression did not correlate with alkaloid grain content under the abiotic stress treatments in the three different sweet cultivars (Frick et al. 2018). In fact, under the different stresses the expression of *LaLDC* and other alkaloid biosynthetic genes appears to decrease relative to control plants. Therefore the mechanisms that increase alkaloid content in cultivars under drought/temperature stress warrants further investigation.

Since alkaloids protect lupin plants from insect herbivory, the expression and alkaloid accumulation from aphid predation were also investigated (Frick et al. 2019). The jasmonate signaling pathway regulates the production of secondary metabolites such as alkaloids and plays a role in defense against aphid predation in legumes (Edwards et al. 2003; Kamphuis et al. 2013; Wasternack and Hause 2013; Wink 1992). Jasmonate responsive genes in narrow-leafed lupin were identified and these included two lipoxygenase genes (*LaLox4-like, LaLox5-like*) and a proteinase inhibitor (*LaPI-like*) gene and these genes as well as alkaloid biosynthetic genes *LaLDC, LaCAO,* and *LaAT* were all upregulated in bitter accession “P27255” following exogenous methyljasmonate treatment. The jasmonate inducible genes were also upregulated following exogenous methyljasmonate application in the sweet cultivar “Tanjil”, but in contrast to the bitter accession “P27255”, the alkaloid biosynthetic genes *LaLDC, LaCAO,* and *LaAT* were not (Frick et al. 2019). Wounding experiments also demonstrated that the jasmonic acid responsive genes were induced in both sweet and bitter lines, but the alkaloid biosynthetic genes were not. In a final experiment using two narrow-leafed lupin cultivars with varying levels of aphid resistance, the response of the jasmonate and alkaloid biosynthesis genes was investigated following green peach aphid (*Myzus persicae*) predation. The two cultivars “Kayla” and “Tallerack” did not exhibit changes in gene expression of the alkaloid biosynthesis genes, nor was a change in grain alkaloid content observed following aphid predation in either cultivar (Frick et al. 2019).

Alkaloid levels in the narrow-leafed lupin grain are controlled by the *iucundus* locus, which has been mapped to a 746 Kb region on pseudochromosome NLL-07 (Hane et al. 2017). To identify candidate genes for *iucundus*, a transcriptomic analysis was conducted by Kroc and colleagues (2019a, b) comparing differentially expressed genes in leaves of high- and low-alkaloid narrow-leafed lupin accessions. The authors identified an APETALA2/ethylene response transcription factor (*RAP2-7*) and a 4-hydroxy-tetrahydro-diplicolinate synthase (*DHDPS*) in the region of interest from 1,489 differentially expressed transcripts in bitter versus sweet accessions. The expression profile of both *RAP2-7* and *DHDPS* were similar to that of known alkaloid biosynthesis genes *LaLDC, LaCAO,* and *LaAT* (Kroc et al. 2019b). When comparing the coding sequence of both candidate proteins, RAP2-7 had two amino acid substitutions between bitter and sweet accession (M48V and S196R), whereas a frameshift leading to a deletion of a guanine distinguished the sweet from bitter accession for the DHDPS protein sequence. Subsequently, Kroc and colleagues investigated the alkaloid content of 199 narrow-leafed lupin accessions and showed that an alkaloid content of greater than 0.9% total alkaloid content of the seed dry weight correlated with the bitter RAP2-7 sequence, whereas a total alkaloid content less than 0.5% correlated with the sweet RAP2-7 sequence thus making this the likely candidate gene controlling alkaloid content in narrow-leafed lupin (Kroc et al. 2019a).

In summary, thanks in part to transcriptomic studies, a picture is starting to emerge of the genes involved in alkaloid biosynthesis and regulation as well as candidate alkaloid transport genes (Frick et al. 2017, 2018; Kroc et al. 2019b; Yang et al. 2017). Investigations using gene expression studies to determine which environmental factors influence alkaloid content in sweet cultivars are also starting to emerge. Wounding and aphid predation in the sweet cultivars assessed to date does not appear to induce genes involved in alkaloid biosynthesis, nor do they increase the overall alkaloid content in the grain. In contrast abiotic stresses such as increased temperature and drought appear to influence alkaloid grain content in a cultivar dependent manner, although expression of alkaloid biosynthesis genes did not correlate with the increased alkaloid content following these abiotic stress treatments. With a changing climate, abiotic stress caused by increased temperature and drought in the lupin growing areas is likely to increase and therefore investigation into the regulation of alkaloid biosynthesis and transport throughout the plant under these stresses warrants further investigation.

5.2.8 Organ Abscission in Yellow Lupin

To date, two parallel sequencing approaches have been used to generate transcriptome data for yellow lupin (Glazinska et al. 2017; Parra-González et al. 2012). The first study used 454-sequencing to generate transcriptome data to generate gene-based molecular markers as described in Sect. 5.2.1 (Parra-González et al. 2012), whereas the second study focuses on organ abscission (Glazinska et al. 2017). Lupins produce inflorescences with numerous flowers, which are initiated when the shoot apex becomes a conical structure rather than the rounded dome seen in the vegetative stage. The number of flowers in an inflorescence varies and up to 40 can be formed on the main stem, yet approximately 90% of flowers shed and only a few set pods. Pod abscission thus has a major influence on seed formation and an economic impact in terms of lupin crop productivity, yet the mechanisms that control this in lupins are not well understood. To gain more insight in pod abscission, Glazinska et al. (2017) generated Illumina RNA sequencing libraries of flowers, pods, and flower pedicels of the yellow lupin cultivar "Taper". These included non-abscising versus abscising flowers and their respective separately collected pedicels from 54-day-old plants, and non-abscising versus abscising pods from 75-day-old plants, collected with their pedicels containing an inactive abscission zone and an active abscission zone, respectively. This yielded between 46,619,042 and 56,784,288 reads per library which were de novo assembled into 219,514 contigs with an average length of 774 bp. Subsequently, differential expression analysis was conducted to investigate the molecular changes during organ abscission, where the transcriptomes of abscised flowers and non-abscised flowers, abscised flower pedicels and non-abscised flower pedicels, and abscised pods and non-abscised pods were compared. This identified 1,343, 2,933, and 1,491 changes in transcripts for flower, flower pedicels, and pods, respectively. Changes in transcript expression for genes associated with cell wall metabolism, metabolic pathways and hormone signaling were identified, where a total of 11 transcripts were validated by qPCR. Changes in 27 transcripts associated with cell wall metabolism were observed, where 22 were upregulated and 5 down-regulated in non-abscising flowers. A significantly higher number of cell wall metabolism genes were differentially regulated in flower pedicels and pods, where 81 were upregulated and 32 down-regulated in non-abscising flower pedicels and 7 were upregulated and 69 down-regulated in non-abscising pods. These changes appear to be a combination of the process of abscission as well as organ development of the organs that did not shed (Glazinska et al. 2017).

Transcriptional changes were also observed in key plant hormone signaling pathways including the auxin, ethylene, and gibberellin signaling pathways. Auxin and ethylene appear to be important regulators of floral abscission in yellow lupin with decrease in expression observed in genes involved in auxin biosynthesis and release of auxin from its conjugates and an increase of auxin transporters and ethylene biosynthesis genes and ethylene response factors (Glazinska et al. 2017). In addition, changes in gibberellin associated transcripts were observed, where transcripts involved in gibberellin biosynthesis were mostly down-regulated across the different abscised samples. Transcripts involved in gibberellin catabolism were upregulated in abscised flowers and flower pedicels and down-regulated in abscised pods. This thus shows that gibberellin plays a role in retaining the pods on the plants.

In conclusion, the yellow lupin abscission transcriptome study made a start in the elucidation of the biological mechanisms that control floral abscission in lupins. While further validation by qPCR and functional characterization will dissect the molecular control of organ abscission, the observed transcriptional changes are aligning with observations made in other crop systems. Future research will lead to insights that will allow improved flower and pod set in lupin cultivars.

5.2.9 SmallRNAs in Narrow-Leafed Lupin

A recent study generated smallRNA sequencing datasets to identify differentially expressed smallRNAs during seed development in narrow-leafed lupin (DeBoer et al. 2019). A total of 43 conserved miRNAs belonging to 16 families, and 13 novel narrow-leafed lupin-specific miRNAs were identified from foliar and root and five seed development stages. Members of the miRNA families miR167, miR399, miR156, miR319, and miR164 were upregulated in lupin seeds, which were confirmed by qRT-PCR for miR156, miR166, miR164, miR1507, and miR396 during five developmental seed stages (DeBoer et al. 2019). To determine whether any of these upregulated miRNAs during seed development regulate processes important for protein content and seed maturation, target predictions were conducted. Forty-nine putative unique targets were identified for the conserved miRNAs of which 40.4% encoded transcription factors, with the other predicted targets involved in various cellular processes. For the novel miRNAs, 20 unique target genes were identified, where in contrast to the conserved miRNAs only 20% of the targets encode for transcription factors. Subsequently, the authors were able to validate the target prediction for two miRNAs, where miR399 mediated cleavage of *Lup029358.1*, a putative Ubiquitin-conjugating E2 enzyme, involved in maintenance of phosphate homeostasis in other plant species and miR159 mediated cleavage of *Lup032338.1*, a putative MYB transcription factor. In addition to the identification of conserved and novel miRNAs in narrow-leafed lupin the authors identified *AGO* and *DCL* genes important for the biogenesis of miRNAs. The highly abundant miRNAs during seed development predominantly targeted transcription factors, some of which have known roles in gene regulation in other crops during seed development and future investigations offer novel opportunities to increase the quality of the lupin grain, as a high protein, gluten-free, non-GMO human food.

5.3 Conclusion

While compared to other legume species the number of gene expression studies in lupins have been limited, the current datasets lay the foundation for lupin crop improvement. This can be achieved through the adoption of gene-based molecular markers in the construction of genetic maps for targeted trait discovery in segregating populations and the continued improvement of lupin genome annotations using the generated transcriptome datasets for the various lupin species. While currently, a genome assembly is only available for narrow-leafed lupin, a genome assembly for white lupin is underway and with the cost of genome sequencing still coming down, additional lupin genome sequences will be generated. Likewise, additional transcriptome datasets will be generated across the lupin species to investigate key agronomic traits. Current transcriptome datasets have generated a gene expression atlas for narrow-leafed lupin (Kamphuis et al. 2015) and a transcript index for both white (O'Rourke et al. 2013; Secco et al. 2014) and yellow lupin (Glazinska et al. 2017). In addition, specific transcriptome studies have investigated specific traits such as disease resistance (to anthracnose; Fischer et al. 2015), alkaloid biosynthesis and regulation (Frick et al. 2018; Kroc et al. 2019b; Yang et al. 2017), seed storage proteins (Foley et al. 2015), cluster root development and phosphorus acquisition, (O'Rourke et al. 2013; Secco et al. 2014; Venuti et al. 2019; Wang et al. 2014) and floral abscission (Glazinska et al. 2017) in lupins. Taken together additional transcriptome studies will further aid lupin crop improvement, providing opportunities for more durable and stable crops resulting in improved yields and grower confidence.

References

Berger J, Buirchell B, Luckett D, Nelson M (2012a) Domestication bottlenecks limit genetic diversity and constrain adaptation in narrow-leafed lupin (*Lupinus angustifolius* L.). Theor Appl Genet 124:637–652

Berger JD, Buirchell B, Luckett DJ, Palta JA, Ludwig C et al (2012b) How has narrow-leafed lupin changed in its 1st 40 years as an industrial, broadacre crop? A G x E-based characterization of yield-related traits in Australian cultivars. Field Crops Res 126:152–164

Bunsupa S, Okada T, Saito K, Yamazaki M (2011) An acyltransferase-like gene obtained by differential gene expression profiles of quinolizidine alkaloid-producing and nonproducing cultivars of *Lupinus angustifolius*. Plant Biotechnol 28:89–94

Bunsupa S, Katayama K, Ikeura E, Oikawa A, Toyooka K et al (2012) Lysine decarboxylase catalyzes the first step of quinolizidine alkaloid biosynthesis and coevolved with alkaloid production in Leguminosae. Plant Cell 24:1202–1216

Cannon SB, McKain MR, Harkess A, Nelson MN, Dash S et al (2015) Multiple polyploidy events in the early radiation of nodulating and nonnodulating legumes. Mol Biol Evol 32:193–210

Cheng L, Bucciarelli B, Shen J, Allan D, Vance CP (2011) Update on lupin cluster roots: update on white lupin cluster root acclimation to phosphorus deficiency. Plant Physiol 156:1025–1032

Cowling WA, Tarr A (2004) Effect of genotype and environment on seed quality in sweet narrow-leafed lupin (*Lupinus angustifolius* L.). Aust J Agric Res 55:745–751

Croxford AE, Rogers T, Caligari PD, Wilkinson MJ (2008) High-resolution melt analysis to identify and map sequence-tagged site anchor points onto linkage maps: a white lupin (*Lupinus albus*) map as an exemplar. New Phytol 180:594–607

DeBoer K, Melser S, Sperschneider J, Kamphuis LG, Garg G et al (2019) Identification and profiling of narrow-leafed lupin (Lupinus angustifolius) microRNAs during seed development. BMC Genomics 20:135

Duranti M, Consonni A, Magni C, Sessa F, Scarafoni A (2008) The major proteins of lupin seed: characterisation and molecular properties for use as functional and nutraceutical ingredients. Trends Food Sci Technol 19:624–633

Edwards O, Ridsdill-Smith T, Berlandier F (2003) Aphids do not avoid resistance in Australian lupin (*Lupinus angustifolius, L. luteus*) varieties. Bull Entomol Res 93:403–411

Fischer K, Dieterich R, Nelson MN, Kamphuis LG, Singh KB et al (2015) Characterization and mapping of *LanrBo*: a locus conferring anthracnose resistance in narrow-leafed lupin (*Lupinus angustifolius* L.). Theor Appl Genet 128:2121–2130

Foley RC, Gao L-L, Spriggs A, Soo LY, Goggin DE et al (2011) Identification and characterisation of seed storage protein transcripts from *Lupinus angustifolius*. BMC Plant Biol 11:59

Foley RC, Jimenez-Lopez JC, Kamphuis LG, Hane JK, Melser S et al (2015) Analysis of conglutin seed storage proteins across lupin species using transcriptomic, protein and comparative genomic approaches. BMC Plant Biol 15:106

Frick KM, Kamphuis LG, Siddique KHM, Singh KB, Foley RC (2017) Quinolizidine alkaloid biosynthesis in lupins and prospects for grain quality improvement. Front Plant Sci 8:87

Frick KM, Foley RC, Kamphuis LG, Siddique KHM, Garg G et al (2018) Characterisation of the genetic factors affecting quinolizidine alkaloid biosynthesis and its response to abiotic stress in narrow-leafed lupin (*Lupinus angustifolius* L.). Plant Cell Environ 41:2155–2168

Frick KM, Foley R, Siddique KHM, Singh KB, Kamphuis LG (2019) The role of jasmonate signalling in quinolizidine alkaloid biosynthesis, wounding and aphid predation response in narrow-leafed lupin. Funct Plant Biol 46:443–454

Gao LL, Hane JK, Kamphuis LG, Foley R, Shi BJ et al (2011) Development of genomic resources for the narrow-leafed lupin (*Lupinus angustifolius*): construction of a bacterial artificial chromosome (BAC) library and BAC-end sequencing. BMC Genomics 12:521

Glazinska P, Wojciechowski W, Kulasek M, Glinkowski W, Marciniak K et al (2017) *De novo* transcriptome profiling of flowers, flower pedicels and pods of *Lupinus luteus* (Yellow Lupine) reveals complex expression changes during organ abscission. Front Plant Sci 8:641

Hane J, Ming Y, Kamphuis LG, Nelson MN, Garg G et al (2017) A comprehensive draft genome sequence for lupin (*Lupinus angustifolius*), an emerging health food: insights into plant–microbe interactions and legume evolution. Plant Biotechnol J 15:318–330

Hayden MJ, Nguyen TM, Waterman A, Chalmers KJ (2008) Multiplex-ready PCR: a new method for multiplexed SSR and SNP genotyping. BMC Genomics 9:80

Kamphuis LG, Zulak K, Gao L-L, Anderson JP, Singh KB (2013) Plant—aphid interactions with a focus on legumes. Funct Plant Biol 40:1271–1284

Kamphuis LG, Hane JK, Nelson MN, Gao L, Atkins CA et al (2015) Transcriptome sequencing of different narrow-leafed lupin tissue types provides a comprehensive uni-gene assembly and extensive gene-based molecular markers. Plant Biotechnol J 13:14–25

Kroc M, Koczyk G, Swiecicki W, Kilian A, Nelson MN (2014) New evidence of ancestral polyploidy in the Genistoid legume *Lupinus angustifolius* L. (narrow-leafed lupin). Theor Appl Genet 127:1237–1249

Kroc M, Czepiel K, Wilczura P, Koczyk G, Święcicki W (2019a) Alkaloid biosynthesis in lupins. In: Third international legume societ conference, Poznan, Poland

Kroc M, Koczyk G, Kamel KA, Czepiel K, Fedorowicz-Strońska O et al (2019b) Transcriptome-derived investigation of biosynthesis of quinolizidine alkaloids in narrow-leafed lupin (*Lupinus angustifolius* L.) highlights candidate genes linked to iucundus locus. Sci Rep 19:2231

Książkiewicz M, Nazzicari N, Yang H, Nelson MN, Renshaw D et al (2017) A high-density consensus linkage map of white lupin highlights synteny with narrow-leafed lupin and provides markers tagging key agronomic traits. Sci Rep 7:15335

McGinn S, Gut IG (2013) DNA sequencing - spanning the generations. Nat Biotechnol 30:366–372

Meng ZB, You XD, Suo D, Chen YL, Tang C et al (2013) Root-derived auxin contributes to the phosphorus-deficiency-induced cluster-root formation in white lupin (*Lupinus albus*). Plant Physiol 148:481–489

Mousavi-Derazmahalleh M, Bayer PE, Nevado B, Hurgobin B, Filatov D et al (2018) Exploring the genetic and adaptive diversity of a pan-Mediterranean crop wild relative: narrow-leafed lupin. Theoret Appl Genet 131:887–901

Nelson MN, Phan HT, Ellwood SR, Moolhuijzen PM, Hane J et al (2006) The first gene-based map of *Lupinus angustifolius* L.-location of domestication genes and conserved synteny with *Medicago truncatula*. Theoret Appl Genet 113:225–238

Nelson MN, Moolhuijzen PM, Boersma JG, Chudy M, Lesniewska K et al (2010) Aligning a new reference genetic map of *Lupinus angustifolius* with the genome sequence of the model legume, *Lotus japonicus*. DNA Res 17:73–83

Nevado B, Atchison GW, Hughes CE, Filatov DA (2016) Widespread adaptive evolution during repeated evolutionary radiations in New World lupins. Nat Commun 7:12384

O'Rourke JA, Yang SS, Miller SS, Bucciarelli B, Liu J et al (2013) An RNA-seq transcriptome analysis of orthophosphate-deficient white lupin reveals novel insights into phosphorus acclimation in plants. Plant Physiol 161:705–724

Parra-González L, Aravena-Abarzua G, Navarro-Navarro C, Udall J, Maughan J et al (2012) Yellow lupin (*Lupinus luteus* L.) transcriptome sequencing: molecular marker development and comparative studies. BMC Genomics 13:425

Phan HT, Ellwood SR, Adhikari K, Nelson MN, Oliver RP (2007) The first genetic and comparative map of white lupin (*Lupinus albus* L.): identification of QTLs for anthracnose resistance and flowering time, and a locus for alkaloid content. DNA Res 14:59–70

Schmutz J, Cannon SB, Schlueter J, Ma J, Mitros T et al (2010) Genome sequence of the palaeopolyploid soybean. Nature 463:178–183

Schmutz J, McClean PE, Mamidi S, Wu GA, Cannon SB et al (2014) A reference genome for common bean and genome-wide analysis of dual domestications. Nat Genet 46:707–713

Secco D, Shou H, Whelan J, Berkowitz O (2014) RNA-seq analysis identifies an intricate regulatory network controlling cluster root development in white lupin. BMC Genomics 15:230

Tian, L, Peel, GJ, Lei, Z, Aziz, N, Dai, X et al (2009) Transcript and proteomic analysis of developing white lupin (*Lupinus albus* L.) roots. BMC Plant Biol 9:1

Varshney RK, Chen W, Li Y, Bharti AK, Saxena RK et al (2012) Draft genome sequence of pigeonpea (*Cajanus cajan*), an orphan legume crop of resource-poor farmers. Nat Biotechnol 30:83–89

Varshney RK, Song C, Saxena RK, Azam S, Yu S et al (2013) Draft genome sequence of chickpea (*Cicer arietinum*) provides a resource for trait improvement. Nat Biotechnol 31:240–246

Venuti S, Zanin L, Marroni F, Franco A, Morgante M et al (2019) Physiological and transcriptomic data highlight common features between iron and phosphorus acquisition mechanisms in white lupin roots. Plant Sci 285:110–121

Wang Z, Straub D, Yang H, Kania A, Shen J et al (2014) The regulatory network of cluster-root function and development in phosphate-deficient white lupin (*Lupinus albus*) identified by transcriptome sequencing. Physiol Plant 151:323–338

Wasternack C, Hause B (2013) Jasmonates: biosynthesis, perception, signal transduction and action in plant stress response, growth and development. An update to the 2007 review in Annals of Botany. Ann Bot 111:1021–1058

Wink M (1992) The role of quinolizidine alkaloids in plant-insect interactions. Insect-Plant Interact 4:131–166

Yang H, Tao Y, Zheng Z, Zhang Q, Zhou G et al (2013) Draft genome sequence, and a sequence-defined genetic linkage map of the legume crop species *Lupinus angustifolius* L. PLoS ONE 8:e64799

Yang T, Nagy I, Mancinotti D, Otterbach SL, Andersen TB et al (2017) Transcript profiling of a bitter variety of narrow-leafed lupin to discover alkaloid biosynthetic genes. J Exp Bot 68:5527–5537

Young ND, Debellé F, Oldroyd GE, Geurts R, Cannon SB et al (2011) The *Medicago* genome provides insight into the evolution of rhizobial symbioses. Nature 480:520–524

Zimmermann J, Voss H, Schwager C, Stegemann J, Ansorge W (1988) Automated Sanger dideoxy sequencing reaction protocol. FEBS Lett 233:432–436

6 Molecular Marker Resources Supporting the Australian Lupin Breeding Program

Michał Książkiewicz and Hua'an Yang

Abstract

Over the last 60 years the Australian lupin industry has emerged to become the largest producer in the world, accounting for 85% of global lupin seed market. This progress was achieved by the rapid domestication process of the narrow-leafed lupin (*Lupinus angustifolius* L.) as a grain legume crop. Narrow-leafed lupin improvement has been based on the identification of donors carrying desirable alleles conferring particular agronomic traits and, subsequently, their orchestrated transfer by classical genetic approaches into domesticated germplasm. These traits include, among others, reduced pod shattering, low alkaloid content, seed water permeability, early flowering, and resistance to diseases caused by pathogenic fungi: anthracnose (*Colletotrichum lupini*) and Phomopsis stem blight (*Diaporthe toxica*). Moreover, some of these traits are related with recessive alleles requiring additional breeding effort. To facilitate selection of desirable genotypes in the progenies and cross derivatives, molecular markers linked to particular trait loci were developed and implemented in Australian breeding program.

6.1 Introduction

During the process of narrow-leafed lupin (*Lupinus angustifolius* L.) domestication numerous advantageous agronomic traits were identified and transferred into current breeding materials and cultivars. Many of these traits are conferred by recessive alleles, requiring additional crossing effort to maintain the homozygote state, whereas some of them depend on two or more unlinked genes. Intensive domestication process of the narrow-leafed lupin, which have occurred during the recent 60 years, has reduced the genetic diversity and constrained adaptation of industrialized germplasm to such extent that further improvement of the species as a crop will require extensive exploitation of wild stocks (Berger et al. 2012). Incorporation of novel potentially valuable alleles from wild or non-adapted accessions into elite germplasm without the aid of molecular selection tools could require several years of pre-breeding crossing efforts to leverage all these traits (Cowling et al. 2009). To address this issue numerous sequence-defined markers matching major domestication traits were developed and subsequently implemented in narrow-leafed lupin breeding program run by the Department of

M. Książkiewicz (✉)
Institute of Plant Genetics Polish Academy of Sciences, Strzeszyńska 34, 60-479 Poznań, Poland
e-mail: mksi@igr.poznan.pl

H. Yang
Department of Agriculture and Food Western Australia, 3 Baron-Hay Court, South Perth 6151, Australia
e-mail: huaan.yang@dpird.wa.gov.au

K. B. Singh et al. (eds.), *The Lupin Genome*, Compendium of Plant Genomes,
https://doi.org/10.1007/978-3-030-21270-4_6

Agriculture and Food, Western Australia (DAFWA) and the Grains Research and Development Corporation (GRDC) (Table 6.1). Recently, DAFWA with two other departments was merged into Department of Primary Industries and Regional Development (DPIRD), whereas the lupin breeding program was fully licensed to the Australian Grain Technologies (AGT) company.

Table 6.1 Marker implemented in the narrow-leafed lupin breeding programs

Trait	Marker	Primers	Detection	Products
Tardus	TaM1 2.1 cM	F: AACAGAGGATTGCAAATC R: ATTGGGTCTCTCTCTCTC	PCR	*ta* 227 bp *Ta* 226 bp or 219 bp
Tardus	TaM2 3.7 cM	F: TTTTGCTAGTCTTTGGATGAGC R: CTCAAACCAGCCTACCAAA	PCR	*ta* 203 bp *Ta* 222 bp or 241 bp
Tardus	TaM3 4.0 cM	F: CATCTTCTTGCTTGCACATA R: GTTGCCAGATTCATTCAGA	CAPS, DraI	*ta* 410 bp and 190 bp *Ta* 410 bp, 100 bp and 90 bp
Tardus	TaLi 1.4 cM	F: GATCTGAAAAGGAATATGAAG R: ATCCTACTAAATCCTGGTACAG	PCR	*ta* 511 bp *Ta* 6 alleles, ~250–320 bp
Lentus	LeM1 2.6 cM	F: TTAACGAACCTACCATTTG R: GGGAACAACAACAAC	PCR	*le* 126 bp *Le* no product
Lentus	LeM2 1.3 cM	F: AGAAAAAGATGAATGCACG R: GTCTAACAACAACAACAAC	PCR	*le* 204 bp *Le* no product
Lentus	LeLi ~6 cM	F: CCCACAGCTAAAATTATACC R: GGCAGGATGTAAGTTTAGG	PCR	*le* 251 bp *Le* 247 bp
Iucundus	IucLi 0.9 cM	F: TCTTAGATGTATGATGAGTATGG R: CCAGGATAAATTAGTTGTGTC	PCR	*iuc* 291 bp *Iuc* 8 alleles, 274–309 bp
Mollis	MoA ~0.0 cM	F: GAAGCATTCGATGAATTC R: TAACATCAACAAGGTGAGAATC	SSCP	*mol* 199 bp (44T, 170A) *Mol* 199 bp (44A, 170 T)
Mollis	MoLi ~0.0 cM	F: TTAAAGCTCCTCCGGGAG R: CTTATACTATTAGACTAACGCC	PCR	*mol* 314 bp *Mol* 281 bp, 294 bp or 300 bp
Ku *Julius*	KuHM1 ~0.0 cM	F: CAAAAACAATAATAACGACAAC R: AGACATACCTTGTATGCGG	PCR	*ku* 287 bp *Ku* 280 bp (*Jul* not tested)
Ku *Julius*	LanFTc1 INDEL1 candidate gene	F: AGCATGCGAGAAAACAACG R: CAGCTTACTCCATAGTTCAAAGCA	PCR	*ku* 2277 bp *Ku* 854 bp *Ku** (P22660) 1069 bp *Jul* no product
Ku *Julius*	LanFTc1 INDEL2 candidate gene	F: TCATATGGCTGCAACCTGAA R: GCTTCCTTCCTTTGCCTTCT	PCR	*ku* 5969 bp *Ku* 4546 bp *Ku** (P22660) 4761 bp *Jul* 757 bp
Lanr1	AntjM1 3.5 cM	F: CATCCTCACATATGAAGC R: CCCATTGTTGTTGTTG	PCR	*lanr1* 273 bp or 270 bp *Lanr1* 276 bp
Lanr1	AntjM2 2.3 cM	F: TCATCTCTAAATCCTATCTCAG R: GTAGAAAATATCATTGCAGAAAAG	PCR	*lanr1* 212 bp *Lanr1* 217 bp
Lanr1	AnSeq3 0.9 cM	F: GAATTCAGAGACAGGACTC R: AGTTTTTTTGGTTATCTAG	SSCP	*lanr1* 93 bp (45C) *Lanr1* 93 bp (45T)
Lanr1	AnSeq4 0.9 cM	F: GAATTCTGATGTGAAACAAC R: CTCCTGGCTGAGCTTTG	SSCP	*lanr1* 93 bp (67G) *Lanr1* 93 bp (67A)
Lanr1	DAFWA213 ~0.0 cM	F: AGTACCTTCATTTGTATGCTCCAC[C/T] R: CTCCCAATTTTGACAGAGAAGAACTGG	Fluidigm	*lanr1* 92 bp 68C *Lanr1* 92 bp 68T

(continued)

Table 6.1 (continued)

Trait	Marker	Primers	Detection	Products
AnMan	AnManM1 5.1 cM	F: TTGAGCTTGGTATATAAACG R: TCGAGCAATAAATGATATG	PCR	*anMan* 216 bp or 218 bp *AnMan* 228 bp
LanrBo	BoSeq196 13.7 cM	F: CACCCAACATTCCAGCTCTG R: CTTCAGCCAAGGAAAAGAAGTC	HRM	Reference line Bo7212
Phr1	Ph258M1 7.8 cM	F: TCCAGACTGACTATATTCTTAG R: CAGGCACATATATCTTTATACC	PCR	*phr1* 303 bp *Phr1* 254 bp
Phr1	Ph258M2 5.7 cM	F: GAACCATTGTAACTAAATCC R: GGGAACAACAACAAC	PCR	*phr1* 206 bp *Phr1* 203 bp
PhtjR	PhtjM4 1.1 cM	F: GAATTCAACCAACGTGG R: GTGGATACAACCTCACTGTC	SSCP	*phtjR* 92 bp (33C) *PhtjR* 92 bp (33A)
PhtjR	PhtjM5 2.1 cM	F: GAATTCCATATGCAATGG R: CTTAATTGTTAATTTGTTATTTGC	SSCP	*phtjR* 90 bp (19C) *PhtjR* 90 bp (19T)
PhtjR	PhtjM7 1.1 cM	F: CTTTTTAGCTTACTTCAATTAGC R: CTAATTCAATGAGCTTCTCTT	SSCP	*phtjR* 88 bp (41G) *PhtjR* 88 bp (41T)
PhtjR	InDel2	F: GATAAAGTATATCTAAATTATGTTTGC R: CTATATTTTGTATCAATTATAACAAATT	PCR	*phtjR* 122 bp *PhtjR* 134 bp
PhtjR	InDel10	F: GTTAAGTGGTAAATTGACTCATG R: GTTTTRCATTCTTGCAAAGATAAAATTAG	PCR	*phtjR* 102 bp *PhtjR* 94 bp

6.2 Methods Applied for Development of Molecular Markers

The studies on marker-assisted selection in narrow-leafed lupin were initiated in the late 1990s, when no information of the genome sequence, nor even the position of agronomic trait loci on the linkage map, was available (Brien et al. 1999). Indeed, the first linkage maps of the species being under development at that time were very simple and carried random amplified DNA polymorphisms (RAPDs) or other arbitrary markers, hardly reproducible among laboratories and unsuitable for molecular selection (Kruszka and Wolko 1999; Wolko 1995). The marker-assisted selection concept required rapid development of large number of reliable markers to increase a chance of finding polymorphic loci matching important agronomic traits improved during the domestication process. To facilitate the production of new markers applicable for future narrow-leafed lupin genotyping, various molecular methods have been harnessed.

The first approach was to target microsatellite simple sequence repeats (SSRs), which are highly polymorphic and abundant in eukaryotic genomes (Litt and Luty 1989). Some PCR-based methods tagging SSRs were already established at that time. Those included a random amplified microsatellite polymorphism (RAMP), being a combination of a SSR-anchor primer technique and RAPD, as well as an amplified fragment length polymorphism (AFLP), based on restriction digestion of genomic DNA and subsequent PCR amplification from primers complementary to adaptors ligated to the sticky ends of restriction fragments (Vos et al. 1995; Wu et al. 1994). AFLP method was sufficient to generate a high number of polymorphic markers during preliminary narrow-leafed lupin screening (Brien et al. 1999), however, their hypothetical subsequent conversion to sequence-specific PCR-based applications was found to be very inconvenient and ineffective (Shan et al. 1999). Therefore, a modification of the existing methods, by replacing one of the AFLP primers with a microsatellite-anchor primer, was proposed as a solution enabling further sequencing of an amplified band. Moreover, additional digestion of DNA with frequently cutting enzyme, *Hae*III, was introduced before PCR amplification step to reduce the number of bands in high-resolution sequencing gel used for product separation (Yang et al. 2001). The method was named microsatellite-anchored fragment length polymorphism (MFLP) and successfully implemented to generate the first linkage map of the species comprising 21 linkage groups (Boersma et al. 2005). Indeed, further exploitation of MFLP approach yielded numerous sequence-specific markers linked to various agronomic traits including pod shattering (Boersma et al. 2007c, 2009; Li et al. 2010, 2012b), alkaloid content (Li et al. 2011), seed water permeability (Boersma et al. 2007a; Li et al. 2012a), vernalization responsiveness (Boersma et al. 2007b), and resistance to diseases caused by pathogenic fungi: anthracnose (Buirchell and Yang 2006; Yang et al. 2004, 2008, 2010; You et al. 2005), Phomopsis stem blight (Yang et al. 2002), and lupin rust (Sweetingham et al. 2006).

To enable comparative mapping and targeting particular genes, including those putatively conferring important agronomic traits, primers were anchored in exon sequences to produce sequence-defined, gene-based markers. Such markers were being developed and introduced on a linkage map of the species, comprising a single linkage group for each of the 20 chromosomes, and successively updated on a case-by-case basis (Kroc et al. 2014; Książkiewicz et al. 2013, 2015; Leśniewska et al. 2011; Narożna et al. 2017; Nelson et al. 2006, 2017; Przysiecka et al. 2015; Wyrwa et al. 2016). This approach provided markers anchored in candidate genes underlying vernalization responsiveness in flowering and low alkaloid content as well as one gene-based marker tightly linked to white flower color locus (Kroc et al. 2019a, 2014; Nelson et al. 2017; Przysiecka et al. 2015; Taylor et al. 2019).

The progress on high-throughput sequencing chemistry and equipment observed during the recent decade opened new opportunities for large-scale DNA fingerprinting and rapid cost-effective marker development. These novel methods, harnessed in narrow-leafed lupin applied research, included restriction-site-associated DNA sequencing (RADseq), Diversity Arrays Technology (DArTseq) as well as transcriptome and whole-genome shotgun sequencing and assembly (Hane et al. 2017; Kamphuis et al. 2015; Mousavi-Derazmahalleh et al. 2018a, b; Yang et al. 2012, 2013a, b, 2015a, b; Zhou et al. 2018). These studies yielded numerous sequence-defined markers co-segregating or tightly linked to several agronomic traits as well as provided valuable information on the contrast in genetic and adaptive diversity between domesticated and untapped narrow-leafed lupin germplasm. DArT sequencing was also used to generate high number of markers, which facilitated linkage map improvement and pseudochromosome assembly refinement (Hane et al. 2017). Moreover, DArTseq was applied for genome-wide association studies targeting phenotypic traits, such as flowering time, flower color, hard/soft seededness, alkaloid status, pod dehiscence, height at maturity, and 100-seed weight. Some associations were found between markers and four traits, however, after applying false discovery rate p-value adjustment of these associations, only two SNPs associated with pod dehiscence remained highly significant (Mousavi-Derazmahalleh et al. 2018a, b).

6.3 Detection Methods

The MFLP method generates a high number of amplified products with a relatively small length difference between particular amplicons (Yang et al. 2001). Therefore, such markers were amplified with γ^{33}P radiolabeled primers and visualized by sequencing gel (5% acrylamide, 7M urea) electrophoresis using Protean II (Bio-Rad) or similar vertical electrophoresis unit (Yang et al. 2002). The same approach was applied for MFLP-derived, PCR-based markers for many years in Australian narrow-leafed lupin breeding programs. Routine implementation of this procedure revealed to be relatively cost-effective, as up to 768 plants could be genotyped using a single 96-well polyacrylamide denaturing sequencing gel and eight multiple loadings in 10 min. intervals (Li et al. 2012b). Marker bands were detected by autoradiography with overnight exposure of the X-ray film to the dried gel (Boersma et al. 2007a).

Gene-based SNP markers used for linkage mapping (Kroc et al. 2014; Książkiewicz et al. 2013, 2015; Narożna et al. 2017; Nelson et al. 2017; Przysiecka et al. 2015; Wyrwa et al. 2016) have been scored mainly by the Cleaved Amplified Polymorphic Sequence (CAPS) (Konieczny and Ausubel 1993) or derived CAPS (dCAPS) (Neff et al. 1998) approaches. Restriction sites and dCAPS primers were identified using dCAPS Finder 2.0 (Neff et al. 2002). Digestion products were separated by simple agarose gel electrophoresis. However, CAPS and dCAPS methods have not been widely implemented into the Australian marker-assisted narrow-leafed lupin selection programs (Boersma et al. 2009).

High-throughput sequencing methods applied to the narrow-leafed lupin genome provided numerous single nucleotide substitution markers, having identical lengths of both alleles. These markers have been resolved using several different methods. Initially, marker-assisted selection exploited single-stranded conformation polymorphisms (SSCP) (Sunnucks et al. 2000) on 6% acrylamide gel using a sequi-gen CT sequencing cell (Bio-Rad) (Yang et al. 2013a). However, it was a tedious and time-consuming approach, hardly scalable to address the principle of molecular breeding, i.e., genotyping of all plant material before the initiation of flowering to enable further crossing of lines carrying desired pattern of alleles.

To tackle the emerging challenge of large-scale genotyping, automated methods of polymorphic PCR product screening have been exploited. Therefore, some RADseq polymorphic sequences derived from NGS platform Illumina Hiseq2000 were transformed to PCR-based markers and subjected to the high resolution melting (HRM) using Light Cycler (Yang et al. 2015a). HRM was also implemented in Poland for the incorporation of sequence-tagged site intergeneric legume markers into the narrow-leafed lupin linkage map, highlighting the applicability of this method for resolving SNPs from high-throughput sequencing data. Nevertheless, non-negligible difficulties in optimization procedure have been encountered for several markers (Kamel et al. 2015). Recently, a Fluidigm system has been adopted to provide high-throughput genotyping platform for selected array of SNP markers, both for linkage map improvement and marker-assisted selection (Hane et al. 2017; Kamphuis et al. 2015; Yang et al. 2013b). Moreover, a KASP assay has been designed for 40 markers and successfully implemented for two narrow-leafed lupin recombinant inbred line mapping populations, Australian "83A:476 × P27255" and Polish "Emir x LAE-1" in Poland (Kozak et al. 2017).

6.4 Traits Subjected to Molecular Selection

6.4.1 Pod Shattering

Pod shattering is a natural adaptation enabling better distribution of the seeds and expansion of the growing area of plants. It became the highly undesired trait in mechanized agriculture, which requires the ability of the crop to retain its seed until pod harvesting at full maturity. Two major genes were subjected to molecular selection, *Lentus* and *Tardus*. The recessive *lentus* (*le*) allele modifies the orientation of the sclerified endocarp in the pod, substantially reducing torsional forces after drying (Gladstones 1967). Additionally, *le* allele changes a pod pigmentation, resulting in a purplish hue of an immature pod and a bright yellowish-brown color on the internal surface of a mature pod. Therefore, a discrimination between *Le* and *le* may be performed by phenotype observation. The recessive *tardus* (*ta*) allele affects the sclerenchyma strips of the dorsal and ventral pod seams, greatly increasing the fusion of two pod halves and moderately impeding their separation during desiccation (Gladstones 1967). Scoring for *Ta*/*ta* is very difficult as this gene does not confer any other visible plant phenotype. Pod shattering observations must be compared with the knowledge on the *Le*/*le* genotypes to discriminate between *Ta* and *ta* alleles (Boersma et al. 2009). Both genes were targeted in Australian breeding program. Recessive alleles of both genes were first combined in the Australian Uniharvest (1971) cultivar providing non-shattering phenotype (Cowling 1999). However, under extreme conditions of drought and high temperatures even genotypes carrying both recessive genes revealed a certain level of pod dehiscence (Święcicki and Święcicki 1995). One MFLP marker linked to the *Tardus* gene (2.1 cM) was converted to a codominant PCR-based TaM1 marker (Boersma et al. 2009). Moreover, the first sequence-based linkage map of the species contained two markers linked to the *Tardus* trait: UWA244 (RFLP, 3.7 cM) and Lup001 (CAPS, 4.0 cM) (Nelson et al. 2006). These markers were converted to a codominant PCR-based TaM2 marker and a CAPS TaM3 marker having restriction sites in both alleles to provide internal control of enzymatic cleavage (Boersma et al. 2009). In the panel of wild and landrace accessions (n = 33) these markers revealed 51–61% of predictive ability of the *Ta* phenotype, being applicable together to ~70% of accessions. To improve the selection process, another PCR-based marker was developed, TaLi, yielding one *ta* allele and six *Ta* alleles. 94% of correlation (n = 150) was observed between *Ta*/*ta* phenotypes and TaLi genotypes (Li et al. 2010). A pair of dominant MFLP markers linked to the *Lentus* gene (located at genetic distances of 1.3 and 2.6 cM from the gene) was converted to the presence (*le*) /absence (*Le*) PCR markers, LeM2 and LeM1, respectively (Boersma et al. 2007c). Validation revealed high number of false-positive scores, narrowing the selection applicability of LeM1 and LeM2 markers to approximately 35% of pod shattering landrace accessions (n = 36). Furthermore, a codominant LeLi PCR-based marker was designed and revealed to be positively correlated with the trait for ~55% of core collection lines (n = 125) (Li et al. 2012b). Whole-genome sequencing effort provided 26 markers within ±1 cM from the *Tardus* as well as 18 SNP markers within ±1 cM from the *Lentus*, including 6 co-segregating, (Hane et al. 2017). Moreover, a new ultra-high-density map carries 23 markers within ±1 cM distance from the *Tardus* trait and 15 markers within ±2 cM distance from the *Lentus* (Zhou et al. 2018). These markers can now be implemented in narrow-leafed lupin breeding programs, but require validation.

6.4.2 Alkaloid Content

Three natural low alkaloid recessive alleles were identified early in narrow-leafed lupin, namely *depressus*, *esculentus,* and *iucundus* (von Sengbusch 1942). These recessive genes influence the total alkaloid content in the grain, decreasing the total seed dry weight alkaloids to ~0.01–0.06%: *depressus* has the highest impact on the alkaloid content, whereas *iucundus* the lowest (Hackbarth and Troll 1956). From these genes only *iucundus* was widely exploited for NLL breeding (Gladstones 1970). Incorporation of this gene into European breeding programs in the initial stage of the narrow-leafed lupin domestication process (years 1928–1947) (Cowling 1999) resulted in a sharply decreased content of seed alkaloids in modern cultivars, even below 0.01% of dry weight (Kamel et al. 2016). The *esculentus* and *depressus* genes were introduced into some former German breeding materials (St. 415 and St. 14, respectively) and further abandoned (Święcicki and Święcicki 1995). To target *iucundus* in Australian breeding, a codominant PCR markers IucLi has been developed (Li et al. 2011). The marker is difficult to score because it has numerous *Iuc* alleles differing from the *iuc* allele by 1–18 bp. However, IucLi revealed 86.4% accuracy in core collection (n = 125) constituting a versatile tool for molecular screening. Linkage map improvements provided LaSSR_025 marker at a distance 0.4 cM from *iucundus* (Hane et al. 2017; Kamphuis et al. 2015) and another 16 markers within ±1 cM distance from the trait (Zhou et al. 2018). The potential applicability of these markers in modern breeding has not yet been confirmed. Recently, a Polish transcriptome-based study yielded a PCR-based dCAPS marker iuc_RAP2-7 matching a candidate gene for *iucundus*, an APETALA2/ethylene response transcription factor (Kroc et al. 2019a, b; see Sect. 5.2.7).

6.4.3 Soft Seededness

Physical dormancy is a common agronomic trait in legumes. It is related with the development of a water-impermeable seed coat (Miao et al. 2001), likely caused by the presence of phenolics and, hypothetically, suberin-impregnated layers of palisade cells as observed in pea and soybean (Smýkal et al. 2014). Recessive allele *mollis* confers water-permeable testa at maturity (Forbes and Wells 1968; Mikolajczyk 1966). Despite the sequencing of the genome (see Sect. 3.3) and mapping flanking markers, no candidate gene underlying *mollis* in narrow-leafed lupin has been identified (Hane et al. 2017; Zhou et al. 2018). As the permeability of seed coat is maternally determined, *mollis* is considered as the most difficult gene for selection by phenotype observation. To enable molecular selection in Australian breeding program, a MFLP marker from the linkage map (Boersma et al. 2005) co-segregating with the *mollis* trait has been converted to the SSCP marker MoA, carrying 2 SNP sites and showing 100% correlation with the trait across wild and domesticated germplasm (n = 76) (Boersma et al. 2007a). As the SSCP procedure is tedious and hardly scalable, another PCR-based marker tightly linked to gene *mollis* has been designed, MoLi, yielding one *mol* and three different *Mol* alleles. The marker MoLi has been successfully validated in 91.3% of lines tested (n = 150) (Li et al. 2012a).

6.4.4 Early Flowering

Wild lupins require a period of low temperature during seed germination and early plant growth to promote flowering (Adhikari et al. 2012; Landers 1995). This phenomenon, known as vernalization responsiveness, is a natural adaptation to temperate climate preventing from flowering in the season with harsh weather, however, it is an undesired trait in agriculture, significantly reducing applicability of this species as a crop in late spring sowings or in warm climate zones. Two natural dominant mutations, both removing vernalization requirement, were found during domestication process, *Ku* in 1961 (Gladstones and Hill 1969) and *Julius* few years later (Mikolajczyk 1966). Based on the phenotyping of particular cross derivatives, *Ku* was hypothesized to be identical with *Julius* (Gladstones et al. 1998; Rahman and Gladstones 1972). Glasshouse tests revealed that the vernalization is essential requirement for normal flowering in most genotypes not containing the *Ku* allele (Landers 1995). *Ku* reduces the vegetation period of spring sowings in the climate of central Europe by 14–16 days as plants undergo partial vernalization in field conditions (Święcicki and Święcicki 1995). *Ku* was widely introduced to breeding programs in Australia and in Europe (Cowling 1999; Stefanova and Buirchell 2010) whereas *Julius* was exploited mainly in Poland (Taylor et al. 2019) and the current knowledge of flowering time control is reviewed in detail in Sect. 9.1. A dominant MFLP marker co-segregating with *Ku* on the linkage map (Boersma et al. 2005) has been converted to a PCR-based, codominant KuHM1 marker (Boersma et al. 2007b). Development of the linkage map with sequence-defined markers (Nelson et al. 2006) provided an opportunity to search for candidate genes by synteny to a sequenced genome, like *Medicago truncatula*. This approach yielded two markers anchored in a candidate gene: dFTc (Kroc et al. 2014) and LanFTc1_INDEL (Nelson et al. 2017). Codominant PCR marker LanFTc1_INDEL amplified 854 bp (*Ku*) and 2,277 bp (*ku*) products and revealed 100% correlation with the *Ku*/*ku* phenotype in wild and domesticated germplasm (n = 216). However it did not amplify any product in few early lines carrying a *Julius* allele. Recently, another INDEL marker to genotype these lines has been developed, flanking a *Julius* 5,162 bp deletion and overlapping also *Ku* LanFTc1_INDEL (Taylor et al. 2019), which allows the differentiation of the *Ku* and the *Julius* promoter sequence variants.

6.4.5 Anthracnose Resistance

Susceptibility to anthracnose, a disease caused by the pathogenic fungus, *Colletotrichum lupini* (Bondar) Nirenberg, Feiler & Hagedorn, is a major issue hampering worldwide lupin cultivation (Nirenberg et al. 2002). The resistance to anthracnose in the narrow-leafed lupin is controlled by several single dominant genes that were discovered in different germplasm resources, namely, *Lanr1* in cv. Tanjil, *AnMan* in cv. Mandelup, and *LanrBo* in the breeding line Bo7212 (Fischer et al. 2015; Ruge-Wehling et al. 2009; Yang et al. 2004, 2008). First, an MFLP-derived PCR marker AntjM1 was generated, having one *Lanr1* (resistant) and two *lanr1* (susceptible) alleles. However, this marker, located 3.5 cM from the *Lanr1* gene, produced many false-positive results due to high recombination rate between AntjM1 and *Lanr1* loci (You et al. 2005). The marker was applied to screen F_2 breeding populations from only 32 and 61 crosses from approximately 400 crosses in 2003 and 2004 (Yang et al. 2015b). Nucleotide-binding site GLPL-A motif has been used to develop the PCR codominant AntjM2 marker. Marker AntjM2 was revealed to have only 3% false-positive rate (n = 53) enabling widespread implementation in Australian lupin breeding program (You et al. 2005). Next, with the aid of the RADseq, two markers, Anseq3 and Anseq4, both flanking *Lanr1* locus by 0.9 cM, were generated. Resistance and susceptibility alleles had the same lengths (93 bp) but differed by SNP variants (Yang et al. 2012). SSCP method was used for marker scoring. Moreover, two SNP markers co-segregating with the *Lanr1*, DAFWA213 and DAFWA5820, were identified in the whole-genome sequencing study (Yang et al. 2013b). DAFWA213 was confirmed to be applicable to marker-assisted selection using Fluidigm system. Another *Lanr1* candidate diagnostic marker, Indel_23269, has been developed recently (Yang et al. 2015b). More recently, the dense reference genetic map narrowed down the region of interest for *Lanr1* to a 388 Kb area on chromosome 11 (NLL-11), where six sequence-based markers co-segregate with *Lanr1* (Hane et al 2017). The presence of resistance alleles in domesticated germplasm resulted in high effectiveness of anthracnose resistance breeding (Fischer et al. 2015; Yang et al. 2012). Development and implementation of truly selective markers into breeding programs greatly contributed to this success, increasing from 50 to 98% the proportion of breeding lines showing disease resistance in field trials with simultaneous reduction by 80% the size of field disease screening trials and working time spent by pathologist and technical officer (Yang et al. 2015b). It means that with the same effort the lupin breeding team has been developing twice as many resistant breeding lines as they used to do before implementation of diagnostic markers into selection procedures.

The two other genetic anthracnose resistance resources, conferred by *AnMan* and *LanrBo* genes, have been exploited to a lesser extent than *Lanr1*. To support Australian breeding, just one codominant PCR marker, AnManM1, has been developed for *AnMan* resistance gene from Mandelup (Yang et al. 2008). The marker is located ~5 cM from a resistance gene and was shown to have ~5% of false-positive score rate in domesticated germplasm (n = 23). Attempts have been also undertaken in Germany (Julius Kühn-Institut and Saatzucht Steinach GmbH) to develop molecular markers tagging the last unexplored major narrow-leafed lupin anthracnose resistance gene, *LanrBo* (Ruge-Wehling et al. 2009). A sequence-defined, HRM marker BoSeq196 has been developed, however, due to large distance from the gene (13.7 cM) its applicability in marker-assisted selection is limited (Fischer et al. 2015).

6.4.6 Phomopsis Stem Blight Resistance

Phomopsis stem blight is caused by a pathogenic fungus, *Diaporthe toxica* Will., Highet, Gams & Sivasith, anamorph *Phomopsis leptostromiformis* (Kühn) Bubák (Jago et al. 1982; Williamson et al. 1994). At the latent stage of stem infection of susceptible plants the fungus produces phomopsins, which are a serious threat for animals grazing on lupin stubble (Cowley et al. 2014). Methods of screening for Phomopsis stem blight rely on observations of percentage surface area on senescent stems covered with lesions (Cowling et al. 1987) or microscopic examination of subcuticular coralloid structure of infected stems (Williamson et al. 1991). Nondestructive glasshouse infection test can be performed by inoculation of lateral branch regenerating from the second main stem node after topping the main stem above this node (Shankar et al. 2002). Resistance in *L. angustifolius* is expressed as a slower saprophytic colonization of host tissue (Shankar et al. 1996). There are at least three different genetic sources of *D. toxica* resistance in *L. angustifolius*, all originating from the Australian lupin collection currently housed at Agriculture Victoria in Horsham. Reference germplasm resources for these genes are breeding line 75A:258 (*Phr1* gene) and cultivars Merrit (*Phr2*) and Tanjil (*PhtjR*). With the use of the molecular fragment length polymorphism (MFLP) technique, markers linked to the hypothetical *Phr1* resistance gene were designed, Ph258M1 (7.8 cM) and Ph258M2 (5.7 cM) (Yang et al. 2002). The markers revealed 10% of false-positive rate during validation step (n = 21). Next generation sequencing of restriction-site-associated DNA fragments was exploited to develop the set of single nucleotide polymorphism (SNP) markers linked to *PhtjR* gene, namely PhtjM4, PhtjM5, and PhtjM7 showing 100% correlation with the *PhtjR* phenotype (n = 26) (Yang et al. 2013a). Recently, whole-genome resequencing approach was harnessed to develop new set of markers tagging *PhtjR* gene, including two INDEL PCR markers, InDel2 and InDel10, proved to be diagnostic for all Australian commercial cultivars (n = 27) (Yang et al. 2015a). 8 co-segregating RADseq markers were identified recently (Zhou et al. 2018).

6.4.7 Lupin Rust Resistance

Lupin rust, caused by pathogenic fungus *Uromyces lupinicolus* Bubák, has been observed on narrow-leafed lupin cultivars and landraces in Europe as early as in the 1920s and 1930s (Klinkowski 1939; Poeverlein 1936). A source of resistance was identified in domesticated germplasm, including cultivars Tallerack and Jindalee. Four MFLP markers were identified as linked to the rust resistance gene from Tallerack and one of these markers was sequenced and converted into a PCR-based marker for future screening if the need arises (Sweetingham et al. 2006).

6.5 Adoption of Molecular Markers in European Breeding Programs

The history of molecular marker-assisted selection in narrow-leafed lupin breeding dates back to years 2002–2004 when the first molecular markers linked to disease resistance loci were published (Yang et al. 2002, 2004). Despite the consecutive development and public release of markers linked to all major domestication traits which are essential for the improvement of this species as a crop worldwide, their implementation into marker-assisted selection was limited only to Australian breeding programs. Recently, some European breeders and scientists have launched initial molecular screening procedures for particular traits. To identify potential sources of anthracnose resistance for Russian and Belarussian breeding programs, four anthracnose resistance markers, AntjM1, AntjM2, Anseq3, and Anseq4, were surveyed and found to be potentially useful for selection of germplasm (n = 50) originating from Belarus, Russia, Poland, and Australia (Grishin et al. 2015). Moreover, the bunch of molecular markers tagging the *mollis* (MoLi, MoA), *tardus* (TaLi, TaM1, TaM2), *lentus* (LeLi, LeM1, LeM2), *Ku* (LanFTc1_INDEL), *Lanr1* (AntjM1, AntjM2, Anseq3, Anseq4), *AnMan* (AnManM1), *Phr1* (Ph258M1, Ph258M2), and *PtjR* (PhtjM4, PhtjM5, PhtjM7, Indel2, Indel10) genes was used to explore the genetic content of selected narrow-leafed lupin lines (n = 202) from the Polish *Lupinus* Gene Bank maintained by Poznan Plant Breeding Ltd. (Bielski et al. 2016; Książkiewicz et al. 2016, 2019). Recently, a marker iuc_RAP2-7 targeting low alkaloid *iucundus* gene was validated in Polish *Lupinus* Gene Bank collection (n = 202) and revealed 100% correlation with bitter phenotype and 97.1% with the sweet one (Kroc et al. 2019a).

6.6 Conclusion

The last 20 years have witnessed progress in applied research targeting domestication genes of narrow-leafed lupin. Exploited techniques and methods followed current trends in molecular sciences at the time. The implementation of MFLP protocol (Yang et al. 2001) provided polymorphic bands linked to all major agronomic traits. These bands were subsequently being converted by Sanger sequencing on one-by-one basis to PCR-based markers (Table 6.1) released from 2002 to 2012 (Boersma et al. 2005, 2007a, b, c; Li et al.2010, 2011, 2012a, b; Lin et al. 2009; Sweetingham et al. 2006; Yang et al. 2002, 2004, 2008, 2010; You et al. 2005). MFLP-derived markers were typically mapped 1–5 cm from the target locus and revealed non-negligible ratios of false-positive and false-negative scores in validation germplasm surveys. However, their use in selection was revealed to be beneficial for a relatively wide range of crosses.

Gene-based markers have been published since 2006 by research groups from several countries, predominantly Australia and Poland. These markers were generated using PCR-based Sanger sequencing to saturate a linkage maps or to localize studied genes without any substantial improvement around domestication loci (Kroc et al. 2014; Książkiewicz et al. 2013, 2015; Leśniewska et al. 2011; Narożna et al. 2017; Nelson et al. 2006, 2010; Przysiecka et al. 2015; Szczepaniak et al. 2018; Wyrwa et al. 2016). However, the combination of genome assembly data and gene expression profiling targeting particular traits provided PCR-based markers anchored directly in candidate genes underlying two important agronomic loci (*Ku*/*Julius* and *iucundus*), constituting a versatile tool for marker-assisted selection (Kroc et al. 2019a; Nelson et al. 2017; Taylor et al. 2019).

The exploitation of high-throughput sequencing technology for germplasm genotyping yielded numerous SNP markers tagging the vast majority of narrow-leafed lupin domestication traits (Hane et al. 2017; Kamphuis et al. 2015; Yang et al. 2012, 2013a, b, 2015a, b; Zhou et al. 2018) and current flanking or co-segregating markers for these traits range from low-throughput SSCP, CAPS/dCAPS, and HRM assays to high-throughput Fluidigm SNP and DArTseq assays. These sequence-based molecular markers can be converted to suit an individual lupin breeding programs genotyping platform/approach of choice. Successful conversion to PCR-based markers from sequencing data, can be achieved by aligning the sequences to a genome or transcriptome assembly, and this has been successfully applied in white lupin for markers linked to early flowering, anthracnose resistance, and low alkaloid content (Książkiewicz et al. 2017; Rychel and Książkiewicz 2019; Rychel et al. 2019).

References

Adhikari KN, Buirchell BJ, Sweetingham MW (2012) Length of vernalization period affects flowering time in three lupin species. Plant Breed 131:631–636. https://doi.org/10.1111/j.1439-0523.2012.01996.x

Berger JD, Buirchell BJ, Luckett DJ, Nelson MN (2012) Domestication bottlenecks limit genetic diversity and constrain adaptation in narrow-leafed lupin (*Lupinus angustifolius* L.). Theor Appl Genet 124:637–652. https://doi.org/10.1007/s00122-011-1736-z

Bielski W, Plewiński P, Rudy E, Rychel S, Książkiewicz M, Naganowska B, Wolko B (2016) Genotyping of *Lupinus angustifolius* L. core collection using sequence-defined markers linked to agronomic traits. In: Paper presented at the Polski Kongres Genetyki, Łódź, Poland, 19–22 September 2016

Boersma JG, Buirchell BJ, Sivasithamparam K, Yang H (2007a) Development of a PCR marker tightly linked to *mollis*, the gene that controls seed dormancy in *Lupinus angustifolius* L. Plant Breed 126:612–616. https://doi.org/10.1111/j.1439-0523.2007.01417.x

Boersma JG, Buirchell BJ, Sivasithamparam K, Yang H (2007b) Development of a sequence-specific PCR marker linked to the *Ku* gene which removes the vernalization requirement in narrow-leafed lupin. Plant Breed 126:306–309. https://doi.org/10.1111/j.1439-0523.2007.01347.x

Boersma JG, Buirchell BJ, Sivasithamparam K, Yang H (2007c) Development of two sequence-specific PCR markers linked to the *le* gene that reduces pod shattering in narrow-leafed lupin (*Lupinus angustifolius* L.). Genet Mol Biol 30:623–629. https://doi.org/10.1590/S1415-47572007000400020

Boersma JG, Nelson MN, Sivasithamparam K, Yang HA (2009) Development of sequence-specific PCR markers linked to the *Tardus* gene that reduces pod shattering in narrow-leafed lupin (*Lupinus angustifolius* L.). Mol Breed 23:259–267. https://doi.org/10.1007/s11032-008-9230-210.1007/s11032-008-9230-2

Boersma JG, Pallotta M, Li C, Buirchell BJ, Sivasithamparam K, Yang H (2005) Construction of a genetic linkage map using MFLP and identification of molecular markers linked to domestication genes in narrow-leafed lupin (*Lupinus angustifolius* L.). Cell Mol Biol Lett 10:331–344

Brien SJ, Cowling WA, Potter RH, Jones RAC, Jones MGK (1999) A molecular marker for early maturity (*Ku*) and marker-assisted breeding of *Lupinus angustifolius*. In: van Santen E, Wink M, Weissmann S, Roemer P (eds) Lupin, an ancient crop for the new millenium. 9th international lupin conference, Klink/Müritz, Germany. International Lupin Association, Canterbury, New Zealand, pp 115–117

Buirchell BJ, Yang H (2006) Breeding narrow-leafed lupins in Western Australia for yield, disease resistance and quality using recurrent selection and molecular markers. In: van Santen E, Hill GD (eds) México, where old and new world lupins meet. 11th international lupin conference, Guadalajara, Mexico, 2005. International Lupin Association, Canterbury, New Zealand, pp 10–13

Cowley R, Luckett DJ, Ash GJ, Harper JDI, Vipin CA, Raman H, Ellwood S (2014) Identification of QTLs associated with resistance to Phomopsis pod blight (*Diaporthe toxica*) in *Lupinus albus*. Breed Sci 64:83–89. https://doi.org/10.1270/jsbbs.64.83

Cowling WA (1999) Pedigrees and characteristics of narrow-leafed lupin cultivars released in Australia from 1967 to 1998. Bull Agric West Aust 4365:4–11. https://doi.org/10.13140/RG.2.1.1441.1600

Cowling WA, Buirchell BJ, Falk DE (2009) A model for incorporating novel alleles from the primary gene pool into elite crop breeding programs while reselecting major genes for domestication or adaptation. Crop Pasture Sci 60:1009–1015. https://doi.org/10.1071/CP08223

Cowling WA, Hamblin J, Wood PM, Gladstones JS (1987) Resistance to Phomopsis stem blight in *Lupinus angustifolius* L. Crop Sci 27:648–652. https://doi.org/10.2135/cropsci1987.0011183X002700040007x

Fischer K et al (2015) Characterization and mapping of *LanrBo*: a locus conferring anthracnose resistance in narrow-leafed lupin (*Lupinus angustifolius* L.). Theor Appl Genet 128:2121–2130. https://doi.org/10.1007/s00122-015-2572-3

Forbes I, Wells HD (1968) Hard and soft seededness in blue lupine, *Lupinus angustifolius* L.: inheritance and phenotype classification. Crop Sci 8:195–197. https://doi.org/10.2135/cropsci1968.0011183X000800020018x

Gladstones J (1967) Selection for economic characters in *Lupinus angustifolius* and *L. digitatus*. Aust J Exp Agric 7:360–366. https://doi.org/10.1071/EA9670360

Gladstones J, Hill G (1969) Selection for economic characters in *Lupinus angustifolius* and *L. digitatus*. 2 Time of flowering. Aust J Exp Agric 9:213–220. https://doi.org/10.1071/EA9690213

Gladstones JS (1970) Lupins as crop plants. Field Crop Abstr 23:26

Gladstones JS, Atkins CA, Hamblin J (1998) Lupins as crop plants: biology, production, and utilization. CAB International, New York

Grishin SY, Zayakin VV, Nam IY, Ageeva PA, Lukashevich MI, Kuptsov NS (2015) Identification of the *Lanr1* gene of resistance to anthracnose of narrow-leafed lupine (*Lupinus angustifolius* L.) using DNA-markers AnSeq3 and AnSeq4 Sel'skokhozyaistvennaya. Biologiya 50:30–36. https://doi.org/10.15389/agrobiology.2015.1.30eng

Hackbarth J, Troll HJ (1956) Die Lupinen als Körnerleguminosen und Futterpflanzen. Handbuch der Pflanzenzüchtung. Verlag Paul Parey, Berlin, pp 1–51

Hane JK et al (2017) A comprehensive draft genome sequence for lupin (*Lupinus angustifolius*), an emerging health food: insights into plant-microbe interactions and legume evolution. Plant Biotechnol J 15:318–330. https://doi.org/10.1111/pbi.12615

Jago MV, Peterson JE, Payne AL, Campbell DG (1982) Lupinosis: response of sheep to different doses of phomopsin. Aust J Exp Biol Med Sci 60:239–251. https://doi.org/10.1038/icb.1982.29

Kamel KA, Kroc M, Swiecicki W (2015) Application of the high resolution melting analysis for genetic mapping of sequence tagged site markers in narrow-leafed lupin (*Lupinus angustifolius* L.). Acta Biochim Pol 62:533–540. https://doi.org/10.18388/abp.2015_977

Kamel KA, Święcicki W, Kaczmarek Z, Barzyk P (2016) Quantitative and qualitative content of alkaloids in seeds of a narrow-leafed lupin (*Lupinus angustifolius* L.) collection. Genet Resour Crop Evol 63:711–719. https://doi.org/10.1007/s10722-015-0278-7

Kamphuis LG, Hane JK, Nelson MN, Gao L, Atkins CA, Singh KB (2015) Transcriptome sequencing of different narrow-leafed lupin tissue types provides a comprehensive uni-gene assembly and extensive gene-based molecular markers. Plant Biotechnol J 13:14–25. https://doi.org/10.1111/pbi.12229

Klinkowski M (1939) Beobachtungen über Krankheiten und Schädlinge iberischer Wildformen von Serradella und Lupine. Z Pflanzenkrankh (Pflanzenpathol) Pflanzenschutz 49:305–321

Konieczny A, Ausubel FM (1993) A procedure for mapping *Arabidopsis* mutations using co-dominant ecotype-specific PCR-based markers. Plant J 4:403–410. https://doi.org/10.1046/j.1365-313X.1993.04020403.x

Kozak B, Książkiewicz M, Plewiński P, Rychel S, Naganowska B, Galek R (2017) Integracja map łubinu wąskolistnego. In: Paper presented at the Znaczenie łubinu dla bioróżnorodności, Wrocław, Poland, 24–25 April 2017

Kroc M, Czepiel K, Wilczura P, Mokrzycka M, Swiecicki W (2019a) Development and validation of a gene-targeted dCAPS marker for marker-assisted selection of low-alkaloid content in seeds of narrow-leafed lupin (*Lupinus angustifolius* L.). Genes 10:428. https://doi.org/10.3390/genes10060428

Kroc M et al (2019b) Transcriptome-derived investigation of biosynthesis of quinolizidine alkaloids in narrow-leafed lupin (*Lupinus angustifolius* L.) highlights candidate genes linked to *iucundus* locus. Sci Rep 9:2231. https://doi.org/10.1038/s41598-018-37701-5

Kroc M, Koczyk G, Święcicki W, Kilian A, Nelson MN (2014) New evidence of ancestral polyploidy in the Genistoid legume *Lupinus angustifolius* L. (narrow-leafed lupin). Theor Appl Genet 127:1237–1249. https://doi.org/10.1007/s00122-014-2294-y

Kruszka K, Wolko B (1999) Linkage maps of morphological and molecular markers in lupin. In: van Santen E, Wink M, Weissmann S, Roemer P (eds) Lupin, an ancient crop for the new millenium. 9th international lupin conference, Klink/Müritz, Germany. International Lupin Association, Canterbury, New Zealand, pp 100–106

Książkiewicz M et al (2016) Profiling resistance of the narrow leafed lupin to pathogenic fungi: *Colletotrichum lupini* and *Diaporthe toxica*. In: Paper presented at the Polski Kongres Genetyki, Łódź, Poland, 19–22 September 2016

Książkiewicz M et al (2019) Validation of *Diaporthe toxica* resistance markers in European *Lupinus angustifolius* germplasm and identification of novel resistance donors for marker-assisted selection. J Appl Genet (in press). https://doi.org/10.1007/s13353-019-00521-y

Książkiewicz M et al (2017) A high-density consensus linkage map of white lupin highlights synteny with narrow-leafed lupin and provides markers tagging key agronomic traits. Sci Rep 7:15335. https://doi.org/10.1038/s41598-017-15625-w

Książkiewicz M et al (2013) Comparative genomics of *Lupinus angustifolius* gene-rich regions: BAC library exploration, genetic mapping and cytogenetics. BMC Genomics 14:79. https://doi.org/10.1186/1471-2164-14-79

Książkiewicz M et al (2015) Remnants of the legume ancestral genome preserved in gene-rich regions: insights from *Lupinus angustifolius* physical, genetic, and comparative mapping. Plant Mol Biol Rep 33:84–101. https://doi.org/10.1007/s11105-014-0730-4

Landers KF (1995) Vernalization responses in narrow-leafed lupin (*Lupinus angustifolius*) genotypes. Aust J Agric Res 46:1011–1025. https://doi.org/10.1071/AR9951011

Leśniewska K, Książkiewicz M, Nelson MN, Mahé F, Aïnouche A, Wolko B, Naganowska B (2011) Assignment of 3 genetic linkage groups to 3 chromosomes of narrow-leafed lupin. J Hered 102:228–236. https://doi.org/10.1093/jhered/esq107

Li X, Renshaw D, Yang H, Yan G (2010) Development of a co-dominant DNA marker tightly linked to gene *tardus* conferring reduced pod shattering in narrow-leafed lupin (*Lupinus angustifolius* L.). Euphytica 176:49–58. https://doi.org/10.1007/s10681-010-0212-1

Li X, Yang H, Buirchell B, Yan G (2011) Development of a DNA marker tightly linked to low-alkaloid gene *iucundus* in narrow-leafed lupin (*Lupinus angustifolius* L.) for marker-assisted selection. Crop Pasture Sci 62:218–224. https://doi.org/10.1071/CP10352

Li X, Buirchell B, Yan G, Yang H (2012a) A molecular marker linked to the *mollis* gene conferring soft-seediness for marker-assisted selection applicable to a wide range of crosses in lupin (*Lupinus angustifolius* L.) breeding. Mol Breed 29:361–370. https://doi.org/10.1007/s11032-011-9552-3

Li X, Yang H, Yan G (2012b) Development of a co-dominant DNA marker linked to the gene *lentus* conferring reduced pod shattering for marker-assisted selection in narrow-leafed lupin (*Lupinus angustifolius*) breeding. Plant Breed 131:540–544. https://doi.org/10.1111/j.1439-0523.2012.01978.x

Lin R et al (2009) Development of a sequence-specific PCR marker linked to the gene "pauper" conferring low-alkaloids in white lupin (*Lupinus albus* L.) for marker assisted selection. Mol Breed 23:153–161. https://doi.org/10.1007/s11032-008-9222-2

Litt M, Luty JA (1989) A hypervariable microsatellite revealed by in vitro amplification of a dinucleotide repeat within the cardiac muscle actin gene. Am J Hum Genet 44:397–401

Miao ZH, Fortune JA, Gallagher J (2001) Anatomical structure and nutritive value of lupin seed coats. Aust J Agric Res 52:985–993. https://doi.org/10.1071/AR00117

Mikolajczyk J (1966) Genetic studies in *Lupinus angustifolius*. 2. Inheritance of some morphological characters in blue lupine. Genetica Polonica 7:153–180

Mousavi-Derazmahalleh M et al (2018a) Exploring the genetic and adaptive diversity of a pan-Mediterranean crop wild relative: narrow-leafed lupin. Theor Appl Genet 131:887–901. https://doi.org/10.1007/s00122-017-3045-7

Mousavi-Derazmahalleh M et al (2018b) The western Mediterranean region provided the founder population of domesticated narrow-leafed lupin. Theor Appl Genet. https://doi.org/10.1007/s00122-018-3171-x

Narożna D, Książkiewicz M, Przysiecka Ł, Króliczak J, Wolko B, Naganowska B, Mądrzak CJ (2017) Legume isoflavone synthase genes have evolved by whole-genome and local duplications yielding transcriptionally active paralogs. Plant Sci 264:149–167. https://doi.org/10.1016/j.plantsci.2017.09.007

Neff MM, Neff JD, Chory J, Pepper AE (1998) dCAPS, a simple technique for the genetic analysis of single nucleotide polymorphisms: experimental applications in *Arabidopsis thaliana* genetics. Plant J 14:387–392. https://doi.org/10.1046/j.1365-313X.1998.00124.x

Neff MM, Turk E, Kalishman M (2002) Web-based primer design for single nucleotide polymorphism analysis. Trends Genet 18:613–615. https://doi.org/10.1016/S0168-9525(02)02820-2

Nelson MN et al (2017) The loss of vernalization requirement in narrow-leafed lupin is associated with a deletion in the promoter and de-repressed expression of a *Flowering Locus T* (*FT*) homologue. New Phytol 213:220–232. https://doi.org/10.1111/nph.14094

Nelson MN et al (2010) Aligning a new reference genetic map of *Lupinus angustifolius* with the genome sequence of the model legume *Lotus japonicus*. DNA Res 17:73–83. https://doi.org/10.1093/dnares/dsq001

Nelson MN et al (2006) The first gene-based map of *Lupinus angustifolius* L.-location of domestication genes and conserved synteny with *Medicago truncatula*. Theor Appl Genet 113:225–238. https://doi.org/10.1007/s00122-006-0288-0

Nirenberg HI, Feiler U, Hagedorn G (2002) Description of *Colletotrichum lupini* comb. nov. in modern terms. Mycologia 94:307–320. https://doi.org/10.2307/3761809

Poeverlein H (1936) Die Verbreitung der süddeutschen Uredineen Berichte der Bayerischen Botanischen Gesellschaft zur Erforschung der Flora 22:86-120

Przysiecka Ł, Książkiewicz M, Wolko B, Naganowska B (2015) Structure, expression profile and phylogenetic inference of chalcone isomerase-like genes from the narrow-leafed lupin (*Lupinus angustifolius* L.) genome. Front Plant Sci 6:268. https://doi.org/10.3389/fpls.2015.00268

Rahman M, Gladstones J (1972) Control of lupin flower initiation by vernalization, photoperiod and temperature under controlled environment. Aust J Exp Agric 12:638–645. https://doi.org/10.1071/EA9720638

Ruge-Wehling B, Dieterich R, Thiele C, Eickmeyer F, Wehling P (2009) Resistance to anthracnose in narrow-leafed lupin (*Lupinus angustifolius* L.): sources of resistanceand development of molecular markers. J kulturpflanzen 61:62–65

Rychel S, Książkiewicz M (2019) Development of gene-based molecular markers tagging low alkaloid pauper locus in white lupin (*Lupinus albus* L.). J Appl Genet 60:269–281. https://doi.org/10.1007/s13353-019-00508-9

Rychel S, Książkiewicz M, Tomaszewska M, Bielski W, Wolko B (2019) *FLOWERING LOCUS T*, *GIGANTEA*, *SEPALLATA* and *FRIGIDA* homologs are candidate genes involved in white lupin (*Lupinus albus* L.) early flowering. Mol Breed 39:43. https://doi.org/10.1007/s11032-019-0952-0d

Shan X, Blake TK, Talbert LE (1999) Conversion of AFLP markers to sequence-specific PCR markers in barley and wheat. Theor Appl Genet 98:1072–1078. https://doi.org/10.1007/s001220051169

Shankar M, Cowling WA, Sweetingham MW (1996) The expression of resistance to latent stem infection by *Diaporthe toxica* in narrow-leafed lupins. Phytopathology 692–697. https://doi.org/10.1094/Phyto-86-692

Shankar M, Sweetingham MW, Cowling WA (2002) Identification of alleles at two loci controlling resistance to Phomopsis stem blight in narrow-leafed lupin (*Lupinus angustifolius* L.). Euphytica 125:35–44. https://doi.org/10.1023/A:1015704728492

Smýkal P, Vernoud V, Blair MW, Soukup A, Thompson RD (2014) The role of the testa during development and in establishment of dormancy of the legume seed. Front Plant Sci 5. https://doi.org/10.3389/fpls.2014.00351

Stefanova KT, Buirchell B (2010) Multiplicative mixed models for genetic gain assessment in lupin breeding. Crop Sci 50:880–891. https://doi.org/10.2135/cropsci2009.07.0402

Sunnucks P, Wilson AC, Beheregaray LB, Zenger K, French J, Taylor AC (2000) SSCP is not so difficult: the application and utility of single-stranded conformation polymorphism in evolutionary biology and molecular ecology. Mol Ecol 9:1699–1710. https://doi.org/10.1046/j.1365-294x.2000.01084.x

Sweetingham M, Yang H, Buirchell BJ, Shea G, Shield I (2006) Resistance to rust in narrow-leafed lupin and development of molecular markers. In: van Santen E, Hill GD (eds) México, where old and new world lupins meet. 11th International Lupin Conference, Guadalajara, Mexico, 2005. International Lupin Association, Canterbury, New Zealand, pp 14–16

Szczepaniak A, Książkiewicz M, Podkowiński J, Czyż KB, Figlerowicz M, Naganowska B (2018) Legume cytosolic and plastid acetyl-coenzyme-A carboxylase genes differ by evolutionary patterns and selection pressure schemes acting before and after whole-genome duplications. Genes 9:563. https://doi.org/10.3390/genes9110563

Święcicki W, Święcicki WK (1995) Domestication and breeding improvement of narrow-leafed lupin (*L. angustifolius* L.). J Appl Genet 36:155–167

Taylor CM et al (2019) INDEL variation in the regulatory region of the major flowering time gene *LanFTc1* is associated with vernalization response and flowering time in narrow-leafed lupin (*Lupinus angustifolius* L.). Plant Cell Environ 42:174–187. https://doi.org/10.1111/pce.13320

von Sengbusch R (1942) Süßlupinen und Öllupinen Die Entstehungsgeschichte einiger neuer Kulturpflanzen. Landwirtsch Jahrb 91:719–880

Vos P et al (1995) AFLP: a new technique for DNA fingerprinting. Nucleic Acids Res 23:4407–4414. https://doi.org/10.1093/nar/23.21.4407

Williamson PM, Highet AS, Gams W, Sivasithamparam K, Cowling WA (1994) *Diaporthe toxica* sp. nov., the cause of lupinosis in sheep. Mycol Res 98:1364–1368. https://doi.org/10.1016/S0953-7562(09)81064-2

Williamson PM, Sivasithamparam K, Cowling WA (1991) Formation of subcuticular coralloid hyphae by *Phomopsis leptostromiformis* upon latent infection of narrow-leafed lupins. Plant Dis 75:1023–1026. https://doi.org/10.1094/PD-75-1023

Wolko B (1995) Markery molekularne w badaniach genetycznych rodzaju Lupinus Hodowla Roślin. Aklimatyzacja i Nasiennictwo 35:3–64

Wu KS, Jones R, Danneberger L, Scolnik PA (1994) Detection of microsatellite polymorphisms without cloning. Nucleic Acids Res 22:3257–3258. https://doi.org/10.1093/nar/22.15.3257

Wyrwa K, Książkiewicz M, Szczepaniak A, Susek K, Podkowiński J, Naganowska B (2016) Integration of *Lupinus angustifolius* L. (narrow-leafed lupin) genome maps and comparative mapping within legumes. Chromosome Res 24:355–378. https://doi.org/10.1007/s10577-016-9526-8

Yang H, Boersma JG, You M, Buirchell BJ, Sweetingham MW (2004) Development and implementation of a sequence-specific PCR marker linked to a gene conferring resistance to anthracnose disease in narrow-leafed lupin (*Lupinus angustifolius* L.). Mol Breed 14:145–151. https://doi.org/10.1023/B:MOLB.0000038003.49638.97

Yang H et al (2015a) Application of whole genome re-sequencing data in the development of diagnostic DNA markers tightly linked to a disease-resistance locus for marker-assisted selection in lupin (*Lupinus angustifolius*). BMC Genomics 16:660. https://doi.org/10.1186/s12864-015-1878-5

Yang H, Li C, Lam HM, Clements J, Yan G, Zhao S (2015b) Sequencing consolidates molecular markers with plant breeding practice. Theor Appl Genet 128:779–795. https://doi.org/10.1007/s00122-015-2499-8

Yang H et al (2010) Development of sequence-specific PCR markers associated with a polygenic controlled trait for marker-assisted selection using a modified selective genotyping strategy: a case study on anthracnose disease resistance in white lupin (*Lupinus albus* L.). Mol Breed 25:239–249. https://doi.org/10.1007/s11032-009-9325-4

Yang H, Renshaw D, Thomas G, Buirchell B, Sweetingham M (2008) A strategy to develop molecular markers applicable to a wide range of crosses for marker assisted selection in plant breeding: a case study on anthracnose disease resistance in lupin (*Lupinus angustifolius* L.). Mol Breed 21:473–483. https://doi.org/10.1007/s11032-007-9146-2

Yang H, Shankar M, Buirchell J, Sweetingham W, Caminero C, Smith C (2002) Development of molecular markers using MFLP linked to a gene conferring resistance to *Diaporthe toxica* in narrow-leafed lupin (*Lupinus angustifolius* L.). Theor Appl Genet 105:265–270. https://doi.org/10.1007/s00122-002-0925-1

Yang H, Sweetingham MW, Cowling WA, Smith PMC (2001) DNA fingerprinting based on microsatellite-anchored fragment length polymorphisms, and isolation of sequence-specific PCR markers in lupin (*Lupinus angustifolius* L.). Mol Breed 7:203–209. https://doi.org/10.1023/A:1011363205557

Yang H, Tao Y, Zheng Z, Li C, Sweetingham MW, Howieson JG (2012) Application of next-generation sequencing for rapid marker development in molecular plant breeding: a case study on anthracnose disease resistance in *Lupinus angustifolius* L. BMC Genomics 13:318. https://doi.org/10.1186/1471-2164-13-318

Yang H et al (2013a) Rapid development of molecular markers by next-generation sequencing linked to a gene conferring phomopsis stem blight disease resistance for marker-assisted selection in lupin (*Lupinus angustifolius* L.) breeding. Theor Appl Genet 126:511–522. https://doi.org/10.1007/s00122-012-1997-1

Yang H et al (2013b) Draft genome sequence, and a sequence-defined genetic linkage map of the legume crop species *Lupinus angustifolius* L. PLoS ONE 8: e64799. https://doi.org/10.1371/journal.pone.0064799

You M, Boersma JG, Buirchell BJ, Sweetingham MW, Siddique KHM, Yang H (2005) A PCR-based molecular marker applicable for marker-assisted selection for anthracnose disease resistance in lupin breeding. Cell Mol Biol Lett 10:123–134

Zhou G et al (2018) Construction of an ultra-high density consensus genetic map, and enhancement of the physical map from genome sequencing in *Lupinus angustifolius*. Theor Appl Genet 131:209–223. https://doi.org/10.1007/s00122-017-2997-y

Chromosomal Structure, History, and Genomic Synteny Relationships in *Lupinus*

7

Steven B. Cannon

Abstract

The genome assembly of narrow-leafed lupin (NLL, *Lupinus angustifolius*) provides critical information for inferring the evolutionary history of NLL relative to other species in the large genistoid clade and to other crop and model legumes. Based on analyses of genomic synteny, phylogenetic reconstructions, gene families, and chromosome counts, it is apparent that the ancestor of all lupin species experienced a whole-genome triplication approximately 20–30 million years ago; and that multiple chromosomal breakages, fusions, and independent duplications have subsequently led to various chromosomal counts in *Lupinus*. In comparison with other sequenced crop and model legume species, substantial genomic synteny is evident, frequently extending tens of megabases. Such synteny is useful information when looking for orthologous loci that may share functions identified in other legume species. At the same time, the early triplication in *Lupinus* and subsequent rearrangements in the genistoid and other legume lineages have scrambled the respective genomes at the chromosomal level; and the evolutionary distance to other crop and model legumes is substantial, with an estimated ~56 million years to the common ancestor.

7.1 Introduction

The genus *Lupinus* is large, diverse, and relatively old, with c. 220–230 species (Lavin et al. 2005), and distinct Old-World and New-World species groups (Hughes and Eastwood 2006; Naganowska et al. 2003). The diversity of species in the genus provides information for better understanding evolution of the *Lupinus* genomes—and then for understanding evolution of *Lupinus* in comparison to the many other crop and model legume species. With genome assemblies available for most such species, including for narrow-leafed lupin (NLL; *Lupinus angustifolius* L.), synteny comparisons also provide maps of correspondences between genome assemblies—at both macro scales (e.g., chromosome regions) and micro scales (e.g., genes and regulatory elements). This chapter will also briefly review online tools that can help researchers access and explore genomic data for NLL and to examine corresponding regions in other well-studied legume species.

7.1.1 Chromosomal Background and NLL Assembly

New-World species include the food legume *L. mutabilis* Sweet (tarwi or perl lupin), and

S. B. Cannon (✉)
USDA-Agricultural Research Service, Corn Insects and Crop Genetics Research Unit, Ames, IA 50011, USA
e-mail: Steven.Cannon@USDA.GOV

K. B. Singh et al. (eds.), *The Lupin Genome*, Compendium of Plant Genomes,
https://doi.org/10.1007/978-3-030-21270-4_7

Old-World domesticated species include *L. albus* L. (white lupin), *L. albus* L. (yellow lupin), and *L. angustifolius* L. Haploid chromosome counts range from 16 to 25 for Old-World species, and from 22 to 26 for New-World species (Naganowska et al. 2003).

The haploid genome size of NLL has been estimated, using flow cytometry, to be ~924 Mb (Kasprzak et al. 2006; Naganowska et al. 2003). The genome assembly reported in Hane et al. (2017) spans 609 Mb in scaffolds larger than 200 bp, with 470 Mb in pseudomolecule assemblies and 139 Mb in remaining, unplaced scaffolds. Scaffolds were placed into pseudomolecules using a dense, high-resolution genetic map of 9,972 markers (Gao et al. 2011; Kamphuis et al. 2015; Kroc et al. 2014). Although the proportion of sequence represented in the pseudomolecule assemblies is only approximately half the estimated total genome size, the assembly captures the large majority (98%) of core eukaryotic genes in the CEGMA gene set (Hane et al. 2017; Parra et al. 2007), and more than 98% of transcriptome assemblies from diverse NLL tissues mapped to the assembly (Hane et al. 2017). These are indications that the 2016 NLL assembly can serve as a useful reference for most NLL genes.

Genomic structural analyses of the 2016 NLL assembly also indicate that the pseudomolecules are sufficiently complete and well-ordered for the assembly to serve as the basis for evaluating the structural history of the genome. A plot of the genetic map compared with the physical assemblies (Fig. 7.1) shows generally smooth ordering of genomic sequence on all chromosomes, as well as features that are common in genetic-by-physical plots in other species: namely, high rates of recombination at chromosome ends (indicated by steep slopes) and regions of low recombination in pericentromeric regions (indicated by shallow slopes).

Of 105 *Lupinus* species with chromosome counts in the Chromosome Counts Database (Rice et al. 2015), the largest number of species have 1 N (haploid) counts of 24 (64 species) or 18 (19 species). There are small numbers of species with other counts: 4@16, 19@18, 2@19, 1@20, 3@21, 64@24, 3@25, 7@26, 2@48. The haploid chromosomal count of 20 for NLL is, therefore, atypical for the genus—and the counts across the genus are generally higher than for other crop and model legumes. For example, the most frequent 1 N count in the *Phaseoleae* tribe is 11. This tribe includes bean (*Phaseolus*) and cowpea (*Vigna*), and soybean (*Glycine*), as well as less commonly known crop species such as winged bean (*Psophocarpus*) and Bambara groundnut (*Vigna subterranea*). Among these, only *Glycine* has a higher count—of 20 chromosomes, resulting from a whole-genome duplication (WGD) ~10 Mya (Schmutz et al. 2010). Similarly, the 1 N counts for many cool-season legume are lower than for *Lupinus*: 8 for *Medicago* and *Cicer* and *Vicia*, 7 for *Lotus* and *Pisum*. Cultivated peanut, in the dalbergioid clade, has 20 chromosomes—but this is a recent allotetraploid, derived from the merger of two 10-chromosome diploid *Arachis* progenitors (Bertioli et al. 2016). What explains the higher chromosome counts in *Lupinus*—and the atypical count in NLL?

7.1.2 Phylogenetic and Synteny Evidence Indicate a Whole-Genome Triplication in *Lupinus*

The first report of polyploidy in NLL was by Kroc et al. (2014), primarily on the basis of comparisons synteny between an NLL genetic map and genome sequence from *Medicago truncatula*. There are now several additional lines of evidence that clarify the timing and the nature of the polyploidy. These indicate that lupin underwent a whole-genome triplication (WGT) within the genistoid clade, approximately 20–30 Mya, on top of the older (~55 Mya) WGD that predated the papilionoid radiation (Cannon et al. 2015). Briefly outlining these lines of evidence:

- First: Synteny comparisons with legumes such as *Lotus japonicus* and *Phaseolus vulgaris* generally show three-to-one synteny relationships for any given region in the

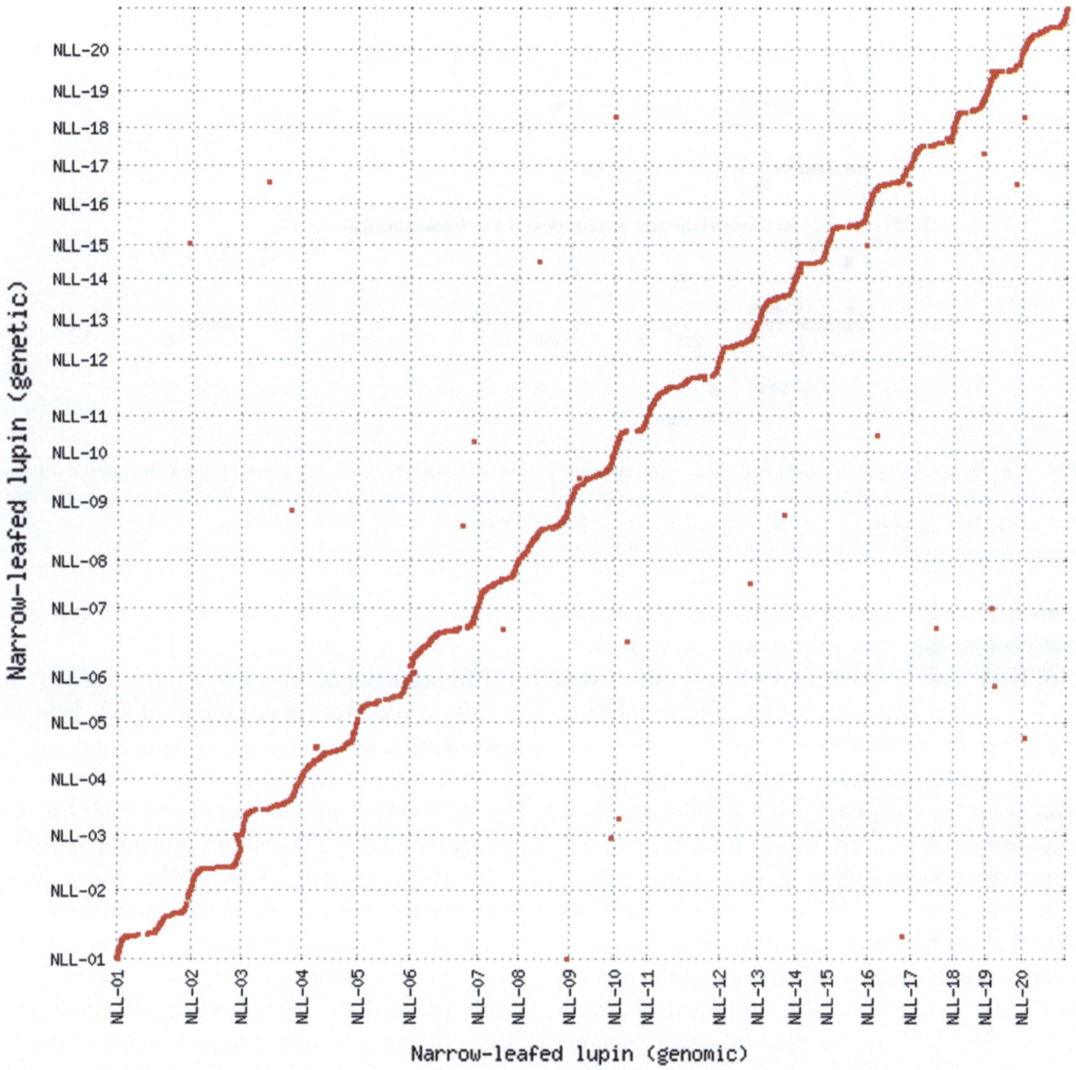

Fig. 7.1 Plot of genetic map (Y-axis) by physical sequence (X-axis) for narrow-leafed lupin (NLL). The genetic map consists of 2959 framework markers from a high-resolution cross (Hane et al. 2017; Kroc et al. 2014). The physical map consists of the chromosomal pseudo-molecule assemblies for NLL

comparison species—or six-to-two relationships if the comparisons are relaxed to include older syntenic matches.

- Second: gene families frequently show triplicated lupin paralogs, deriving after the split with other (non-genistoid) legume species.
- Third: chromosome counts are consistent with tripling from 9 chromosomes in the basal genistoid species, to 27, followed by moderate reduction to ~20–25 chromosomes.

Synteny evidence of a WGT is illustrated in Fig. 7.2, which shows comparisons of NLL chromosome 13 with the common-bean genome (*Phaseolus vulgaris*) and with the rest of the NLL genome, from the Legume Information System (https://legumeinfo.org; Dash et al. 2016). Colored blocks indicate conserved synteny (runs of genes that are generally in the same order in both genomic regions). In the comparison with *Phaseolus*, most NLL regions have

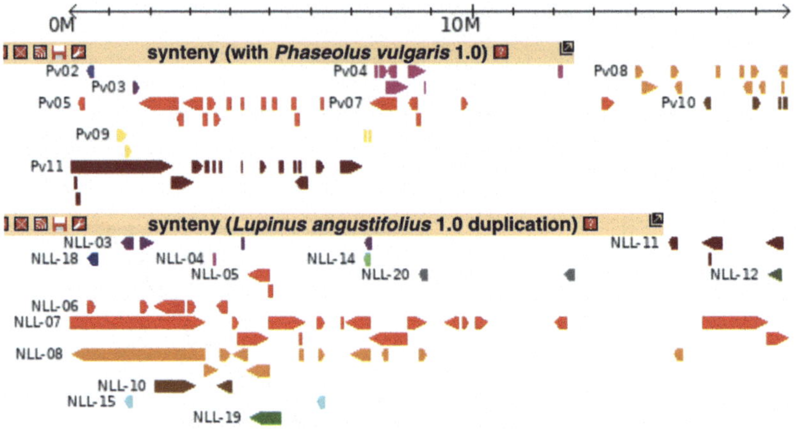

Fig. 7.2 Genomic synteny between NLL chromosome 13 and common bean (*Phaseolus vulgaris*) and with the rest of the NLL genome. Colored blocks of a given color represent synteny (blocks of genes that occur in the same order in both genomic regions). Pointed ends in the blocks indicate the relative direction of the alignments in the two regions

correspondences with two *Phaseolus* regions. In the comparison with NLL, most NLL regions also have correspondences with two other NLL regions (often large), as well as additional fragmented correspondences.

The correspondence with two regions in *Phaseolus* is consistent with a whole-genome duplication shortly predating the origin of the Papilionoideae subfamily. Because the genistoid clade diverged early in the Papilionoideae from the clade containing the Phaseoleae and the cool-season legume crops, the evolutionary distances are similar from NLL to either duplicate in *Phaseolus*, so there are generally two syntenic regions for any NLL region. For example, in Fig. 7.2, the first half of NLL chromosome 13 (NLL-13) matches Pv05 (red) and Pv11 (brown)—albeit with fragmentation.

The correspondences of NLL to itself are also consistent both with a genistoid WGT and with the pre-papilionoid WGD. The expected pattern for a WGT is for the reference chromosome (in this case, NLL-13) to match two other chromosomes, for a total of three corresponding chromosomes. For example, the left side of NLL-13 matches NLL-07 (red) and NLL-08 (orange). Evidence of the pre-papilionoid WGD is seen in the other, more fragmented matches—for example, NLL-03, -06, and -10, from the left side of the reference chromosome, NLL-13.

The second line of evidence for a WGT in *Lupinus* comes from gene families. Many legume gene families show clusters of paralogous genes in NLL. One such gene family is shown in Fig. 7.3—which provides evidence both for the pre-papilionoid WTD and the genistoid WGT.

In the two clusters of three NLL genes, each gene comes from a different chromosome or scaffold, indicating that these gene duplications result from a genome-wide event rather than local duplications. Note two nearly complete clades of the represented legume species, deriving from the pre-papilionoid WGD, and an additional gene duplication affecting Glycine and a WGT affecting *Lupinus*.

The third line of evidence for a WGT in *Lupinus* is the combined phylogenetic and chromosomal count data in *Lupinus* and other genistoid species.

A selection of species from the large genistoid clade, combined with the modal 1 N chromosome counts for genera (Fig. 7.4), illustrates that the most parsimonious basal chromosome count for the clade is 9, but following the split of *Dilchius* from the sister clade containing *Lupinus*, the most frequent 1 N chromosome counts are 24

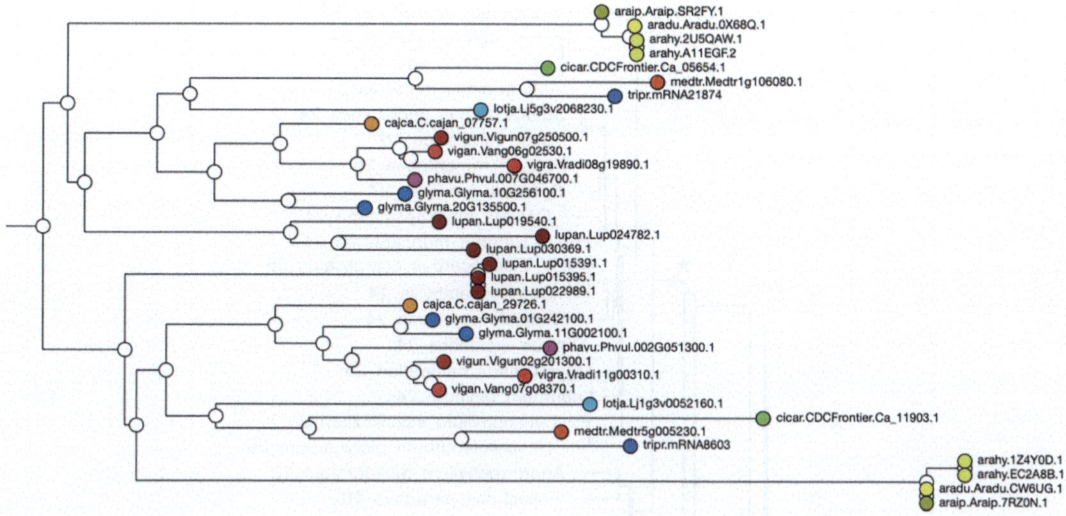

Fig. 7.3 Legume gene family L_36MFSW (an uncharacterized protein) from LegumeInfo, https://legumeinfo.org/search/phylotree. Genes are colored by species. Five-letter prefixes are composed of the three first letters of the genus and the first two letters of the species epithet, e.g., "glyma" for *Glycine max* or "lupan" for *Lupinus angustifolius*

(8 genera), 25 (3 genera), and 26 (2 genera). This is consistent with a genome triplication, increasing the 1 N count from 9 to 27, at around 20 Mya, based on the legume phylogenetic timings of (Lavin et al. 2005). The phylogenetic tree is calculated based on an alignment of the maturase K (matK), from (LPWG 2017).

Following the WGT in the genistoid clade, a series of further genomic rearrangements, splits, and fusions have resulted in varying chromosome counts in *Lupinus*. Examples of these changes can be seen in Fig. 7.5, which shows the comparison of a white lupin (WL) genetic map (Ksiazkiewicz et al. 2017) by the NLL genomic sequence. The 25 WL linkage groups (representing the 25 WL haploid chromosomes) generally correspond with the 20 NLL chromosomes fairly directly, but with various fusions, splits, and rearrangements. For example, NLL-01 corresponds with WL-12 and WL-13, albeit with apparent rearrangements on both sides. NLL-02 corresponds with WL-04, albeit with an apparent large inversion in one of the species.

7.2 Online Resources for Lupin Genetic and Genomic Research

Several online repositories have tools that should be useful to lupin researchers. The Lupin Genome Portal at https://www.lupinexpress.org has genomic sequences for NLL, including sequence download, genetic map viewer, and BLAST search tools. The Legume Information System at https://legumeinfo.org (Dash et al. 2016) also has genome viewers (GBrowse and JBrowse; see Fig. 7.2 for example), gene and genome sequences for download, a gene family viewer, BLAST search, an annotation service for annotating submitted sequences, and a microsynteny viewer. The microsynteny viewer ("Context Viewer"), accessed via https://legumeinfo.org/lis_context_viewer, displays corresponding gene sets from available sequenced legume genomes. The Gene Tree Viewer (see Fig. 7.3 for example), the Context Viewer, and BLAST sequence search all provide methods that enable a

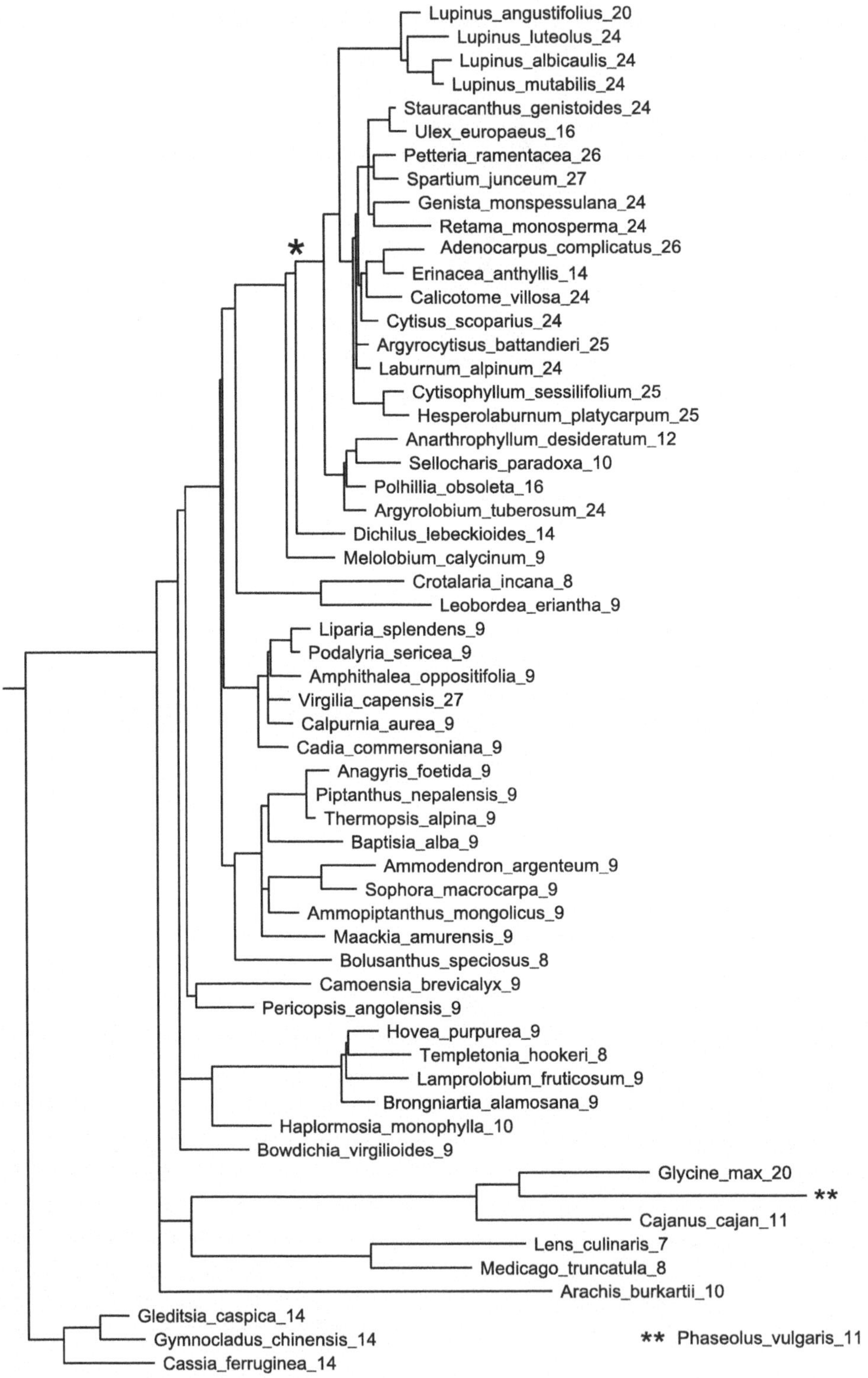

Fig. 7.4 Species tree, calculated from matK sequences (LPWG 2017), for species in the genistoid clade and selected other legume species. Numbers following each species are the haploid chromosome counts typical for that genus, from the Chromosome Counts Database (Rice et al. 2015). Large asterisk marks hypothesized transition from 9 chromosomes to 27. The tree is rooted with sequences from the Caesalpinioideae subfamily

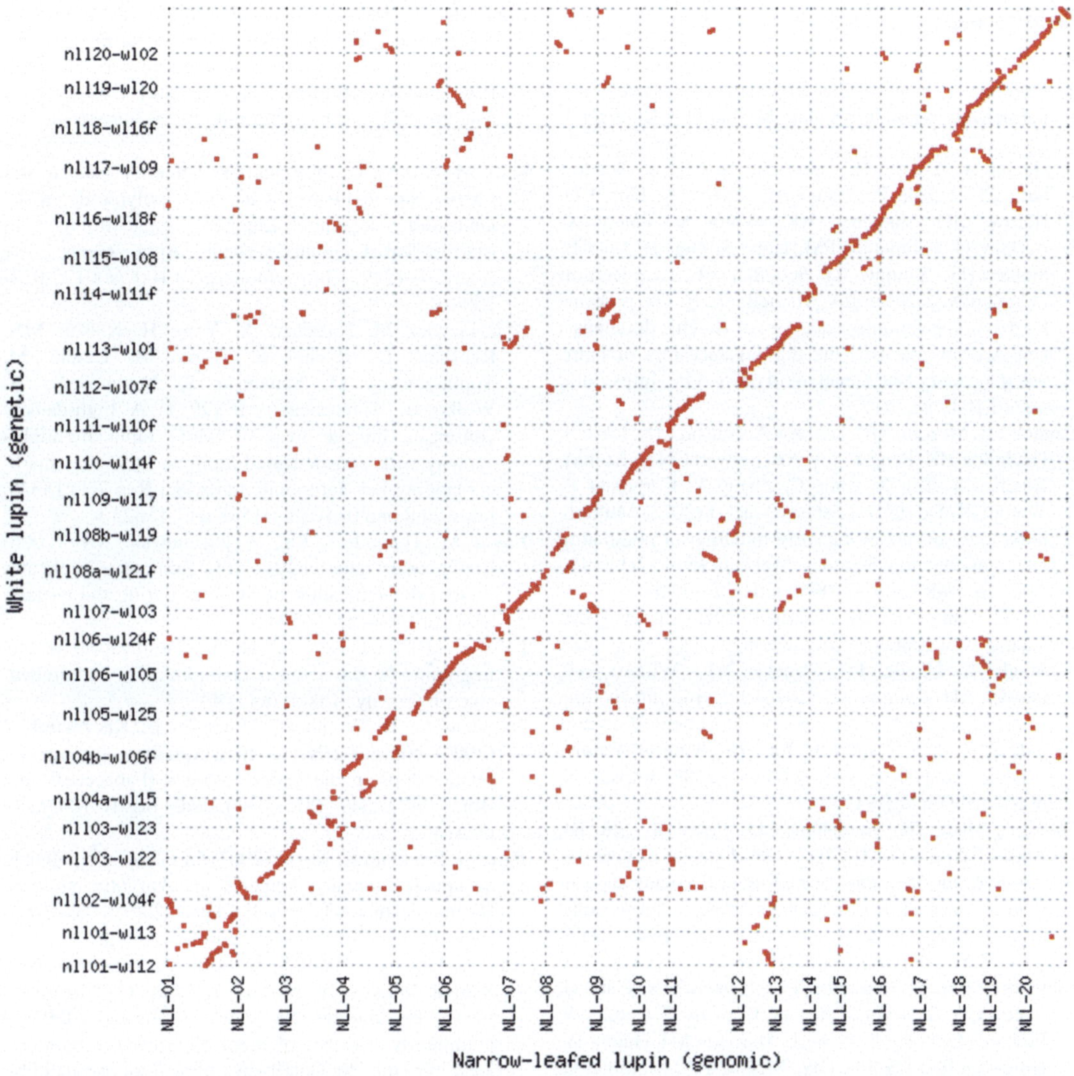

Fig. 7.5 Plot of white lupin (WL) genetic map (Y-axis) by narrow-leafed lupin (NLL) physical sequence (X-axis). The genetic map consists of 3624 framework markers in a high-resolution cross (Ksiazkiewicz et al. 2017). Linkage groups in the WL genetic map have been reordered to correspond (approximately) with the order of the NLL genome. The WL linkage group names have this format: nll02-wl04f, where "nll02" indicates correspondence with NLL chromosome 2, "wl04" is the WL linkage group, and the terminal "f" indicates that the linkage group has been flipped for correspondence with respect to NLL

researcher to identify genes and regions that correspond across many legume species. This can be useful for investigating orthologs in other legumes, starting from lupin sequences with suspected function, and for discovering lupin sequences with homology to genes in other legumes for which functions have been determined.

References

Bertioli DJ, Cannon SB, Froenicke L, Huang G, Farmer AD, Cannon EK, Liu X, Gao D, Clevenger J, Dash S, Ren L, Moretzsohn MC, Shirasawa K, Huang W, Vidigal B, Abernathy B, Chu Y, Niederhuth CE, Umale P, Araujo AC, Kozik A, Kim KD, Burow MD, Varshney RK, Wang X, Zhang X, Barkley N, Guimaraes PM, Isobe S, Guo B, Liao B, Stalker HT, Schmitz RJ, Scheffler BE, Leal-Bertioli SC, Xun X, Jackson SA, Michelmore R, Ozias-Akins P (2016) The genome sequences of Arachis duranensis and Arachis ipaensis, the diploid ancestors of cultivated peanut. Nat Genet 48(4):438–446. https://doi.org/10.1038/ng.3517

Cannon SB, McKain MR, Harkess A, Nelson MN, Dash S, Deyholos MK, Peng Y, Joyce B, Stewart CN Jr, Rolf M, Kutchan T, Tan X, Chen C, Zhang Y, Carpenter E, Wong GK, Doyle JJ, Leebens-Mack J (2015) Multiple polyploidy events in the early radiation of nodulating and nonnodulating legumes. Mol Biol Evol 32(1):193–210. https://doi.org/10.1093/molbev/msu296

Dash S, Campbell JD, Cannon EK, Cleary AM, Huang W, Kalberer SR, Karingula V, Rice AG, Singh J, Umale PE, Weeks NT, Wilkey AP, Farmer AD, Cannon SB (2016) Legume information system (LegumeInfo.org): a key component of a set of federated data resources for the legume family. Nucleic Acids Res 44(D1):D1181–1188. https://doi.org/10.1093/nar/gkv1159

Gao LL, Hane JK, Kamphuis LG, Foley R, Shi BJ, Atkins CA, Singh KB (2011) Development of genomic resources for the narrow-leafed lupin (Lupinus angustifolius): construction of a bacterial artificial chromosome (BAC) library and BAC-end sequencing. BMC Genomics 12:521. https://doi.org/10.1186/1471-2164-12-521

Hane JK, Ming Y, Kamphuis LG, Nelson MN, Garg G, Atkins CA, Bayer PE, Bravo A, Bringans S, Cannon S, Edwards D, Foley R, Gao LL, Harrison MJ, Huang W, Hurgobin B, Li S, Liu CW, McGrath A, Morahan G, Murray J, Weller J, Jian J, Singh KB (2017) A comprehensive draft genome sequence for lupin (Lupinus angustifolius), an emerging health food: insights into plant-microbe interactions and legume evolution. Plant Biotechnol J 15(3):318–330. https://doi.org/10.1111/pbi.12615

Hughes C, Eastwood R (2006) Island radiation on a continental scale: exceptional rates of plant diversification after uplift of the Andes. Proc Natl Acad Sci USA 103(27):10334–10339. https://doi.org/10.1073/pnas.0601928103

Kamphuis LG, Hane JK, Nelson MN, Gao L, Atkins CA, Singh KB (2015) Transcriptome sequencing of different narrow-leafed lupin tissue types provides a comprehensive uni-gene assembly and extensive gene-based molecular markers. Plant Biotechnol J 13 (1):14–25. https://doi.org/10.1111/pbi.12229

Kasprzak A, Safar J, Janda J, Dolezel J, Wolko B, Naganowska B (2006) The bacterial artificial chromosome (BAC) library of the narrow-leafed lupin (Lupinus angustifolius L.). Cell Mol Biol Lett 11 (3):396–407. https://doi.org/10.2478/s11658-006-0033-3

Kroc M, Koczyk G, Swiecicki W, Kilian A, Nelson MN (2014) New evidence of ancestral polyploidy in the Genistoid legume Lupinus angustifolius L. (narrow-leafed lupin). Theor Appl Genet 127 (5):1237–1249. https://doi.org/10.1007/s00122-014-2294-y

Ksiazkiewicz M, Nazzicari N, Yang H, Nelson MN, Renshaw D, Rychel S, Ferrari B, Carelli M, Tomaszewska M, Stawinski S, Naganowska B, Wolko B, Annicchiarico P (2017) A high-density consensus linkage map of white lupin highlights synteny with narrow-leafed lupin and provides markers tagging key agronomic traits. Sci Rep 7(1):15335. https://doi.org/10.1038/s41598-017-15625-w

Lavin M, Herendeen PS, Wojciechowski MF (2005) Evolutionary rates analysis of Leguminosae implicates a rapid diversification of lineages during the tertiary. Syst Biol 54(4):575–594

LPWG (2017) A new subfamily classification of the Leguminosae based on a taxonomically comprehensive phylogeny. Taxon 66(1):44–77

Naganowska B, Wolko B, Sliwinska E, Kaczmarek Z (2003) Nuclear DNA content variation and species relationships in the genus Lupinus (Fabaceae). Ann Bot 92(3):349–355. https://doi.org/10.1093/aob/mcg145

Parra G, Bradnam K, Korf I (2007) CEGMA: a pipeline to accurately annotate core genes in eukaryotic genomes. Bioinformatics (Oxford, England) 23(9):1061–1067. https://doi.org/10.1093/bioinformatics/btm071

Rice A, Glick L, Abadi S, Einhorn M, Kopelman NM, Salman-Minkov A, Mayzel J, Chay O, Mayrose I (2015) The chromosome counts database (CCDB)—a community resource of plant chromosome numbers. New Phytol 206(1):19–26. https://doi.org/10.1111/nph.13191

Schmutz J, Cannon SB, Schlueter J, Ma J, Mitros T, Nelson W, Hyten DL, Song Q, Thelen JJ, Cheng J, Xu D, Hellsten U, May GD, Yu Y, Sakurai T, Umezawa T, Bhattacharyya MK, Sandhu D, Valliyodan B, Lindquist E, Peto M, Grant D, Shu S, Goodstein D, Barry K, Futrell-Griggs M, Abernathy B, Du J, Tian Z, Zhu L, Gill N, Joshi T, Libault M, Sethuraman A, Zhang XC, Shinozaki K, Nguyen HT, Wing RA, Cregan P, Specht J, Grimwood J, Rokhsar D, Stacey G, Shoemaker RC, Jackson SA (2010) Genome sequence of the palaeopolyploid soybean. Nature 463(7278):178–183. https://doi.org/10.1038/nature08670

How Have Narrow-Leafed Lupin Genomic Resources Enhanced Our Understanding of Lupin Domestication?

8

Jemma L. Taylor, Gabriella De Angelis and Matthew N. Nelson

Abstract

Lupins provide an insightful model for plant domestication with five species domesticated over a wide range of time and geography. The most intensively studied species is narrow-leafed lupin, a twentieth-century domesticate where the addition of each successive domestication trait was documented in the scientific literature. Foundational to the advances made in our understanding of lupin domestication was the availability of excellent genetic resources: Well-annotated wild seed collections, published pedigrees of Australian narrow-leafed lupin cultivars and a suite of wild × domesticated cross populations. Rapid developments in genomic technologies culminating in the reference genome for narrow-leafed lupin have greatly increased our understanding of the origins of domesticated lupins, how diversity has been profoundly affected and the molecular control of domestication genes. This chapter provides an overview of our current understanding of lupin domestication and how this knowledge can equip lupin breeders to create more diverse and productive cultivars.

8.1 Background

The legume genus *Lupinus* is special in many respects, not least its extraordinary history of domestication of several species across a wide range of time and geography. *Lupinus* encompasses around 275 species distributed across the Mediterranean region and North Africa ('Old World' lupins) and the Americas ('New World' lupins) (Hughes and Eastwood 2006). While Old World lupins represent just 13 of those 275 species, four Old World species can be considered fully domesticated, while several others show signs of historic cultivation and selection (Swiecicki and Swiecicki 1995). The oldest fully domesticated species is white lupin (*L. albus* L.). There is clear evidence that white lupin was cultivated in Egypt by 300 BC and possibly as early as 2000–1000 BC (Wolko et al. 2011). Three further Old World species are more modern domesticates: Narrow-leafed lupin (*L. angustifolius* L.), yellow lupin (*L. luteus* L.) and West Australian blue lupin or sandplain lupin (*L. cosentinii* Guss.; often mistaken for *L. digitatus* Forsk.). Despite the majority of *Lupinus* species being from the New World, just one—Andean

J. L. Taylor · G. De Angelis · M. N. Nelson (✉)
Royal Botanic Gardens, Kew, Ardingly, UK
e-mail: matthew.nelson@csiro.au

J. L. Taylor
e-mail: j.taylor2@kew.org

M. N. Nelson
The University of Western Australia, Perth, Australia

CSIRO Agriculture & Food, Perth, Australia

K. B. Singh et al. (eds.), *The Lupin Genome*, Compendium of Plant Genomes,
https://doi.org/10.1007/978-3-030-21270-4_8

lupin (*L. mutabilis* Sweet)—has been domesticated, likely 3000–4000 years ago.

Today, the most widely cultivated species are *L. angustifolius* and *L. albus*, while *L. luteus* and *L. mutabilis* are niche crops, and *L. cosentinii* is not currently cultivated to our knowledge. The production of lupin seeds as an agricultural product still occurs mainly in Australia but also in parts of Europe, Africa and South America. Although production has fluctuated over the last 20 years, over a million tonnes are produced every year. In 2017, the largest producers were Australia (1,031,425 t), Poland (168,678 t) and the Russian Federation (161,680 tonnes) (FAO 2017).

It is easy to understand why lupins attracted the attention of early farmer and hunter-gatherers: their seeds are large and highly nutritious with protein contents of around 30–45%, comparable to soybean (Lucas et al. 2015). Wild types and landraces are high in bitter quinolizidine alkaloids but humans quickly learned to remove most of the alkaloids by soaking and rinsing in water. As a snack food some residual bitterness in the lupin seeds provides a pleasant, distinctive flavour, which remains popular in Spain ('altramuces'), Italy ('lupini'), Ethiopia ('gibto'), Egypt and Sudan ('termes') and South America ('tarwi' or 'chocho'). Naturally low alkaloid, 'sweet' cultivars have been developed, which now represent most of the lupins cultivated worldwide. Sweet cultivars are grown primarily for grain for animal feed but increasingly as a healthy adjunct to the human diet in breads and pastries, or to provide a gluten-free alternative to wheat flour (Gresta et al. 2017).

The *Lupinus* genus has been the subject of genomic studies since the late 1990s (Wolko et al. 2011). Most extensive genomic research has focused on the most widely cultivated species *L. angustifolius*, which culminated in the publication of a high-quality reference genome (Hane et al. 2017). This chapter explores how the genomic resources in *L. angustifolius* are enabling a greater understanding of lupin domestication, which is becoming an increasingly insightful model for crop domestication and species evolution more generally.

8.2 Multiple Lupin Domestications Spanning Time and Space

Domestication can be defined as the taming of wild plants and animals to become more productive for humans, enabling the development of trade specializations and burgeoning human populations (Diamond 2002). It involved progressively accumulating domestication traits that made the plants increasingly more useful and productive to people. The founding father of the discipline, Nikolai Vavilov, described this as the 'homologous series in inherited variation' (Vavilov 1951) and which is now known as the 'domestication syndrome' (Hammer 1984). These traits included reduced fruit dehiscence, increased apical dominance, removal of seed dormancy, altered time of flowering and maturity, and reduced bitter compounds in seeds (Doebley et al. 2006). The domestication of lupin occurred several times throughout human history and across wide geographical regions (Gladstones 1998).

8.2.1 Ancient Lupin Domestication

The first records that suggest lupin had been adopted and adapted for use within human culture are from Greek and Roman texts. However, it is believed that lupins had been cultivated around the Mediterranean much earlier, having spread from the place of initial cultivation, Egypt (Gladstones 1974). Archaeological remains of *L. albus* have been found in Greece and Cyprus dating from around the Bronze Age (Zohary et al. 2012). It is believed that it was white lupin that the ancient Greek writers Hippocrates (400–356 B.C) and Theophrastus (372–288 B.C) record, discussing soil type and harvesting requirements for the crop. The Roman writer, the elder Cato (243–149 B.C) referred to its use as a cattle feed and as a green manure, and the poet Virgil writes of its use in crop rotation with cereals (Hondelmann 1984). It was at some point during this early cultivation that the initial domestication of white lupin would have occurred, selecting the permeable seed coats to aid even germination and non-shattering pods to reduce wastage during harvest (Gladstones 1970). The history of lupin use and domestication in the 'New World' is harder to follow as there are fewer records. The early cultivation of *L. mutabilis* in the Andes of South America has been dated to around 700 B.C. (Hondelmann 1984). Later, it was the Incas who used lupin extensively in crop rotations until the Spanish conquest in the early sixteenth century (Wolko et al. 2011). Similar traits as those in *L. albus* would have been selected for in the early domestication and adoption of *L. mutabilis* into South American agriculture, and there are no true wild lines remaining without these domestication traits (Eastwood and Hughes 2008).

Lupin cultivation and domestication underwent a renaissance in the eighteenth century. This was by royal decree in Prussia as a means of soil improvement using *L. albus*. This species did not thrive in the Northern European climate and was replaced successfully with *L. luteus*, which was used for seed production for animal feed as well as soil improvement in crop rotations (Wolko et al. 2011; Hondelmann 1984). *L. angustifolius* was also introduced to Europe over the following 100 years and along with *L. luteus* was taken up by farmers in Northern Europe as both species had good frost tolerance and suitable maturity timing compared to *L. albus* (Wolko et al. 2011).

8.2.2 Modern Lupin Domestication

The driver for the modern era domestication of lupins was to find sweet varieties which were low in alkaloids (Hondelmann 1984). Up to this point, all lupins were bitter and the consumption of seed was only possible after soaking the seeds for a period of time in water. If sweet varieties could be developed, then it could open up a greater use of the crop for animal feed as well as for humans without the risk of toxicity. The first recorded discovery of sweet plants for both *L. luteus* and *L. angustifolius* was in the late 1920s by German scientist and plant breeder Dr. Reinhold von Sengbusch. This was only possible after the development of a simple, high-throughput assay to detect the presence of alkaloids (Hondelmann 1984; Gladstones 1970). The identification of sweet types of *L. albus* was subsequently achieved and this, along with breeding for early maturity, was carried out in 1930–1940s in northern Europe leading to varieties such as Nahrquell being released post-war in West Germany (Gladstones 1970).

At the same time, other key seed traits—indehiscence, water permeability (soft seededness) and white colouring—were included in the selection and proved successful in *L. luteus*. Lupin breeding also began in Poland in the 1930s, focused mainly on *L. luteus* and *L. angustifolius* but it was not until after the Second World War that interest for lupins grew, particularly in the Mediterranean, Australasia and South Africa. In the 1950s, a breeding programme was established in Western Australia for *L. angustifolius* where the full domestication of this species was achieved by incorporating domestication genes from several sources (Gladstones 1977) (Table 8.1). A *L. angustifolius* breeding programme was also set-up in the USA in the 1940s and continued to the 1960s with advances made in disease resistance, particularly to anthracnose and grey leaf mould. These varieties and knowledge were then combined into the Australian breeding programme (Gladstones 1977). Yield improvements then became the focus for the breeding efforts as the domestication process had been completed.

L. mutabilis was another lupin species for which von Sengbusch produced sweet types in the 1930s, as other domestication traits such as non-shattering were already in place. Mutation breeding for sweetness using ethyl methanesulfonate (EMS) was also attempted later on, lowering alkaloid levels to around 0.2–0.3% (Clements et al. 2008; Williams et al. 1984). However, it was breeding work based on a natural mutant by von Baer and Gross in Chile that led to the production of an extremely low-alkaloid cultivar, Inti (Gross et al. 1988). Breeding was then continued in Western Australia from 1999, focusing on flowering time and male sterility. The Australian Lupin Collection containing a number

Table 8.1 Key domestication genes in narrow-leafed lupin

Domestication trait	Gene name	Dominant or recessive	Description	Chromosome, interval size[a]	Origin and reference
Low alkaloid	*iucundus*	Recessive	Reduced level of quinolizidine alkaloids in the plant. Possibly controlled by *RAP2-7* gene (Kroc et al. 2019)	NLL-07, 746 Kb	Discovered by Von Sengbusch in 1928 (Hackbarth 1957; Von Sengbusch 1942)
Soft seededness	*mollis*	Recessive	Water permeable seed allowing immediate germination	NLL-17, 119.5 Kb	Unknown origin in 1930 (Mikolajczyk 1966; Forbes and Wells 1968)
White flowers and seeds	*leucospermus*	Recessive	Anthocyanin pigments are suppressed leading to white flowers and seeds, and no purple colouring in leaves and stems	NLL-03, 907.1 Kb	Natural variant (Hallqvist 1921; Hackbarth and Troll 1959)
Non-shattering pods	*tardus*	Recessive	Pod seams are fused together, reducing shattering	NLL-01, 517.6 Kb	Discovered in 1960 (Gladstones 1967)
Non-shattering pods	*lentus*	Recessive	Endocarp cells in the pod walls lose their parallel orientation, reducing shattering	NLL-08, 387.1 Kb	Discovered in 1960 (Gladstones 1967)
Early flowering	*Ku*	Dominant	Loss of vernalization requirement for flowering leading to early flowering in warmer environments. Controlled by a *Flowering Locus T* (*FT*) homologue (Nelson et al. 2017)	NLL-10, 413 Kb	Discovered by Gladstones in 1961 (Gladstones and Hill 1969)

[a]Hane et al. (2017)

of *L. mutabilis* accessions with differing characteristics provided additional traits, which could be combined into breeding programmes for the continual improvement of *L. mutabilis* cultivars (Clements et al. 2008).

L. cosentinii domestication and breeding was undertaken by Gladstones in the 1950s, around a century after it had been initially introduced to the country for flour production. It was a good choice for domestication as it had naturalized well into the Western Australian environment and thrived on infertile, sandy soils as well as having some drought tolerance (Gladstones 1970). By this time, it was used mainly for soil improvement and sheep feed. Domestication traits that were targeted for *L. cosentinii* improvement were low-alkaloid seed (from artificial mutagenesis), non-shattering pods, early flowering and soft seededness, from natural mutations (Gladstones and Francis 1965; Gladstones and Hill 1969; Gladstones 1958, 1967). All of these were incorporated into the cultivar 'Erregulla'. However, it was not widely taken up due to problems with deformed seeds and reduced seed filling (Cowling et al. 1998). While this species is not currently grown to any appreciable extent, it may provide a useful legume rotation crop in a drying climate, as there is anecdotal evidence of drought tolerance in this species (Gladstones 1970).

8.3 Genetic and Genomic Resources Supporting Domestication Research in Lupin

Lupins provide an excellent model for understanding domestication genes due to the multiple independent domestications across wide spatial (Europe, South America and Australia) and temporal (from 4000 to 50 years ago) ranges. Two crucial features supporting domestication studies are the availability of extensive and well-annotated seed collections made for *Lupinus* species, and the relatively small diploid genomes (2C = 1.16–2.44 pg, equivalent to around 600–1200 Mbp per haploid genome; (Naganowska et al. 2003)), which makes them tractable to genomic analyses.

8.3.1 Genetic Resources

Lupin breeders and researchers are blessed with excellent germplasm resources, especially for the domesticated species, *L. angustifolius*, *L. albus*, *L. luteus* and *L. mutabilis*. The value of seed collections is not only related to the number of accessions (estimated to be over 36,000 in the largest 40 collections (Wolko et al. 2011)), but also to the geographic spread and annotated passport data (which is very good for a large proportion of accessions). The largest and best characterized collection is located in Perth, Australia, the majority of which is currently being transferred to the Australian Grain Genebank in Horsham, Australia for long-term conservation (Sally Norton, pers. comm.). These international collections cover both Old World and New World species, wild and landrace types as well as many breeding lines. These seed collections are an invaluable resource for understanding plant domestication as well as a source of genetic variation for important agronomic traits such as abiotic and biotic stress tolerance (Berger et al. 2017).

Recombinant inbred line (RIL) populations have been created for *L. angustifolius*, *L. albus* and *L. luteus* (Berger et al. 2013), which are valuable for investigating the genetic basis of domestication traits. RIL populations are produced by crossing contrasting parental lines to produce an F_1 hybrid, which is self-pollinated to produce a large F_2 population. Each F_2 individual is then subjected to inbreeding by a process of single seed descent to generate an inbred population (typically F_8 generation) in which traits of interest have segregated. The value of RIL populations is the capacity to generate unlimited seed for replicated phenotyping and sharing with research collaborators. The main reference RIL populations for *L. angustifolius*, *L. albus* and *L. luteus* were generated from wide crosses at the Department of Primary Industries and Regional Development (DPIRD, Perth, Australia) (Wolko et al. 2011; Berger et al. 2013). All three populations segregate for domestication traits as they were generated by crossing a domesticated parent with a wild (*L. angustifolius* and *L. luteus*) or partially domesticated landrace (*L. albus*) parent.

The *L. angustifolius* RIL population developed from a cross between 83A:476 (an Australian breeding line) and P27255 (a Moroccan wild accession) has been particularly instrumental for understanding lupin domestication, through the provision of genetic maps to locate domestication genes (Boersma et al. 2005; Nelson et al. 2006, 2010; Kroc et al. 2014; Kamphuis et al. 2015; Zhou et al. 2018) and ultimately as the genetic backbone for the first lupin reference genome (Hane et al. 2017), a key resource for domestication gene discovery. The *L. albus* and *L. luteus* RIL populations are now being used to further our understanding of domestication in those species (Matthew N. Nelson et al. unpublished data).

8.3.2 Genomic Resources

As the most widely grown lupin species, genomic resources are most advanced for *L. angustifolius*. Starting from humble beginnings with protein isozyme markers (Wolko and Weeden 1989), genomic resources for *L. angustifolius* have grown in scale and complexity as technology has evolved. Transcriptomic (that is, sets of expressed gene sequences) resource development began with cloning and sequencing genes expressed in seed tissues (Nelson et al. 2006), then exploded in scale with the advent of next generation sequencing (NGS) platforms, resulting in comprehensive transcriptomes for seed, leaf, flower, pod, stem and root organs (Kamphuis et al. 2015; Foley et al. 2011, 2015; Cannon et al. 2015; Kroc et al. 2019; Yang et al. 2017). For a detailed review of lupin transcriptome studies, see Chap. 5. Two genomic bacterial artificial chromosomes (BAC) libraries based on cultivars Sonet (from Poland; (Kasprzak et al. 2006)) and Tanjil (from Australia; (Gao et al. 2011)) proved to be useful tools for gene discovery before the availability of whole genome surveys (based on Tanjil; (Kamphuis et al. 2015, Yang et al. 2013)) and then comprehensive Tanjil genome assemblies (Zhou et al. 2018; Hane et al. 2017). The Tanjil genome assembly is currently being improved with long sequence-read technology, and a pan-genome is also being developed that will represent species-wide genome diversity through incorporating portions of the *L. angustifolius* genome that are absent in Tanjil but present in domesticated and wild accessions (Karam B. Singh, pers. comm.; Sect. 3.6.1).

While not yet as comprehensive as for *L. angustifolius*, genomic resources have also rapidly developed for other lupin species (see Chaps. 3 and 5). Indeed, the first lupin transcriptomic resources were generated to explore *L. albus* cluster roots (Uhde-Stone et al. 2003), which was followed up later with richer next generation datasets (O'Rourke et al. 2013; Wang et al. 2014; Secco et al. 2014). Other *L. albus* transcriptomes were produced to explore seed storage proteins and for genetic marker development (Foley et al. 2015; Książkiewicz et al. 2017). Transcriptome sequences were used for marker discovery and exploring organ abscission in yellow lupin (Parra-González et al. 2012; Glazinska et al. 2017), and its chloroplast genome was sequenced (Martin et al. 2014). In the broadest sampling reported yet, Nevado et al. (2016) sequenced transcriptomes from 55 New World lupin species in order to understand adaptive evolution of rapidly speciating lupins in the New World. One of those species—*L. polyphyllus*—had been sequenced earlier by Cannon et al. (2015) as part of the 1000 Plants (Leebens-Mack et al. 2019).

High-quality reference genomes are being prepared using long sequence-read technologies and optical mapping for *L. albus* (Hufnagel et al. 2019) and *L. luteus* (Joshua Udall, pers. comm.). These are expected to be as useful as the *L. angustifolius* genome has already proven to be. Taken together, these genomic resources provide powerful tools for lupin domestication gene discovery.

8.4 Lupin Domestication Gene Discovery

8.4.1 *Lupinus angustifolius*

Identifying the genes controlling domestication traits is important for basic understanding of plant evolution but also for improving crops through plant breeding. One of the key constraints in accessing trait diversity in wild relatives for breeding purposes is the poor agronomic performance of early generations of progeny from crosses between breeding lines and wild relatives (Cowling et al. 2009). The availability of diagnostic molecular markers based on domestication genes transforms the speed and efficiency of the conversion of wild material into a suitable domesticated background in which the agronomic value of wild alleles can be measured. Understanding how domestication genes operate may also help to refine and improve the domestication genes themselves, either through prospecting for natural allelic diversity in those genes or by biotechnological intervention through transgenics or targeted mutation. For example, a modified set of phenology genes could be used to expand the adaptation of crops to new or changing climatic regions (Mousavi-Derazmahalleh et al. 2019; Taylor et al. 2019).

Our most advanced understanding of lupin domestication genes comes from studies of *L. angustifolius*. This twentieth-century domestication is special in that each event was recorded at the time in the scientific literature (Table 8.1). There are five domestication traits controlled by six major genes: soft seededness (*mollis*), seed indehiscence (*lentus* and *tardus*), low alkaloid (*iucundus*), early flowering through removal of vernalization requirement (*Ku* and *Julius*) and white flower/green vegetative organ pigmentation (*leucospermus*) as a marker for domestication (wild types having blue flowers and a purple or red tinge throughout the vegetative organs) (Nelson et al. 2006; Taylor et al. 2019). Several studies have mapped each of the six domestication genes to the 83A:476 x P27255 reference genetic map with increasing resolution as marker technology improved and the size of the RIL population expanded (Boersma et al. 2005, 2009; Hane et al. 2017; Kamphuis et al. 2015; Kroc et al. 2014; Nelson et al. 2006, 2010; Zhou et al. 2018). These studies shed light on the chromosomal location of domestication genes but fell short of identifying the causal genes underlying the domestication traits.

The first clue about the identity of a lupin domestication gene was found by Kroc et al. (2014). Alignment of the *L. angustifolius* genetic map to the genome of the model legume *Medicago truncatula* revealed a cluster of three homologues of the flowering time gene, *FT*, on Chromosome 7 of *M. truncatula* at the equivalent map position as the *Ku* locus in *L. angustifolius*. Kroc et al. (2014) developed *FT* gene-based markers and mapped them back into *L. angustifolius*. One of the *FT* markers (FTc) mapped precisely to the *Ku* locus. This lead was followed up by Nelson et al. (2017) who were able to confirm that the *FT* homologue *LanFTc1* not only mapped perfectly to the *Ku* locus but a 1.4 kb deletion in its promoter region was perfectly correlated with vernalization responsiveness in a panel of 216 wild and domesticated accessions of *L. angustifolius*. Of four *FT* homologues found in the *L. angustifolius*, only *LanFTc1* showed elevated gene expression across a range of organ types in response to vernalization treatment in the vernalization responsive accession P27255. Taylor et al. (2019) were then able to demonstrate conclusively that the 1.4 kb deletion was the causal variant responsible for the loss of vernalization responsiveness, presumably due to a loss of regulatory sequence(s) that represses *LanFTc1* expression. This explanation was supported by the discovery of a smaller, partly overlapping 1.2 kb deletion in the *LanFTc1* promoter region of a wild accession from Israel that showed an intermediate flowering time phenotype. This discovery offers the exciting prospect of an allelic series of *LanFTc1* that can be used as a simple breeding tool for targeting lupin varieties to specific climatic regions and flowering time in lupins (see Chap. 9 for more detailed discussion).

The low-alkaloid trait is another important target for gene identification for breeding in all lupin crop species. Quinolizidine alkaloids (QA) are responsible for the bitterness found in the *Lupinus* genus but the specific QAs differ between each of the species. In *L. angustifolius*, the major QA is lupanine (Frick et al. 2017; Wink et al. 1995). While several low-alkaloid genes have been discovered in *L. angustifolius*, only one is used in cultivars: *iucundus* (Table 8.1). It was first discovered by Von Sengbusch in 1928 and found to be a single recessive gene (Hackbarth 1957; Von Sengbusch 1942). However, it was not until the development of genetic tools that the *iucundus* region could be explored and the alkaloid biosynthesis pathway further understood. Li et al. (2011) identified markers linked to *iucundus,* which could be used for marker-assisted selection in wild × domesticated introgressive crossing programmes for broadening genetic diversity in breeding pools. Even more useful would be a perfectly predictive marker based on the causal gene mutation underlying *iucundus*. The *iucundus* gene was mapped to a 746 Kb region on chromosome NLL-07 (Hane et al. 2017). Kroc et al. (2019) used a transcriptomic approach to identify a strong candidate gene for *iucundus* in this interval: *RAP2-7*, an ethylene responsive transcription factor. A less promising candidate gene in the same region could not be fully ruled out: *DHDPS*, a 4-hydroxytetrahydrodipicolinate synthase gene. Further validation work will be required to confirm the causal mutation underlying *iucundus*. Three other alkaloid biosynthesis genes genetically unlinked to *iucundus* have been identified in *L. angustifolius*. The first step in the alkaloid biosynthesis pathway was found to be a lysine decarboxylase (*LDC*; Bunsupa et al. (2012)) and more recently the second step was identified as a copper amine oxidase (*CAO*; Yang et al. (2017)). A third gene, which role is not yet fully understood, is an acyl transferase (*LaAT*; Bunsupa et al. (2011)). The quinolizidine biosynthetic pathway has yet to be elucidated in any species, but most progress achieved to date has been in *L. angustifolius,* which functions as a model for other species. In this regard there has been recent progress made on the genetic factors affecting the biosynthetic pathway and how it responds to some biotic (Frick et al. 2019) and abiotic stresses (Frick et al. 2018) in *L. angustifolius*.

8.4.2 *Lupinus albus*

The reference mapping population for *L. albus* (Kiev Mutant x P27174 RIL population) segregates for just two domestication traits: low alkaloid (controlled by the *pauper* locus) and early flowering. The first genetic map for *L. albus* mapped at low resolution the *pauper* locus and two quantitative trait loci (QTL) for flowering time (Phan et al. 2007). This map was modestly improved by Vipin et al. (2013) although this did not provide more insight into the domestication traits. More recently, Książkiewicz et al. (2017) used genotyping by sequencing (GBS) in the same RIL population to generate a much improved, high-resolution map. They located *pauper* in a well-defined interval on linkage group ALB18 and identified a candidate gene residing in that region —*LaAT*, a gene previously identified in *L. angustifolius* by Bunsupa et al. (2011). The two flowering time QTL previously identified by Phan et al. (2007) were confirmed and furthermore were demonstrated to be involved in the vernalization responsive (Książkiewicz et al. 2017). One of these QTL may respond to the previously reported *brevis* locus (Gladstones 1970) but this has yet to be confirmed. An additional weak QTL was identified, which was not vernalization related. The improved *L. albus* genetic map was aligned to the *L. angustifolius* reference genome but interestingly none of the mapped loci corresponded to the positions of the equivalent *iucundus* and *Ku* loci in *L. angustifolius*. However, the *L. angustifolius* genome provided candidate genes for the flowering time QTLs, as subsequently described by Rychel et al. (2019). This example illustrates both the value and limitations of the *L. angustifolius* reference genome sequence for domestication gene research in other legume species. The reference genome for *L. albus* based on the cultivar Amiga (Hufnagel et al. 2019) is helping identify the genes underlying *pauper* and other low-alkaloid mutant loci in an international collaboration between France, Denmark, Poland, UK and Australia (Nelson et al. unpublished data).

8.4.3 *Lupinus luteus*

The reference RIL population for *L. luteus* was developed from a cross between the Australian cultivar Wodjil (a selection from the Polish cultivar Teo) (French et al. 2001) and P28213 (wild accession from the Azores) (Iqbal et al. 2019). This population segregates for the complete suite of domestication traits: soft seededness, seed indehiscence, vernalization responsiveness in flowering, alkaloid content and flower colour (yellow versus orange). The first map for *L. luteus* was recently released (Iqbal et al. 2019) and analysis of domestication traits is underway (see Chap. 11). Domestication gene discovery will be greatly facilitated by the availability of a reference genome, which is currently under development (Joshua Udall, pers. comm.).

8.4.4 Other Lupin Species

To our knowledge, little progress has been made in other lupin species to identify domestication genes. Foundational resources such as RIL populations should be developed between wild and domesticated accessions of both *L. mutabilis* and *L. cosentinii*. Mining of available transcriptomic datasets (see above) may provide some initial leads to follow-up in more comprehensive experiments.

8.5 Genetic Consequences of Domestication on Genome Diversity

The domestication of grain crops involves a series of population bottlenecks as new domestication alleles undergo extreme selection pressure (Doebley et al. 2006). This leads to a reduction in genetic diversity, which takes time to recover through gene flow from wild populations and spontaneous mutations. It is therefore to be expected that the genetic diversity of a young, twentieth-century domesticate such as *L. angustifolius* will have very depleted diversity compared to its wild ancestors. This was indeed found to be the case in a diversity analysis of 1,248 wild and 95 domesticated accessions using low-resolution Diversity Arrays Technology (DArT) genotyping (Berger et al. 2012). Figure 8.1 graphically illustrates the small portion of diversity captured in Australian and European cultivars compared to wild accessions collected across the Mediterranean Basin. This highlighted the need to understand where useful genetic diversity can be found among wild accessions (Berger et al. 2013).

A detailed analysis of 142 wild accessions using high-resolution single nucleotide polymorphism (SNP) genotyping revealed that accessions from the western Mediterranean region were more diverse and that there had been an historic eastward migration during which there was a shift in phenological adaptation to warmer, lower rainfall environments (Mousavi-Derazmahalleh et al. 2018a). This provides valuable guidance for lupin breeders to identify untapped sources of genetic and adaptive diversity for lupin improvement. Mousavi-Derazmahalleh et al. (2018b) went further to demonstrate that the western Mediterranean region provided the founder populations for the domestication of lupin, which had been suspected previously based on morphological observations (Gladstones 1998). Another important finding was the much higher linkage disequilibrium evidence in domesticated compared to wild accessions,

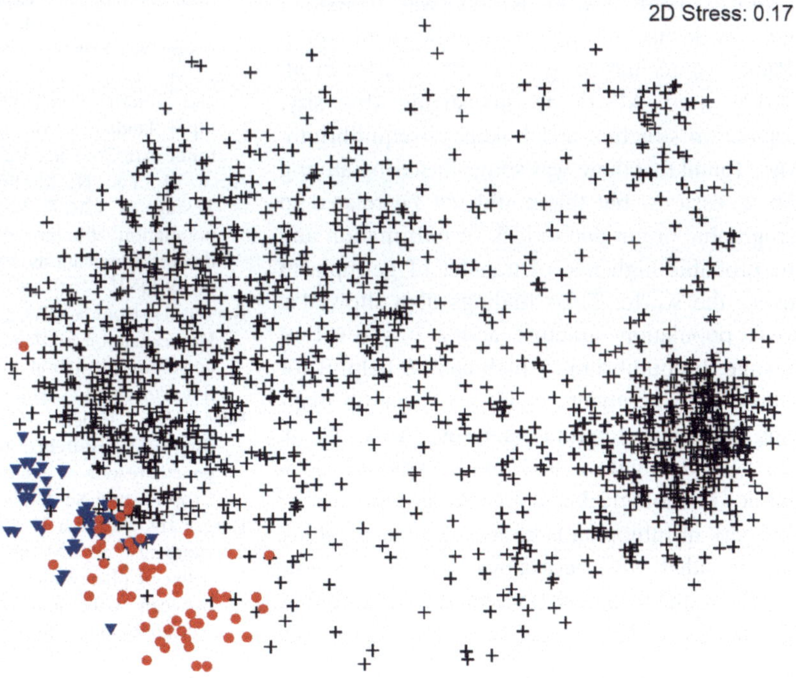

Fig. 8.1 Domesticated cultivars of *L. angustifolius* contain a small proportion of species diversity. This multidimensional scaling plot was based on diversity measured at 137 DArT marker loci in 1,248 wild (black crosses) and 95 domesticated (Australian varieties represented by red circles and European varieties represented by blue triangles) accessions. Redrawn from data presented by Berger et al. (2012)

meaning that plant breeder efforts to accumulate beneficial alleles will be hampered by unwanted linkage to unfavourable alleles (Mousavi-Derazmahalleh et al. 2018b). Only by introducing wild diversity into breeding programmes will such unwanted linkages be broken up over time. Interestingly, a search for footprints of selection around domestication trait loci proved inconclusive, which may have been due to the recentness of *L. angustifolius* domestication.

Less is known about the impact of domestication on the genome diversity of other lupin crop species. Gilbert et al. (1999) investigated the genetic diversity present in 40 *L. albus* accessions using ISSR-PCR. The small sample size, the repeatability of the marker technology limitations and lack of useful passport information accompanying accessions severely limited the conclusions that could be drawn from this study. In a more comprehensive study of 94 landrace and cultivar accessions, Raman et al. (2008, 2014) found that *L. albus* landraces clustered separately from modern cultivars and that within landraces, Ethiopian landraces were the most distinct. Annicchiarico et al. (2010) investigated agronomic and phenological diversity in a more globally representative collection of *L. albus* landraces. Current work is underway to extend this work using high-resolution genotyping (Paolo Annicchiarico, pers. comm.). Iqbal et al. (2012) used AFLPs to investigate diversity, population structure and linkage disequilibrium. They found that there was some clustering among the accessions, but this could not be related to geographic origin due to lack of information and the probable high rate of transfer of germplasm across the world. Their findings also showed a weak population structure and a low level of linkage disequilibrium, which can be helpful for follow on experiments such as association mapping. In more focused analyses, Atnaf et al. (2015) and Atnaf et al. (2017), explored agronomic, phenological and low-resolution marker diversity in Ethiopian landraces. So far, no study has included wild accessions, which can now only be found in Greece (known as *graecus* types; (Gladstones 1998)). Currently, we are investigating molecular and phenological diversity in a large global collection of wild, landrace and cultivar accessions from 15 countries, which we believe will provide insights into the origin and genetic consequences of *L. albus* domestication (M. Nelson, unpublished data).

8.6 Closing Remarks

The genomics revolution has provided powerful new tools to answer basic questions about crop domestication, an insightful model for species evolution. The reference genome sequence of *L. angustifolius* provides a valuable resource for identifying domestication genes and understanding the effects of domestication on genome-wide diversity in lupin crop species. These discoveries provide the knowledge and the genetic tools needed by lupin breeders and pre-breeders to introduce much-needed genetic and adaptive diversity into lupin crops.

References

Annicchiarico P, Harzic N, Carroni AM (2010) Adaptation, diversity, and exploitation of global white lupin (*Lupinus albus* L.) landrace genetic resources. Field Crops Res 119:114–124

Atnaf M, Tesfaye K, Dagne K, Wegari D (2015) Extent and pattern of genetic diversity in Ethiopian white lupin landraces for agronomical and phenological traits. Afr Crop Sci J 23:327–341

Atnaf M, Yao N, Martina K, Dagne K, Wegary D, Tesfaye K (2017) Molecular genetic diversity and population structure of Ethiopian white lupin landraces: implications for breeding and conservation. PLoS ONE 12:e0188696

Berger JD, Buirchell BJ, Luckett DJ, Nelson MN (2012) Domestication bottlenecks limit genetic diversity and constrain adaptation in narrow-leafed lupin (*Lupinus angustifolius* L.). Theor Appl Genet 124:637–652

Berger JD, Clements JC, Nelson MN, Kamphuis LG, Singh KB, Buirchell B (2013) The essential role of genetic resources in narrow-leafed lupin improvement. Crop Pasture Sci 64:361–373

Berger JD, Shrestha D, Ludwig C (2017) Reproductive strategies in Mediterranean legumes: trade-offs between phenology, seed size and vigor within and between wild and domesticated *Lupinus* species collected along aridity gradients. Front Plant Sci 8:548

Boersma JG, Pallotta M, Li C, Buirchell BJ, Sivasithamparam K, Yang H (2005) Construction of a genetic linkage map using MFLP and identification of molecular markers linked to domestication genes in narrow-leafed lupin (*Lupinus angustifolius* L.). Cell Mol Biol Lett 10:331–344

Boersma J, Nelson M, Sivasithamparam K, Yang H (2009) Development of sequence-specific PCR markers linked to the *Tardus* gene that reduces pod shattering in narrow-leafed lupin (*Lupinus angustifolius* L.). Mol Breed 23:259–267

Bunsupa S, Okada T, Saito K, Yamazaki M (2011) An acyltransferase-like gene obtained by differential gene expression profiles of quinolizidine alkaloid-producing and nonproducing cultivars of *Lupinus angustifolius*. Plant Biotechnol 28:89–94

Bunsupa S, Katayama K, Ikeura E, Oikawa A, Toyooka K, Saito K, Yamazaki M (2012) Lysine decarboxylase catalyzes the first step of quinolizidine alkaloid biosynthesis and coevolved with alkaloid production in Leguminosae. Plant Cell 24:1202–1216

Cannon SB, McKain MR, Harkess A, Nelson MN, Dash S, Deyholos MK, Peng Y, Joyce B, Stewart CN, Rolf M, Kutchan T, Tan X, Chen C, Zhang Y, Carpenter E, Wong GK-S, Doyle JJ, Leebens-Mack J (2015) Multiple polyploidy events in the early radiation of nodulating and nonnodulating legumes. Mol Biol Evol 32:193–210

Clements J, Sweetingham M, Smith L, Francis G, Thomas G, Sipsas S (2008) Crop improvement in *Lupinus mutabilis* for Australian agriculture-progress and prospects. Lupins for health and wealth. In: Proceedings of the 12th international lupin conference, Fremantle, Western Australia. International Lupin Association, pp 244–250, 14–18 Sept 2008

Cowling WA, Huyghe C, Święcicki W (1998) Lupin breeding. In: Gladstones JS, Atkins CA, Hamblin J (eds) Lupins as crop plants: biology, production, and utilization. CAB International, Wallingford, UK

Cowling WA, Buirchell BJ, Falk DE (2009) A model for introducing novel genetic diversity from wild relatives into elite crop populations. Crop Pasture Sci 60:1009–1015

Diamond J (2002) Evolution, consequences and future of plant and animal domestication. Nature 418:700–707

Doebley JF, Gaut BS, Smith BD (2006) The molecular genetics of crop domestication. Cell 127:1309–1321

Eastwood RJ, Hughes CE (2008) Origins of domestication of *Lupinus mutabilis* in the Andes. Lupins for health and wealth. In: Proceedings of the 12th international lupin conference, pp 14–18

FAO (2017) FAOSTAT: crop production data. Food and Agriculture Organisation of the United Nations. https://www.fao.org/faostat/en/#data/QC. Accessed Dec 2018

Foley RC, Gao LL, Spriggs A, Soo LY, Goggin DE, Smith PM, Atkins CA, Singh KB (2011) Identification and characterisation of seed storage protein transcripts from *Lupinus angustifolius*. BMC Plant Biol 11:59

Foley RC, Jimenez-Lopez JC, Kamphuis LG, Hane JK, Melser S, Singh KB (2015) Analysis of conglutin seed storage proteins across lupin species using transcriptomic, protein and comparative genomic approaches. BMC Plant Biol 15:106

Forbes I, Wells HD (1968) Hard and soft seededness in Blue Lupine, *Lupinus angustifolius* L.: inheritance and phenotype classification 1. Crop Sci 8:195–197

French R, Sweetingham M, Shea G (2001) A comparison of the adaptation of yellow lupin (*Lupinus luteus* L.) and narrow-leafed lupin (*L. angustifolius* L.) to acid sandplain soils in low rainfall agricultural areas of Western Australia. Aust J Agric Res 52:945–954

Frick KM, Kamphuis LG, Siddique KH, Singh KB, Foley RC (2017) Quinolizidine alkaloid biosynthesis in lupins and prospects for grain quality improvement. Front Plant Sci 8:87

Frick KM, Foley RC, Kamphuis LG, Siddique KHM, Garg G, Singh KB (2018) Characterisation of the genetic factors affecting quinolizidine alkaloid biosynthesis and its response to abiotic stress in narrow-leafed lupin (*Lupinus angustifolius* L.). Plant Cell Environ 41:2155–2168

Frick KM, Foley R, Siddique KHM, Singh KB, Kamphuis LG (2019) The role of jasmonate signalling in quinolizidine alkaloid biosynthesis, wounding and aphid predation response in narrow-leafed lupin. Funct Plant Biol 46:443–454

Gao LL, Hane JK, Kamphuis LG, Foley R, Shi BJ, Atkins CA, Singh KB (2011) Development of genomic resources for the narrow-leafed lupin (*Lupinus angustifolius*): construction of a bacterial artificial chromosome (BAC) library and BAC-end sequencing. BMC Genomics 12:521

Gilbert J, Lewis R, Wilkinson M, Caligari P (1999) Developing an appropriate strategy to assess genetic variability in plant germplasm collections. Theor Appl Genet 98:1125–1131

Gladstones J (1958) Induction of mutation in the west Australian blue lupin (*Lupinus digitatus* Forsk.) by X-irradiation. Aust J Agric Res 9:473–482

Gladstones J (1967) Selection for economic characters in *Lupinus angustifolius* and *L. digitatus*. Aust J Exp Agric 7:360–366

Gladstones JS (1970) Lupins as crop plants. Field Crop Abstr 23:123–148

Gladstones JS (1974) Lupins of the Mediterranean region and Africa. Western Australia Department of Agriculture. Technical Bulletin

Gladstones JS (1977) The narrow-leafed lupin in Western Australia (*Lupinus angustifolius* L.). Western Australian Department of Agriculture

Gladstones J (1998) Distribution, origin, taxonomy, history and importance. In: Gladstones JS, Atkins CA, Hamblin J (eds) Lupins as crop plants: biology, production and utilization. CAB International, Wallingford, UK

Gladstones J, Francis C (1965) Low-alkaloid mutants of *Lupinus digitatus* Forsk. Nature 207:553

Gladstones J, Hill G (1969) Selection for economic characters in *Lupinus angustifolius* and *L. digitatus*. 2. Time of flowering. Aust J Exp Agric 9:213–220

Glazinska P, Wojciechowski W, Kulasek M, Glinkowski W, Marciniak K, Klajn N, Kesy J, Kopcewicz J (2017) De novo Transcriptome profiling of flowers, flower pedicels and pods of *Lupinus luteus* (yellow lupine) reveals complex expression changes during organ abscission. Front Plant Sci 8:641

Gresta F, Wink M, Prins U, Abberton M, Capraro J, Scarafoni A, Hill G (2017) Lupins in European cropping systems. In: Murphy-Bokern D, Stoddard FL, Watson CA (eds) Legumes in cropping systems. The Centre for Agriculture and Bioscience International

Gross R, von Baer E, Koch F, Marquard R, Trugo L, Wink M (1988) Chemical composition of a new variety of the Andean lupin (*Lupinus mutabilis* cv. Inti) with low-alkaloid content. J Food Compos Anal 1:353–361

Hackbarth J (1957) Die Gene der Lupinenarten. II. Schmalblättrige Lupine (*Lupinus angustifolius* L.). Z Pflanzenzüchl 37:81–95

Hackbarth J, Troll H-J (1959) Lupinen als Körnerleguminosen und Futterpflanzen. Handbuch der Pflanzenzüchtung 2:1–51

Hallqvist C (1921) The inheritance of the flower coloour and the seed colour in *Lupinus angustifolius*. Hereditas 2:299–363

Hammer K (1984) Das Domestikationssyndrom. Die Kulturpflanze 32:11–34

Hane JK, Ming Y, Kamphuis LG, Nelson MN, Garg G, Atkins CA, Bayer PE, Bravo A, Bringans S, Cannon S, Edwards D, Foley R, Gao L-L, Harrison MJ, Huang W, Hurgobin B, Li S, Liu C-W, McGrath A, Morahan G, Murray J, Weller J, Jian J, Singh KB (2017) A comprehensive draft genome sequence for lupin (Lupinus angustifolius), an emerging health food: insights into plant–microbe interactions and legume evolution. Plant Biotechnol J 15:318–330

Hondelmann W (1984) The lupin—ancient and modern crop plant. Theor Appl Genet 68:1–9

Hufnagel B, Marques A, Soriano A, Marquès L, Divol F, Doumas P, Sallet E, Mancinotti D, Carrere S, Marande W, Arribat S, Keller J, Huneau C, Blein T, Aime D, Laguerre M, Taylor J, Schubert V, Nelson M, Geu-Flores F, Crespi M, Gallardo-Guerrero K, Delaux P-M, Salse J, Bergès H, Guyot R, Gouzy J, Péret B (2019) Genome sequence of the cluster root forming white lupin. bioRxiv:708917

Hughes C, Eastwood R (2006) Island radiation on a continental scale: exceptional rates of plant diversification after uplift of the Andes. Proc Natl Acad Sci 103:10334–10339

Iqbal MJ, Mamidi S, Ahsan R, Kianian SF, Coyne CJ, Hamama AA, Narina SS, Bhardwaj HL (2012) Population structure and linkage disequilibrium in *Lupinus albus* L. germplasm and its implication for association mapping. Theor Appl Genet 125:517–530

Iqbal M, Huynh M, Udall J, Kilian A, Adhikari K, Berger J, Erskine W, Nelson MN (2019) The first genetic map for yellow lupin enables genetic dissection of adaptation traits in an orphan grain legume crop. BMC Genetics 20:68

Kamphuis LG, Hane JK, Nelson MN, Gao L, Atkins CA, Singh KB (2015) Transcriptome sequencing of different narrow-leafed lupin tissue types provides a comprehensive uni-gene assembly and extensive gene-based molecular markers. Plant Biotechnol J 13:14–25

Kasprzak A, Safár J, Janda J, Dolezel J, Wolko B, Naganowska B (2006) The bacterial artificial chromosome (BAC) library of the narrow-leafed lupin (*Lupinus angustifolius* L.). Cell Mol Biol Lett 11:396–407

Kroc M, Koczyk G, Święcicki W, Kilian A, Nelson MN (2014) New evidence of ancestral polyploidy in the Genistoid legume *Lupinus angustifolius* L. (narrow-leafed lupin). Theor Appl Genet 127:1237–1249

Kroc M, Koczyk G, Kamel KA, Czepiel K, Fedorowicz-Strońska O, Krajewski P, Kosińska J, Podkowiński J, Wilczura P, Święcicki W (2019) Transcriptome-derived investigation of biosynthesis of quinolizidine alkaloids in narrow-leafed lupin (*Lupinus angustifolius* L.) highlights candidate genes linked to *iucundus* locus. Sci Rep 9:2231

Książkiewicz M, Nazzicari N, Yang HA, Nelson MN, Renshaw D, Rychel S, Ferrari B, Carelli M, Tomaszewska M, Stawiński S, Naganowska B, Wolko B, Annicchiarico P (2017) A high-density consensus linkage map of white lupin highlights synteny with narrow-leafed lupin and provides markers tagging key agronomic traits. Sci Rep 7:15335

Leebens-Mack JH, Barker MS, Carpenter EJ, Deyholos MK, Gitzendanner MA, Graham SW, Grosse I, Li Z, Melkonian M, Mirarab S, Porsch M, Quint M, Rensing SA, Soltis DE, Soltis PS, Stevenson DW, Ullrich KK, Wickett NJ, DeGironimo L, Edger PP, Jordon-Thaden IE, Joya S, Liu T, Melkonian B, Miles NW, Pokorny L, Quigley C, Thomas P, Villarreal JC, Augustin MM, Barrett MD, Baucom RS, Beerling DJ, Benstein RM, Biffin E, Brockington SF, Burge DO, Burris JN, Burris KP, Burtet-Sarramegna V, Caicedo AL, Cannon SB, Çebi Z, Chang Y, Chater C, Cheeseman JM, Chen T, Clarke ND, Clayton H, Covshoff S, Crandall-Stotler BJ, Cross H, dePamphilis CW, Der JP, Determann R, Dickson RC, Di Stilio VS, Ellis S, Fast E, Feja N, Field KJ, Filatov DA, Finnegan PM, Floyd SK, Fogliani B, García N, Gâteblé G, Godden GT, Goh F, Greiner S, Harkess A, Heaney JM, Helliwell KE, Heyduk K, Hibberd JM, Hodel RGJ, Hollingsworth PM, Johnson MTJ, Jost R, Joyce B, Kapralov MV, Kazamia E, Kellogg EA, Koch MA, Von Konrat M, Könyves K, Kutchan TM, Lam V, Larsson A, Leitch AR, Lentz R, Li F-W, Lowe AJ, Ludwig M, Manos PS, Mavrodiev E, McCormick MK, McKain M, McLellan T, McNeal JR, Miller RE, Nelson MN, Peng Y, Ralph P, Real D, Riggins CW, Ruhsam M,

Sage RF, Sakai AK, Scascitella M, Schilling EE, Schlösser E-M, Sederoff H, Servick S, Sessa EB, Shaw AJ, Shaw SW, Sigel EM, Skema C, Smith AG, Smithson A, Stewart CN, Stinchcombe JR, Szövényi P, Tate JA, Tiebel H, Trapnell D, Villegente M, Wang C-N, Weller SG, Wenzel M, Weststrand S, Westwood JH, Whigham DF, Wu S, Wulff AS, Yang Y, Zhu D, Zhuang C, Zuidof J, Chase MW, Pires JC, Rothfels CJ, Yu J, Chen C, Chen L, Cheng S, Li J, Li R, Li X, Lu H, Ou Y, Sun X, Tan X, Tang J, Tian Z, Wang F, Wang J, Wei X, Xu X, Yan Z, Yang F, Zhong X, Zhou F, Zhu Y, Zhang Y, Ayyampalayam S, Barkman TJ, Nguyen N-p, Matasci N, Nelson DR, Sayyari E, Wafula EK, Walls RL, Warnow T, An H, Arrigo N, Baniaga AE, Galuska S, Jorgensen SA, Kidder TI, Kong H, Lu-Irving P, Marx HE, Qi X, Reardon CR, Sutherland BL, Tiley GP, Welles SR, Yu R, Zhan S, Gramzow L, Theißen G, Wong GK-S (2019) One thousand plant transcriptomes and the phylogenomics of green plants. One thousand plant transcriptomes initiative. Nature 574:679–685

Li X, Yang H, Buirchell B, Yan G (2011) Development of a DNA marker tightly linked to low-alkaloid gene iucundus in narrow-leafed lupin (*Lupinus angustifolius* L.) for marker-assisted selection. Crop Pasture Sci 62:218–224

Lucas MM, Stoddard FL, Annicchiarico P, Frías J, Martínez-Villaluenga C, Sussmann D, Duranti M, Seger A, Zander PM, Pueyo JJ (2015) The future of lupin as a protein crop in Europe. Front Plant Sci 6:705

Martin GE, Rousseau-Gueutin M, Cordonnier S, Lima O, Michon-Coudouel S, Naquin D, de Carvalho JF, Aïnouche M, Salmon A, Aïnouche A (2014) The first complete chloroplast genome of the Genistoid legume *Lupinus luteus*: evidence for a novel major lineage-specific rearrangement and new insights regarding plastome evolution in the legume family. Ann Bot 113:1197–1210

Mikolajczyk J (1966) Genetic studies in *Lupinus angustifolius*. Part III. Inheritance of the alkaloid content, seed hardness and length of the growing season in blue lupin. Genet Pol 7:181–196

Mousavi-Derazmahalleh M, Bayer PE, Nevado B, Hurgobin B, Filatov D, Kilian A, Kamphuis LG, Singh KB, Berger JD, Hane JK, Edwards D, Erskine W, Nelson MN (2018a) Exploring the genetic and adaptive diversity of a pan-mediterranean crop wild relative: narrow-leafed lupin. Theor Appl Genet 131:887–901

Mousavi-Derazmahalleh M, Nevado B, Bayer PE, Filatov DA, Hane JK, Edwards D, Erskine W, Nelson MN (2018b) The western Mediterranean region provided the founder population of domesticated narrow-leafed lupin. Theor Appl Genet 131:2543–2554

Mousavi-Derazmahalleh M, Bayer PE, Hane JK, Valliyodan B, Nguyen HT, Nelson MN, Erskine W, Varshney RK, Papa R, Edwards D (2019) Adapting legume crops to climate change using genomic approaches. Plant Cell Environ 42:6–19

Naganowska B, Wolko B, Sliwińska E, Kaczmarek Z (2003) Nuclear DNA content variation and species relationships in the genus *Lupinus* (Fabaceae). Ann Bot 92:349–355

Nelson M, Phan H, Ellwood S, Moolhuijzen P, Hane J, Williams A, O'Lone C, Fosu-Nyarko J, Scobie M, Cakir M, Jones M, Bellgard M, Ksiazkiewicz M, Wolko B, Barker S, Oliver R, Cowling W (2006) The first gene-based map of *Lupinus angustifolius* L.-location of domestication genes and conserved synteny with *Medicago truncatula*. Theor Appl Genet 113:225–238

Nelson MN, Moolhuijzen PM, Boersma JG, Chudy M, Lesniewska K, Bellgard M, Oliver RP, Swiecicki W, Wolko B, Cowling WA, Ellwood SR (2010) Aligning a new reference genetic map of *Lupinus angustifolius* with the genome sequence of the model legume, *Lotus japonicus*. DNA Res 17:73–83

Nelson MN, Książkiewicz M, Rychel S, Besharat N, Taylor CM, Wyrwa K, Jost R, Erskine W, Cowling WA, Berger JD, Batley J, Weller JL, Naganowska B, Wolko B (2017) The loss of vernalization requirement in narrow-leafed lupin is associated with a deletion in the promoter and de-repressed expression of a *Flowering Locus T* (*FT*) homologue. New Phytol 213:220–232

Nevado B, Atchison GW, Hughes CE, Filatov DA (2016) Widespread adaptive evolution during repeated evolutionary radiations in New World lupins. Nat Commun 7:12384

O'Rourke JA, Yang SS, Miller SS, Bucciarelli B, Liu J, Rydeen A, Bozsoki Z, Uhde-Stone C, Tu ZJ, Allan D, Gronwald JW, Vance CP (2013) An RNA-Seq transcriptome analysis of orthophosphate-deficient white lupin reveals novel insights into phosphorus acclimation in plants. Plant Physiol 161:705–724

Parra-González LB, Aravena-Abarzúa GA, Navarro-Navarro CS, Udall J, Maughan J, Peterson LM, Salvo-Garrido HE, Maureira-Butler IJ (2012) Yellow lupin (*Lupinus luteus* L.) transcriptome sequencing: molecular marker development and comparative studies. BMC Genomics 13:425

Phan HTT, Ellwood SR, Adhikari K, Nelson MN, Oliver RP (2007) The first genetic and comparative map of white lupin (*Lupinus albus* L.): identification of QTLs for anthracnose resistance and flowering time, and a locus for alkaloid content. DNA Res 14:59–70

Raman R, Luckett DJ, Raman H (2008) Estimation of genetic diversity in albus lupin (*Lupinus albus* L.) using DArT and genic markers. International Lupin Association, Canterbury

Raman R, Cowley RB, Raman H, Luckett DJ (2014) Analyses using SSR and DArT molecular markers reveal that Ethiopian accessions of white lupin (*Lupinus albus* L.) represent a unique genepool. Open J Genet 4:87

Rychel S, Książkiewicz M, Tomaszewska M, Bielski W, Wolko B (2019) *FLOWERING LOCUS T, GIGANTEA, SEPALLATA*, and *FRIGIDA* homologs are candidate genes involved in white lupin (*Lupinus albus* L.) early flowering. Mol Breed 39:43

Secco D, Shou H, Whelan J, Berkowitz O (2014) RNA-seq analysis identifies an intricate regulatory network controlling cluster root development in white lupin. BMC Genomics 15:230

Swiecicki W, Swiecicki W (1995) Domestication and breeding improvement of narrow-leafed lupin (*L. angustifolius* L.). J Appl Genet 2:155–167

Taylor CM, Kamphuis LG, Zhang W, Garg G, Berger JD, Mousavi-Derazmahalleh M, Bayer PE, Edwards D, Singh KB, Cowling WA, Nelson MN (2019) INDEL variation in the regulatory region of the major flowering time gene *LanFTc1* is associated with vernalization response and flowering time in narrow-leafed lupin (*Lupinus angustifolius* L.). Plant Cell Environ 42:174–187

Uhde-Stone C, Gilbert G, Johnson JM, Litjens R, Zinn KE, Temple SJ, Vance CP, Allan DL (2003) Acclimation of white lupin to phosphorus deficiency involves enhanced expression of genes related to organic acid metabolism. Plant Soil 248:99–116

Vavilov NI (1951) The origin, variation, immunity and breeding of cultivated plants. Chron Bot 13:1–366

Vipin CA, Luckett DJ, Harper JD, Ash GJ, Kilian A, Ellwood SR, Phan HT, Raman H (2013) Construction of integrated linkage map of a recombinant inbred line population of white lupin (*Lupinus albus* L.). Breed Sci 63:292–300

von Sengbusch R (1942) Sweet lupins and oil lupins. The history of the origin of some new crop plants. Landwirtschaftlic Jahrb 91:719–880

Wang Z, Straub D, Yang H, Kania A, Shen J, Ludewig U, Neumann G (2014) The regulatory network of cluster-root function and development in phosphate-deficient white lupin (*Lupinus albus*) identified by transcriptome sequencing. Physiol Plant 151:323–338

Williams W, Harrison JEM, Jayasekera S (1984) Genetical control of alkaloid production in *Lupinus mutabilis* and the effect of a mutant allele *Mutal* isolated following chemical mutagenesis. Euphytica 33:811–817

Wink M, Meißner C, Witte L (1995) Patterns of quinolizidine alkaloids in 56 species of the genus *Lupinus*. Phytochemistry 38:139–153

Wolko B, Weeden N (1989) Estimation of *Lupinus* genome polyploidy on the basis of isozymic loci number. Genet Pol 30

Wolko B, Clements JC, Naganowska B, Nelson MN, Yang H (2011) *Lupinus*. In: Kole C (ed) Wild crop relatives: genomic and breeding resources. Springer, Heidelberg

Yang H, Tao Y, Zheng Z, Zhang Q, Zhou G, Sweetingham MW, Howieson JG, Li C (2013) Draft genome sequence, and a sequence-defined genetic linkage map of the legume crop species *Lupinus angustifolius* L. PLoS ONE 8:e64799

Yang T, Nagy I, Mancinotti D, Otterbach SL, Andersen TB, Motawia MS, Asp T, Geu-Flores F (2017) Transcript profiling of a bitter variety of narrow-leafed lupin to discover alkaloid biosynthetic genes. J Exp Bot 68:5527–5537

Zhou G, Jian J, Wang P, Li C, Tao Y, Li X, Renshaw D, Clements J, Sweetingham M, Yang H (2018) Construction of an ultra-high density consensus genetic map, and enhancement of the physical map from genome sequencing in *Lupinus angustifolius*. Theor Appl Genet 131:209–223

Zohary D, Hopf M, Weiss E (2012) Domestication of plants in the old world: the origin and spread of domesticated plants in Southwest Asia, Europe, and the Mediterranean basin. Oxford University Press on Demand

9 Genomic Applications and Resources to Dissect Flowering Time Control in Narrow-Leafed Lupin

Candy M. Taylor, Lars G. Kamphuis, Wallace A. Cowling, Jens D. Berger and Matthew N. Nelson

Abstract

Flowering time is a highly influential phenological trait for crop adaptation, and in the case of narrow-leafed lupin (*Lupinus angustifolius* L.), has been one of the most economically significant traits for crop production in both Australia and Europe. Given the importance of this trait, understanding the genetic basis of flowering time has become an important goal for pre-breeding. In this chapter, we report the current achievements made to dissect the control of flowering in narrow-leafed lupin using a variety of genetic and genomic approaches, and discuss how new and emerging resources will continue to shape our understanding of these complex genetic regulatory networks.

C. M. Taylor · W. A. Cowling
UWA School of Agriculture and Environment, The University of Western Australia, Perth, WA 6009, Australia
e-mail: candy.taylor@research.uwa.edu.au

W. A. Cowling
e-mail: wallace.cowling@uwa.edu.au

L. G. Kamphuis · J. D. Berger · M. N. Nelson (✉)
Commonwealth Scientific and Industrial Research Organisation, Agriculture & Food, Floreat, WA 6014, Australia
e-mail: matthew.nelson@csiro.au

L. G. Kamphuis
e-mail: lars.kamphuis@csiro.au

J. D. Berger
e-mail: jens.berger@csiro.au

L. G. Kamphuis
Centre for Crop and Disease Management, Curtin University, Bentley, WA 6102, Australia

L. G. Kamphuis · W. A. Cowling · M. N. Nelson
The UWA Institute of Agriculture, The University of Western Australia, Perth, WA 6000, Australia

M. N. Nelson
Natural Capital and Plant Health, Kew Royal Botanic Gardens, Wakehurst Place, Ardingly, Haywards Heath, Sussex RH17 6TN, UK

9.1 Genetic Mapping and Accounting for the Phenotypic Contribution of Flowering Time Loci

Since the early 2000s, the extensive collections of germplasm described in Sect. 2.3 have been utilised in several studies exploring narrow-leafed lupin (*Lupinus angustifolius* L.) genetics and genomics. The predominance of narrow-leafed lupin in genetic mapping studies reflects its relatively important agricultural status relative to other lupin species, particularly within Australia. Below, we outline the progress made as the first step to dissect and better explore the regulation of flowering time: mapping and quantifying the phenotypic effect of major genes and other QTLs for flowering time using bi-parental and association mapping populations.

K. B. Singh et al. (eds.), *The Lupin Genome*, Compendium of Plant Genomes,
https://doi.org/10.1007/978-3-030-21270-4_9

9.1.1 Bi-parental Mapping Populations

The most widely used bi-parental population for linkage mapping in narrow-leafed lupin was derived from crossing 83A:476 (maternal parent), an Australian breeding line, and P27255 (paternal parent), a wild type from Morocco. Development of the population began in 1997 with the intention of mapping and deriving markers for several key domestication genes (*iucundus*, low seed alkaloid content; *tardus* and *lentus*, reduced pod shattering; *mollis*, water-permeable seed coat; and *leucospermus*, colour indicator on the flowers, seeds and cotyledons), and the *Ku* early flowering time locus (Nelson et al. 2006), which has been of paramount importance for crop adaptation in Australia (see Chap. 2). The F_2 generation was initially used to construct a partial linkage map using Amplified Fragment Length Polymorphism (AFLP) markers (Brien et al. 1999). However, the first fully completed linkage map (Boersma et al. 2005) and all subsequent revisions (Hane et al. 2017; Kamphuis et al. 2015; Kroc et al. 2014; Nelson et al. 2010, 2006; Zhou et al. 2018) have been assembled using the F_8 or F_9 populations of Recombinant Inbred Lines (RILs) developed by single seed descent from the F_2 generation.

The location of *Ku* has been refined with each progressive revision of the narrow-leafed lupin linkage map based on the 83A:476 x P27255 RIL populations. Initially, it was assigned to LG17, one of 21 linkage groups in the first completed map, which was constructed using 522 mostly dominant Microsatellite-Anchored Fragment Length Polymorphism (MFLP) markers (Boersma et al. 2005). The first gene-based map, which incorporated 359 predominantly codominant Sequence-Tagged Site (STS) markers across 20 linkage groups, later resolved *Ku* within LG01 (Nelson et al. 2006). A more comprehensive linkage map, including markers from previous map versions and further saturated with 200 new mostly gene-based STS markers, saw the assignment of *Ku* change to NLL-10 following the absorption of LG01, LG19 and a small cluster into a single linkage group (Nelson et al. 2010). The location of *Ku* has remained on NLL-10 in two further revised linkage maps (Kamphuis et al. 2015; Kroc et al. 2014), which have added an additional 353 and 275 diverse markers, respectively. More recently, the location of *Ku* was delimitated to a 413 Kb region between molecular markers LaDArT_SNP01240 and LaSNP_499 following the addition of 469 Fluidigm SNP assay and 8,666 DArTseq markers during the construction of a physical map for narrow-leafed lupin based on the Tanjil variety (Hane et al 2017). It is important to note that marker density is no longer the limiting factor in an age of plentiful molecular markers, but instead the size of the mapping population. So far, Hane et al. (2017) provide the best available resolution having used 153 RILs compared to other studies that used as few as 87 RILs (e.g. Zhou et al. 2018).

To date, only one study has yet examined the phenotypic contribution of *Ku* in the 83A:476 x P27255 F_8 RIL population. Boersma et al. (2008) demonstrated that the allele possessed both an additive effect, reducing time to flowering by approximately 10.5 days, and a small GxE effect of approximately ± 0.6 days in two consecutive winter-sown field experiments in Western Australia. Importantly, *Ku* was shown to account for roughly 81% of total phenotypic variation in that study. This outcome validates conclusions that *Ku* and the vernalisation response dominate the regulation of flowering time in narrow-leafed lupin cultivars (Rahman and Gladstones 1972). Not surprisingly, a second QTL study in preparation has also found *Ku* to have an enormous contribution to phenotypic variation in the same population in a number of controlled environments with variable ambient temperatures and photoperiods, explaining over 95% of flowering time variation in non-vernalising conditions (Matthew N. Nelson unpublished data).

In addition to *Ku*, the 83A:476 x P27255 F_8 RIL population has also enabled as many as nine other minor flowering time QTLs to be mapped across six different linkage groups (Boersma et al. 2008). In contrast to *Ku*, these loci are relatively minor in nature, individually influencing flowering time by less than 2 days and explaining less than 2.5% of the total phenotypic variation. These results indicate the importance of *Ku* in the domesticated gene pool and underline the importance of adding new variation from wild germplasm in order to expand flowering times and the target environment range of the crop.

A new bi-parental mapping population is currently in development to map the *efl* locus for intermediate flowering time (Candy Taylor, personal comm.). This population was derived from crosses between Chittick (maternal parent), the original variety carrying the *efl* mutation, and Geebung (paternal parent), an Australian variety with the *ku* allele (Cowling 1999). Two hundred F_2 individuals were used to rapidly broad-map *efl* via a bulked segregant analysis approach incorporating whole-genome re-sequencing data, termed QTLseq (Zhou et al. 2018). Approximately 180 F_6 RILs have simultaneously been developed using an in vivo Accelerated Single Seed Descent (aSSD) method that optimises light quality and spectrum for rapid generation cycling (Croser et al. 2016). This latter population of RILs is currently being used to fine-map *efl* via linkage mapping with a high density of DArTseq markers.

Lastly, a third bi-parental RIL population has been developed between Australian varieties, Marri (*ku*) and Quilinock (*Ku*), to map QTLs for thermal responsiveness (Berger et al. 2013). No phenotypic analyses or genetic mapping with this population have yet been reported.

9.1.2 Association Mapping Populations

As has been demonstrated using the 83A:476 x P27255 F_8 and F_9 narrow-leafed lupin RIL populations, linkage mapping in bi-parental populations is useful for locating major loci associated with domestication and/or broader adaptation, like the *Ku* locus. However, this mapping approach also has limitations that prevent it from being a highly efficient method for broadly surveying genetic variation for a particular trait within a species. Firstly, the development of bi-parental populations, such as F_2, backcross, RIL, Near Isogenic Line (NIL) or Doubled Haploid (DH) populations, involves a significant investment of time and other resources. For example, it took 6 years to progress from the initial crossing between 83A:476 and P27255 to the harvest of the F_8 RIL seed (Nelson et al. 2006). This factor alone often inhibits the development of multiple populations, unless a rapid generation or speed breeding system, such as aSSD, can be utilised. Secondly, though more importantly, the capacity to detect QTLs for a trait depends upon both the phenotypic and genetic variation of the parents, which represents only a small fraction of the total species diversity (Huang and Han 2014). Also, bi-parental populations are generally restricted to two alleles per locus (or up to four alleles in an outcrossing population strategy), whereas the species may contain several functional alleles per locus. Thirdly, bi-parental populations lack genetic resolution due to limited opportunity for meiotic recombination.

An alternative genetic mapping method which overcomes these limitations and makes it possible to efficiently conduct broader surveys of genetic variation for quantitative traits is genome-wide association mapping (Brachi et al. 2011; Huang and Han 2014; Korte and Farlow 2013). Here, populations of unrestricted size comprising diverse individuals of various origins are assembled from germplasm collections, meaning that a greater breadth of the total species diversity can be assessed. Given that the no new major loci for flowering time have been identified in narrow-leafed lupin for several decades (see Chap. 2, Table 2.2), and that our genetically diverse international collections of wild germplasm remain largely unexplored and unmined for flowering time diversity, association mapping is a useful approach for identifying new loci for breeding in future.

Recently, three marker-trait association analyses were conducted in narrow-leafed lupin. All three analyses benefited from the extensive germplasm resources at the Australian Lupin Collection (South Perth, Australia). In the first study, Chen et al. (2016b) used 191 DArT (array-based) markers to screen 111 wild accessions that were phenotyped for root trait diversity, identifying several statistically significant associations. It should be noted, however, that few of these markers had known genomic locations and that the proportion of phenotypic variation explained by marker associations was modest. In a more comprehensive, genome-wide association study, Mousavi-Derazmahalleh et al. (2018a) assembled a panel of 142 wild accessions originating from 11 Mediterranean countries, including Algeria, Morocco, Portugal, Spain, France, Italy, Greece, Cyprus, Turkey, Israel and Syria. The association study combined 11,690 SNP markers derived from DArTseq genotyping by sequencing technology and anchored to the Tanjil reference genome (Hane et al. 2017) and used phenotypes for flowering time that were produced during the field evaluations by Gladstones and Crosbie (1979) and Clements and Cowling (1994). In addition, Mousavi-Derazmahalleh et al. (2018a) incorporated into their analysis climatic data from WorldClim, which corresponded to the geographical coordinates for germplasm collection sites. Unfortunately, no significant associations between markers and flowering time were evident. This outcome may potentially be attributed to the phenotypic data used in the study, as flowering times were scored for subsets of germplasm over several years in different field environments, and this may reduce the power to detect QTLs. However, two markers from linkage groups NLL-07 and NLL-05 were found to be significantly associated with precipitation in the driest month/quarter of the year and mean annual temperature, respectively. As both of these climatic variables are strong drivers for adaptive phenological diversity, this result may indicate areas of the genome facing evolutionary pressure in wild populations of narrow-leafed lupin and which may contain other phenology-associated traits, as demonstrated by (Berger et al. 2017). Lastly, in a third study, Mousavi-Derazmahalleh et al. (2018b) incorporated an additional 85 domesticated accessions to their existing panel of 142 wild accessions. Several associations between SNP markers derived from DArTseq genotyping and flowering times in the Australian Lupin Collection were found to be at the threshold of significance. Potentially, controlled environment studies of flowering time in narrow-leafed lupin, with controlled vernalisation, may increase the significance of such associations.

An expanded Genome-Wide Association Study (GWAS) to identify QTLs for flowering time in narrow-leafed lupin is currently in progress (Candy Taylor, unpublished data). Over 350 accessions are included in the panel, with additional wild types and domesticated material from the Australian Lupin Collection (South Perth, Australia), the Plant Breeding Station (Wiatrowo, Poland) and selected European seed providers supplementing the original panels of Mousavi-Derazmahalleh et al. (2018a, b). Phenotyping was conducted in a controlled environment with strong and mild vernalisation treatments to identify vernalisation responsive QTLs, and thermal time to flowering QTLs, respectively. DArTseq genotyping has also been completed and the research is now in the association analysis phase. The aims for the study are: (1) to detect previously unidentified QTLs for flowering time, including intermediate and late flowering variation, which has been suggested to be more adaptive and profitable for long-season environments (see Sect. 2.2) (Chen et al. 2016a, 2017); and (2) to determine whether any alleles other than *Ku*, *Jul* and *efl* are involved in mediating the vernalisation response in narrow-leafed lupin, which would have implications both for breeding lupins and our understanding of the evolution of this pathway within the genus.

9.2 Identifying Candidate Genes for Major Flowering Time Loci and Understanding the Evolution of Genetic Networks for Flowering Time Regulation in *Lupinus*

Knowledge of the genes within the genetic pathways for floral initiation in narrow-leafed lupin would help improve understanding of how this important trait is regulated and may be manipulated to improve crop adaptation.

Additionally, the identification of the genes involved in flowering time pathways has important implications for understanding the evolution of this important adaptive trait within the *Lupinus* genus and the Papilionoideae subfamily of legumes. Below, we outline genetic and genomic approaches and resources used to identify candidate genes for major flowering time loci in narrow-leafed lupin, and to study gene families known to be involved in floral regulation of model species. Additionally, we discuss the prospects of future candidate gene discovery with continual improvement and availability of various genetic and genomic tools.

9.2.1 Identification of Candidate and Causal Genes for Flowering Time Loci

9.2.1.1 Involvement of the *FT* Gene Family in the Narrow-Leafed Lupin Vernalisation Pathway

In *Arabidopsis thaliana*, *FLOWERING LOCUS T* (*FT*) is an important member of the phosphatidyl ethanolamine-binding protein (PEBP) domain gene family, integrating signals from the vernalisation, photoperiod and autonomous pathways, and at the appropriate time, promoting floral initiation by upregulating floral meristem identity genes (Kim et al. 2009; Turck et al. 2008). Although the PEBP domain family is broadly distributed within the Tree of Life, including within the Plant Kingdom, the *FT* group is exclusive to and widespread among angiosperms (Klintenäs et al. 2012; Wickland and Hanzawa 2015). Likewise, *FT* is a floral integrator and produces a mobile signal known as 'florigen' which promotes flowering, and is well conserved among flowering plants (Turck et al. 2008). Variations to the expression and functionality of *FT* homologues are known to result in phenological diversity in numerous lineages (Laurie et al. 2011; Ogiso-Tanaka et al. 2013; Schwartz et al. 2009).

Recently, an *FT* homologue named *LanFTc1* was identified as a candidate gene for the *Ku* locus, which is a key component in the vernalisation response pathway in narrow-leafed lupin. The initial identification was made by Kroc et al. (2014), who, in an effort to further saturate the reference genetic map for the species, used Expressed Tag Sequences (ESTs) from legumes and gene sequence information from Arabidopsis to design new markers for flowering related homologues. These included: *FTa* and *FTc* homologues of the floral integrator gene, *FT* (Kim et al. 2009); *SOC1*, a second *Arabidopsis* floral integrator gene (Moon et al. 2003); *TFL1*, a transcriptional repressor of genes activated by *FT* in *Arabidopsis* (Hanano and Goto 2011); *VIN3*, a gene involved in establishing repression of *FLOWERING LOCUS C* (*FLC*) during vernalisation and subsequent de-repression of *FT* (Sung and Amasino 2004); *VIP3*, a protein complex forming scaffold that also inhibits *FLC* (Zhang et al. 2003); and *VRN1*, a third repressor of *FLC* (Levy et al. 2002). Linkage mapping in the 83A:476 x P27255 F_8 RIL population revealed that the *FTc*-derived marker mapped to the same position as *Ku* on linkage group NLL-10. This same region was also shown through comparative analysis to have extensive marker collinearity with a proportion of *M. truncatula* linkage group Mt-7 containing three *FT* homologues, one of which is involved in creating an early flowering phenotype in the absence of vernalisation in *M. truncatula* (Jaudal et al. 2013; Laurie et al. 2011).

Convincing evidence to validate *LanFTc1* as the causal gene for not only *Ku*, but also the European early flowering time gene *Julius* (*Jul*), has subsequently been gathered by Nelson et al. (2017) and Taylor et al. (2019) using a variety of complementary genetic and genomic approaches and resources. The full-length *LanFTc1* sequence was initially obtained from the cultivar Sonet (*Ku*) Bacterial Artificial Chromosome (BAC) clone library (Kasprzak et al. 2006) and used as the reference to assemble *LanFTc1* in the 83A:476 (*Ku*) and P27255 (*ku*) RIL parents using short paired-end whole-genome sequence reads and overlapping targeted Sanger sequence reads (Nelson et al. 2017). Alignment of these sequences with those from the Sonet BAC library

(Kasprzak et al. 2006) and Tanjil genome survey sequence and unigene assembly (Gao et al. 2011; Kamphuis et al. 2015) revealed a number of polymorphisms, including a very distinctive 1,423 bp deletion in the promoter of *LanFTc1* in accessions with the *Ku* genotype. Short-read sequencing data were later generated for an additional 41 genetically diverse narrow-leafed lupins and mapped to the *LanFTc1* reference sequences from the P27255 wild type (Nelson et al. 2017) and Tanjil reference genome assembly (Hane et al. 2017; Sect. 3.3), which revealed two additional deletions of 1,208 bp and 5,162 bp overlapping the same region of the promoter (Taylor et al. 2019). PCR-based polymorphism length genetic markers similarly identified the 5,162 bp deletion genotype in several European varieties containing the *Jul* allele for early flowering, including the original *Jul* variety, named Krasnolistny (Mikołajczyk 1966).

A combination of phenotypic and gene expression analyses was used to demonstrate that the series of deletions in the promoter is significantly associated with reduced response to vernalisation, earlier flowering, and de-repressed expression of *LanFTc1* in the absence of vernalisation (Nelson et al. 2017; Taylor et al. 2019). The 1,423 bp and 5,162 bp deletions representing the *Ku* and *Jul* alleles resulted in similar levels of expression of *LanFTc1* and flowering times, which together with genetic sequence data, strongly supports the hypothesis that they are functionally equivalent alleles of the same gene (Rahman and Gladstones 1972). Meanwhile, the 1,208 bp deletion facilitated an intermediate level of expression during early growth, but which rapidly rose to levels similar to the *Ku* and *Jul* alleles prior to floral initiation. This resulted in an intermediate flowering time phenotype, including a moderate reduction in vernalisation requirement relative to the *ku* wild type (Taylor et al. 2019), both of which may be useful in breeding. Similar analyses in a subset of 28 F_8 RILs were used to confirm that the gene expression profiles segregating between *Ku* and *ku* genotypes result from the 1,423 bp deletion which affects *cis*-regulation of *LanFTc1* (Nelson et al. 2017). The 1,208 bp deletion provides novel and useful flowering diversity and provides encouragement that other, yet undiscovered, novel allelic diversity exists for flowering time in narrow-leafed lupin.

Lastly, linkage and association analyses have strengthened the case that *LanFTc1* underlies the *Ku* and *Jul* early flowering time alleles, and that the series of deletions in the promoter region of *LanFTc1* are the causal mutations. Nelson et al. (2017) were able to demonstrate a rapid decay of linkage disequilibrium within 75 Kb upstream and downstream of *LanFTc1*, and that the only two other genes within the immediate genomic region do not appear to have plausible roles in the floral initiation pathway. Complementing this finding, the 1,423 bp deletion and all three deletion genotypes scored as a single multiallelic polymorphism were the only genetic sequence variants within 7 Kb upstream and 2 Kb downstream of *LanFTc1* that were both highly significantly associated and perfectly linked with vernalisation responsiveness and flowering time in 42 genetically diverse narrow-leafed lupins (Taylor et al. 2019).

9.3 The Evolution of Flowering Time Genetic Control in *Lupinus* and the Fabaceae

9.3.1 A MADS-Box Transcription Factor Clade Key to the Vernalisation Pathway in the Brassicaceae is Absent in *Lupinus*

The MADS-box gene family consists of a large group of transcriptional regulators for developmental processes in the animal, fungi and plant kingdoms (De Bodt et al. 2003). Common to all MADS-box genes is a sequence of approximately 180 bp in length known as the MADS-box domain (De Bodt et al. 2003). This domain is present within the DNA-binding domain towards the N-terminus of the protein and is essential for binding to regulatory regions of targeted genes at conserved motif sequences, $CC(A/T)_6GG$, called CArG boxes (CC-A-rich-GG) (Theissen et al.

2000). Additionally, the MADS-box domain has roles in dimerization and binding of accessory factors (Shore and Sharrocks 1995). Although MADS-box genes have a variety of developmental roles in plants, ranging from root growth (Tapia-López et al. 2008; Yu et al. 2014) to fruit development (Gu et al. 1998; Liljegren et al. 2000), many of the most commonly known MADS-box genes are those involved in the determination of flowering time and floral meristem and organ identity (De Bodt et al. 2003).

Analysis of the draft Tanjil genome assembly (Hane et al. 2017) revealed that many of the MADS-box genes involved in flowering time coordination in other dicotyledonous species are also present in narrow-leafed lupin (Table 9.1). Some of these genes include, but are not limited to, homologues of: (1) *SOC1*, involved in integrating signals from the photoperiod, vernalisation, autonomous and gibberellin-dependent pathways in *Arabidopsis* (Moon et al. 2003), and found on linkage group NLL-13 and an unmapped scaffold in narrow-leafed lupin; (2) numerous *AGAMOUS-LIKE* (*AGL*) repressor genes, such as *AGL15* and *AGL18*, which act upstream of *FT* (Adamczyk et al. 2007), and *AGL* activators, including *AGL19*, involved in vernalisation responsiveness (Schönrock et al. 2006); and (3) *SVP*, a repressor of flowering involved in

Table 9.1 A list of prominent MADS-box genes in dicotyledon species indicated as being present or absent in narrow-leafed lupin (*Lupinus angustifolius* L.), adapted from Hane et al. (2017) to include narrow-leafed lupin scaffold and pseudochromosome coordinates

Gene	*Medicago truncatula* ID	*Lupinus angustifolius* ID	Narrow-leafed lupin scaffold coordinates	Pseudochromosome coordinates
AG	*2g017865, 8g087860*	*Lup005280*	Scaffold_136: 699182–701468	NLL-16: 5448773–5451059
		Lup011389	Scaffold_213: 247859–249474	NLL-04: 5376700–5378315
AGL15	*0003s0590*	*Lup015658*	Scaffold_280: 586046–589775	NLL-02: 5896396–5900125
		Lup017483	Scaffold_30_7: 2998082–3001083	NLL-13: 258393–261394
AGL18	*1g053070*	*Lup016817*	Scaffold_3_48: 141213–145185	NLL-08: 13196454–13200426
AGL19/14	*4g102530*	*Lup020546*	Scaffold_377: 208011–225295	NLL-06: 13786882–13804166
AGL6/13	*7g075850, 8g033270*	*Lup011089*	Scaffold_21_21: 79616–82322	NLL-09: 13200016–13202722
		Lup014750	Scaffold_276_3: 29702–33375	Not on pseudochromosome
AGL71/72	Not present	Not present		
ANR1 clade	*5g031000, 2g009890*	*Lup009010*	Scaffold_180: 684618–692667	NLL-19: 17467146–17475195
ANR1 clade		*Lup020683*	Scaffold_38: 973205–973929	NLL-18: 12850825–12851549
AP1	*8g066260, 5g046790*	*Lup006876*	Scaffold_152: 923071–927491	NLL-02: 14347554–14351974
		Lup021855	Scaffold_412: 191853–193923	NLL-11: 7189302–7191372
		Lup024348	Scaffold_491: 59796–64337	NLL-05: 20548584–20553125

(continued)

Table 9.1 (continued)

Gene	*Medicago truncatula* ID	*Lupinus angustifolius* ID	Narrow-leafed lupin scaffold coordinates	Pseudochromosome coordinates
AP3	*5g021270, 3g113030*	*Lup022019*	Scaffold_42_79: 187564–189087	NLL-17: 1459212–1460735
		Lup022831	Scaffold_44_2: 693863–695336	NLL-06: 1396226–1397699
FLC clade	Not present	Not present		
FULa/b	*4g109830, 2g461760*	*Lup018485*	Scaffold_32_11: 918478–921988	NLL-01: 1047437–1050947
FULc	*7g016630*	*Lup023253*	Scaffold_454: 139468–152826	NLL-09: 1327366–1340724
		Lup029962	Scaffold_75_88: 306596–311518	NLL-08: 11058137–11063059
PI	*3g088615, 1g029670*	*Lup011754*	Scaffold_22_1: 214383–217313	NLL-15: 16651639–16654569
		Lup011756	Scaffold_22_2: 11763–13420	Not on pseudochromosome
		Lup019933	Scaffold_357_9: 248391–250123	NLL-15: 16640236–16641968
SEP1/2	*7g016600, 6g015975*	*Lup001397*	Scaffold_105_5: 410921–415660	NLL-20: 19794209–19798948
		Lup029963	Scaffold_75_88: 317393–322687	NLL-08: 11068934–11074228
SEP3/4	*8g097090, 3g084980, 4g109810*	*Lup007613*	Scaffold_161_23: 431548–442587	NLL-01: 30909796–30920835
		Lup018484	Scaffold_32_11: 898871–901239	NLL-01: 1027830–1030198
		Lup021854	Scaffold_412: 162178–183861	NLL-11: 7199364–7221047
		Lup024347	Scaffold_491: 18533–27384	NLL-05: 20507321–20516172
		Lup026304	Scaffold_56: 1800372–1802053	NLL-12: 18239214–18240895
SHP	*3g452380*	*Lup029632*	Scaffold_73: 1513925–1516118	NLL-11: 610866–613059
		Lup019040	Scaffold_33: 2371132–2373649	NLL-02: 23765342–23767859
SOC1	*7g075870, 8g033250*	*Lup014751*	Scaffold_276_3: 44505–54633	Not on pseudochromosome
		Lup024911	Scaffold_50_31: 959650–967364	NLL-13: 14317859–14325573
STK	*3g005530*	*Lup017516*	Scaffold_30_7: 3225105–3237172	NLL-13: 22304–34371
SVPa	*5g032150, 5g032520, 4g093970*	*Lup015985*	Scaffold_29_1: 956508–959816	NLL-06: 4741768–4745076
		Lup025549	Scaffold_53: 1390130–1393883	NLL-17: 4688394–4692147

(continued)

Table 9.1 (continued)

Gene	*Medicago truncatula* ID	*Lupinus angustifolius* ID	Narrow-leafed lupin scaffold coordinates	Pseudochromosome coordinates
SVPc	*5g066180*	*Lup012854*	Scaffold_233_1: 261215–265246	NLL-11: 3209056–3213087
		Lup032504	Scaffold_94_15: 191335–195356	NLL-09: 18775616–18779637
TT16	*1g038300*	*Lup010795*	Scaffold_203_1: 312996–320875	NLL-15: 2970837–2978716
		Lup018771	Scaffold_329: 70703–72785	NLL-14: 523113–525195

ambient temperature signalling and interaction with *FLC* (Kim et al. 2009; Lee et al. 2007), found on linkage groups NLL-06, NLL-09, and NLL-17 in narrow-leafed lupin.

Notably, representation of the *FLC* clade is absent in narrow-leafed lupin based on searches for homologous genes in the genome assembly (Hane et al. 2017). In the Brassicaceae, *FLC* represses flowering (Michaels and Amasino 1999) by inhibiting transcription of floral integrator genes, including *FT* and *SOC1*. While this repression can be alleviated through the autonomous pathway in late developmental stages (Adrian et al. 2009), *FLC* is primarily silenced through exposure to vernalising conditions, which result in epigenetic modification of the chromatin surrounding *FLC*.

Hecht et al. (2005, 2011) have similarly reported the absence of *FLC*-like genes in numerous species from the Hologalegina (galegoid) clade, which comprises cool-season legumes, including the *Lotus*, *Medicago*, *Pisum*, *Vicia*, *Cicer*, *Lens*, *Trifoloium* and *Melilotus* genera (Cardoso et al. 2013; Wojciechowski et al. 2004). Within the Papilionoideae, the largest of the six Fabaceae subfamilies (Azani et al. 2017), *FLC* has only yet been reported in soybean (*Glycine max* L.) (Jung et al. 2012). However, this orthologue is thought to be non-functional and present only in the capacity as an ancestral remnant (Weller and Ortega 2015). This seems likely to be the case given phenology is principally modified by the photoperiod pathway in soybean (Weller and Ortega 2015). Overall, the absence of *FLC*-like genes in the narrow-leafed lupin genome adds to mounting evidence that, similar to the Poaceae (Blümel et al. 2015), the vernalisation response in the Fabaceae differs to that in the model *Arabidopsis*, and that the vernalisation pathway has likely evolved independently in parallel among different angiosperm families (Ream et al. 2012).

9.3.1.1 Divergence of the Phosphatidyl Ethano lamine-Binding Protein (PEBP) Family Within the *Lupinus* Genus and from Other Legumes

The PEBP gene family is an extremely old group of proteins widely conserved among the tree of life, and though involved in numerous biological processes, is commonly found to act in a regulatory capacity for growth and differentiation (Karlgren et al. 2011). Within higher plants, duplication and divergence have resulted in three clades evolving within the PEBP family, including the *MOTHER OF FT and TFL1* (*MFT*)-like, *FT*-like, and *TERMINAL FLOWER1* (*TFL1*)-like clades (Chardon and Damerval 2005; Kobayashi et al. 1999). Having been found in basal plants, such as moss, *MFT*-like is believed to be the oldest and ancestral clade (Hedman et al. 2009), and has roles in seed and embryo development. Following the divergence of the lycophytes, duplication of an *MFT*-like gene is thought to have led to an *FT*/*TFL1*-like group (Hedman et al. 2009), marking the first stage of PEBP family evolution within the plant kingdom. Phylogenetic analysis between PEBP genes in gymnosperms and angiosperms suggests that further duplication of the *FT/TFL1*-like group was then responsible for the subsequent development of distinctive *FT*-like and *TFL1*-like clades after divergence of the angiosperms (Klintenäs et al. 2012; Wickland and Hanzawa 2015). Although sharing a remarkable level of amino-acid sequence identity, residues at positions 134 and 138 have been critical to differentiating the roles of *FT* and *TFL* genes (Klintenäs et al. 2012), and while *TFL1*-like genes have taken on roles of repressing floral development, *FT*-like genes have instead become indispensable in promoting the transition from vegetative to floral growth.

Whole-genome and tandem duplication events have continued to shape the PEBP family in angiosperm lineages, including the Fabaceae family, and the recent availability of the narrow-leafed lupin genetic and genomic resources has enabled substantial progress in resolving the phylogeny and expansion of PEBP genes, including *FT*, within legumes. Previously, phylogenetic analysis had resolved three *FT* homologue clades within model organisms of the Fabaceae: *FTa*, *FTb* and *FTc* (Hecht et al. 2011). While the total number of *FT* homologues varies widely among species (Książkiewicz et al. 2016), and as few as three homologues have been found in *Lotus japonicus* (Yamashino et al. 2013), an average of six *FT* homologues have been identified per species in 11 legumes studied to date (Table 9.2). In general, the three *FT* clades are represented in all legume species. Until recently, the only exception to this generalisation was *FTc*, which was shown by Yamashino et al. (2013) to be absent in *Lotus japonicus*, and more recently by Książkiewicz et al. (2016) in pigeon pea (*Cajanus cajan*). However, utilising the genome survey sequence and unigene assembly resources for narrow-leafed lupin (Kamphuis et al. 2015), Nelson et al. (2017) have also shown an absence of the *FTb* clade for the first time in an agriculturally important legume species. Micro-synteny analyses by (Hane et al. 2017) also confirmed this to be the case, and have additionally highlighted the absence of representation of the *FTa1* subclade of *FTa* homologue in narrow-leafed lupin (Table 9.3).

In addition to variation in representation of the PEBP family in terms of *FT* clade presence/absence, Nelson et al. (2017) were also able to demonstrate an unusual copy number for the *FTc* clade in narrow-leafed lupin. Typically, a singular *FTc* homologue is observed in most legume species. However, two copies are present in both soybean and narrow-leafed lupin (Table 9.2). While the PEBP family in soybean has undergone multiple Whole-Genome

Table 9.2 A summary of the number of *FT* homologues identified in 11 legume species, adapted from Książkiewicz et al. (2016)

Species	*FTa*	*FTb*	*FTc*	Total *FT*	Reference
Arachis duranensis	3	4	1	8	Książkiewicz et al. (2016)
Arachis ipaensis	4	4	1	9	Książkiewicz et al. (2016)
Cajanus cajan	2	2	0	4	Książkiewicz et al. (2016)
Cicer arietinum	3	1	1	5	Książkiewicz et al. (2016)
Glycine max	4	4	2	10	Kong et al. (2010), Książkiewicz et al. (2016)
Lotus japonicas	1	2	0	3	Yamashino et al. (2013), Książkiewicz et al. (2016)
Lupinus angustifolius	2	0	2	4	Książkiewicz et al. (2016), Nelson et al. (2017)
Medicago truncatula	3	2	1	6	Laurie et al. (2011), Książkiewicz et al. (2016)
Phaseolus vulgaris	2	2	1	5	Książkiewicz et al. (2016)
Pisum sativum	2	2	1	5	Hecht et al. (2011)
Vigna radiate	2	4	1	7	Książkiewicz et al. (2016)
Average	2.5	2.5	1.0	6.0	

Table 9.3 A list of phosphatidyl ethanolamine-binding protein (PEBP) genes in dicotyledonous species indicated as being present or absent in narrow-leafed lupin (*Lupinus angustifolius* L.), adapted from Hane et al. (2017) to include narrow-leafed lupin scaffold and pseudochromosome coordinates

Gene	*Medicago truncatula* ID	*Lupinus angustifolius* ID	Narrow-leafed lupin scaffold coordinates	Pseudochromosome coordinates
BFT	*0020s120*	*Lup020111*	Scaffold_36: 2152852–2153927	NLL-07: 8911179–8912254
FTa1/2	*7g084970, 7g085020*	Not present		
FTa3	*6g033040*	*Lup021189*	Scaffold_398: 267518–270764	NLL-20: 21506464–21509710
		[a]	Scaffold_135: 877169–874226	NLL-08: 12283657–12280714
FTb	*7g066630, 7g066690*	Not present		
FTc	*7g085040*	*Lup005674*	Scaffold_14_1: 3314711–3317999	NLL-17: 19284328–19287616
		Lup015264	Scaffold_276_44: 10823–18144	NLL-10: 8023843–8031164
MFT	*8g106840*	*Lup021954*	Scaffold_42_76: 333256–335047	NLL-06: 9683689–9685480
TFL1	*1g060190, 2g086270, 7g104460*	*Lup001307*	Scaffold_104_5: 981187–982389	NLL-06: 29611068–29612270
		Lup019608	Scaffold_347: 151124–153204	NLL-05: 24312971–24315051
		Lup025777	Scaffold_54_99: 3824–4144	Not on pseudochromosome
		Lup026028	Scaffold_557: 19888–21452	NLL-19: 5733078–5734642

[a]One of two *FTa3* homologues in narrow-leafed lupin was identified by Książkiewicz et al. (2016) and Nelson et al. (2017).

Duplication (WGD) events in addition to recent tandem duplication (Wang et al. 2015) and the second *FTc* homologue in this species is highly likely to have arisen during the most recent WGD event roughly 11–13 million years ago (mya) (Hane et al. 2017; Schmutz et al. 2010), two independent macro- and micro-syntenic analyses with several Papilionoideae legumes suggest that the second *FTc* homologue in narrow-leafed lupin was derived through a much smaller scale duplication event (Hane et al. 2017; Książkiewicz et al. 2016). In both of these studies, the genomic region containing *LanFTc1* was found to show conservation and collinearity with regions containing *FTc* homologues in other legume species (Fig. 9.1). However, the genomic region containing *LanFTc2* had a distinctly different syntenic pattern, and instead shared a high degree of collinearity with separate regions lacking any *FT* representation, additionally indicating that this is the more recently derived homologue in narrow-leafed lupin (Książkiewicz et al. 2016; Hane et al. 2017).

As previously discussed, convincing genetic linkage mapping, gDNA sequencing and gene expression data have recently implicated the *LanFTc1* homologue in the vernalisation signally pathway and as the causal gene for the *Ku* early flowering time locus in narrow-leafed lupin (Nelson et al. 2017; Taylor et al. 2019). This has been an important discovery within the Papilionoideae, shedding light on the evolution of functional roles of the three *FT* clades in this subfamily of legumes. Until now, the vernalisation pathway was largely thought to be mediated by the *FTa* clade in cool-season, long-day

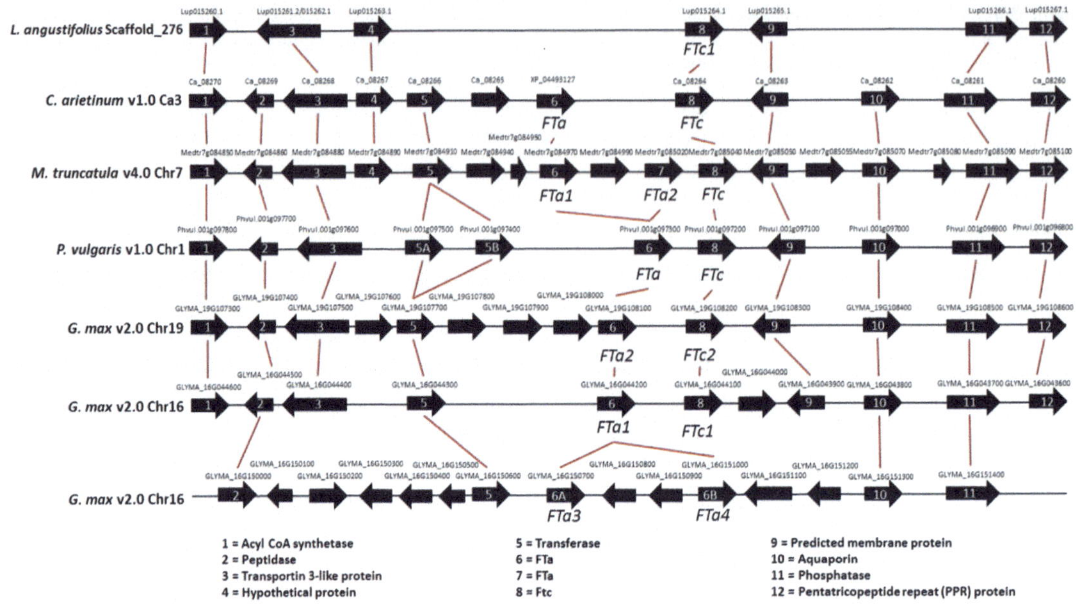

Fig. 9.1 A schematic representation of the shared collinearity of *FTc1/2* and *FTa1/2* homologues in narrow-leafed lupin (*L. angustifolius*) barrel medic (*M. truncatula*), chickpea (*C. arietinum*), common bean (*Phaseolus vulgaris*), and soybean (*G. max*). *Source* Hane et al. (2017)

legumes based on findings in *M. truncatula* (Laurie et al. 2011). Additionally, this discovery provides evidence that the *FTc* clade, which is known to have the most divergent coding sequence relative to most *FT* genes (Weller and Ortega 2015), has taken on diverse functions throughout the Papilionoideae lineage, similar to *FTa* (Laurie et al. 2011). Previously, an *FTc* homologue (*GmFT5a*/*GmFTc1*) has been shown to be important for responsiveness to short-day inductive photoperiods in soybean (*G. max*) (Kong et al. 2010), and while the specific roles are still yet to be determined in *M. truncatula* and pea (*Pisum sativum*), the singular *FTc* homologues in these galegoid species appear to cross regulate *FT* or other integrators upstream of floral meristem identity genes (Hecht et al. 2011; Laurie et al. 2011).

In addition to the genetic and genomic resources for narrow-leafed lupin, the recent release of the revised white lupin genetic map (Książkiewicz et al. 2017) has also enabled fascinating insight into PEBP family divergence within the *Lupinus* genus. Previously, phylogenetic analysis between narrow-leafed lupin and *L. polyphyllus* indicated that the WGT event specific to the Genisteae tribe within the Papilionoideae subfamily is estimated to have occurred prior to the divergence of the New and Old World lupins (Cannon et al. 2015). This result may have given reason to believe that the PEBP family may be relatively similar between the Old and New World lupin groups. However, there is also substantial research to indicate that not only do the two geographic groupings of lupins vary considerably in terms of genome size and chromosome number, but that significant variation also exists within these two groups (Wolko et al. 2011). Therefore, it seems reasonable to assume that significant diversification of the PEBP family and flowering mechanisms would have occurred within the *Lupinus* genus.

Insights gained from synteny analysis between narrow-leafed lupin and white lupin support this latter hypothesis. Although all five QTLs for vernalisation responsive early flowering in white lupin are located in regions with shared collinearity to narrow-leafed lupin, none of these QTLs are found in a region syntenic to that of NLL-10 containing the *Ku* locus in narrow-leafed lupin (Książkiewicz et al. 2017). Furthermore, whereas an *FTc* homologue (*LanFTc1*) underlies the main locus for vernalisation responsive early flowering in narrow-leafed lupin (*Ku*), the *FTa* homologue (corresponding to *LanFTa1* in narrow-leafed lupin; (Hane et al. 2017; Nelson et al. 2017)) has instead been identified as a candidate for the *brev* early flowering time locus in white lupin (Książkiewicz et al. 2017). Interestingly, an earlier analysis of phloem exudate in white lupin revealed the presence of an *FTc* homologue, suggesting a role for FTc in floral initiation signalling in this species (Rodriguez-Medina et al. 2011). Taken together, these lines of evidence clearly illustrate that PEBP family has continued to evolve within the *Lupinus* genus and suggests that the roles of the *FT* clades in integrating exogenous signals for flowering have diverged. Similarly, evolution of the PEBP family within a legume genus is also evident in *Arachis duranensis* and *A. ipaensis*, the progenitors of cultivated peanut (Bertioli et al. 2016), which have differing copy numbers for the *FTa* clade and *TFL1* PEBP family groups (Książkiewicz et al. 2016). Despite this divergent *FT* clade evolution among Old World lupins, it is interesting to note that flowering time varies similarly across rainfall gradients in these species (Berger et al. 2017), suggesting that similar selection pressures have impacted upon divergent genomic regions.

At this stage, our greatest understanding of the PEBP family in the Fabaceae surrounds the *FT* homologues, which integrate signals from various responses, including the vernalisation pathway. However, it is still largely unclear which genes upstream of *FT* are involved in transmitting these signals in this family and through which mechanisms they do so, particularly in *Lupinus*. In *Arabidopsis*, the FLC and SHORT VEGETATIVE PHASE (SVP) proteins are well documented to inhibit *FT* expression prior to vernalisation by binding to motifs within the first intron of *FT* (Gregis et al. 2013; Helliwell et al. 2006; Searle and Coupland 2004; Tao et al. 2012). However, *FLC* is broadly absent in the Papilionoideae legumes (Hecht et al. 2005, 2011), including narrow-leafed lupin (Hane et al. 2017), and so an alternative gene must instead be repressing *FT*-like genes until the arrival of vernalising conditions.

In soybean, a rather unique mechanism for regulation of *FT* homologues has been observed. Through gene expression analyses, Zhai et al. (2014) provide evidence that an *FTb* homologue (*GmFT4*) inhibits flowering time via antagonistic repression of two *FT* homologues, *GmFT2a* and *GmFT5a*/*GmFTc1*, which promote flowering time under inductive short-day photoperiods (Kong et al. 2010). A similar mechanism has been famously reported only once before in sugar beet (*Beta vulgaris* ssp. *vulgaris*) by Pin et al. (2010), indicating the novelty of this mechanism. As gene expression profiles generated through quantitative PCR (qPCR) have revealed that none of the four *FT* homologues are down regulated with vernalisation treatment in narrow-leafed lupin (Nelson et al. 2017), it is highly unlikely that another *FT* homologue is repressing *LanFTc1* in the absence of vernalisation.

Recently, a *VERNALISATION2* (*VRN2*) homologue has been identified in *M. truncatula* as a repressor of the vernalisation and photoperiod responsive *FT* homologue, *MtFTa1* (Jaudal et al. 2016). In *Arabidopsis*, VRN2 is involved in stably repressing *FLC* transcription following vernalisation (Gendall et al. 2001) by forming a complex with POLYCOMB-GROUP REPRESSIVE COMPLEX (PRC2) (Berry and Dean 2015). Although it is still unclear exactly how MtVRN2 functions, it appears to repress upregulation of *MtFTa1* mediated through the vernalisation pathway, and *Mtvrn2* mutants are able to flower early in long days without fulfilling vernalisation (Jaudal et al. 2016). Potentially, a similar mechanism could exist in the *Lupinus* genus, whereby *FT* homologues are directly regulated by repressor proteins rather than indirectly through suppression of repressors, like FLC. The availability of the narrow-leafed lupin draft reference genome (Hane et al. 2017) provides a great resource to further explore this possibility.

Lastly, the upcoming release of a narrow-leafed lupin pan-genome assembly (Gagan Garg, Lars G. Kamphuis and Karam B. Singh, unpublished data) has recently provided a unique opportunity to further explore regulation of the PEBP family, specifically *LanFTc1*. Screening for polymorphisms in this pan-genome panel recently revealed two independent deletions of 1,208 bp and 5,162 bp in the promoter region of *LanFTc1* (Taylor et al. 2019), both of which overlap the location of 1,423 bp deletion associated with the *Ku* allele (Nelson et al. 2017). Similar to the 1,423 bp deletion, the 5,162 bp deletion was associated with early, vernalisation-insensitive flowering and de-repressed expression of *LanFTc1* in the absence of vernalisation. Meanwhile, the 1,208 bp deletion was associated with an intermediate flowering time, response to vernalisation and level of relative *LanFTc1* gene expression in the absence of vernalisation as a young seedling. As these deletions overlap the 1,423 bp deletion in the 5′ regulatory region of *LanFTc1*, this series has refined which parts of the promoter are critical for mediating the vernalisation response at the molecular level. However, it remains unclear how this is achieved and further research, such as methylation studies or genome editing for example by the CRISPR/Cas-9 system (Bortesi and Fischer 2015), will be required to elucidate whether this critical region facilitates promoter element proximity or contains binding sites for transcription factors or proteins related to chromatin modification in the wild-type. Release of the pan-genome assembly will allow new opportunities to explore the regulation and potential involvement of members of the PEBP family and others in flowering time control in narrow-leafed lupin.

9.3.1.2 Representation of the Basic Region/Leucine Zipper (bZIP) Transcription Factors Involved in Floral Integration in Narrow-Leafed Lupin Indicate Earlier Evolution of this Gene Family in the Fabaceae

A gene family closely connected to the PEBP family in angiosperms through their role in floral integration is the basic region/leucine zipper (bZIP) transcription factor family. This group of genes are characterised by the DNA-binding domain of their encoded proteins, which comprise two structures: firstly, a basic region responsible for nuclear localization and DNA contact; and secondly, a region containing leucine and other hydrophobic amino acids that result in an amphipathic α-helix secondary structure, which is essential for the interaction of two bZIP proteins to form a 'zipper' complex for DNA binding (Jakoby et al. 2002). One of the most widely known bZIP proteins is *FD* (sometimes referred to as *FLOWERING LOCUS D*, e.g. (Turck et al. 2008)), which is preferentially expressed in the shoot apex of *Arabidopsis* (Abe et al. 2005). Here, FD is targeted by FT proteins and forms an FT/FD complex, which is then involved in induction of floral meristem identity genes, including *APETALA1* (*AP1*) (Abe et al. 2005; Wigge et al. 2005). Similarly, FD is also targeted within the shoot apex by TFL1, forming a TFL1/FD complex, which contrastingly works to maintain vegetative meristematic growth by repressing the upregulation of *AP1* and other floral identity genes (Hanano and Goto 2011).

The draft Tanjil reference genome (Hane et al. 2017) has revealed that narrow-leafed lupin possesses a single *FD* homologue (Table 9.4). Within the galegoid and phaseoloid tribes of legumes, three subclades of *FD* genes have recently been identified: *FDa*, *FDb* and *FDc* (Sussmilch et al. 2015). *FDc* homologues have so far only been found in phaseoloid species (Sussmilch et al. 2015), including soybean and common bean, indicating that this subclade arose from a duplication event in the lineage ancestral to the phaseoloid tribe. Contrastingly, the *FDa* and *FDb* subclades are present in species from both the phaseoloid and galegoid (Sussmilch et al. 2015), suggesting that these two subclades evolved from a duplication event prior to the divergence of the two tribes, which is estimated to have occurred approximately 54 mya (Lavin et al. 2005). However, as the *FD* homologue present in the narrow-leafed lupin genome corresponds to the *FDa* clade, it provides new evidence that at least the *FDa* subclade arose a lot

Table 9.4 A list of basic region/leucine zipper (bZIP) transcription factor genes involved in floral integration in dicotyledon species indicated as being present or absent in narrow-leafed lupin (*Lupinus angustifolius* L.), adapted from Hane et al. (2017) to include narrow-leafed lupin scaffold and pseudochromosome coordinates

Gene	*Medicago truncatula* ID	*Lupinus angustifolius* ID	Narrow-leafed lupin scaffold coordinates	Pseudochromosome coordinates
FDa	*5g022780*	*Lup018024*	Scaffold_31_17: 522527–523786	NLL-18: 15951182–15952441
FDb	*8g075130*	Not present		
LFY	*3g098560*	*Lup006312*	Scaffold_164_51: 13037–16573	NLL-01: 25488325–25491861
		Lup012198	Scaffold_226_15: 146454–147680	NLL-11: 19627585–19628811
		Lup027481a	Scaffold_61_29: 605238–610275	NLL-12: 14564590–14569627

earlier within the Papilionoide subfamily lineage, prior to the divergence of the genistoid tribe estimated to have occurred roughly 56 mya (Lavin et al. 2005).

9.3.1.3 Homologous Genes Related to Light Detection, Signalling and Response to the Photoperiod Pathway in Narrow-Leafed Lupin

The detection of light and an internal circadian clock, plus the interaction between the two, contribute to the photoperiodic signalling pathway for floral promotion in *Arabidopsis* (Andrés and Coupland 2012; Song et al. 2013). In this model species, light perception relies on a number of different gene families and genes to detect different wavelengths in the electromagnetic spectrum, including: the *PHYTOCHROME* (*PHY*) family, for detection of red/far-red light and daylength (Fankhauser 2001); the *CRYPTOCHROME* (*CRY*) and *PHOTOTROPIN* (*PHO*) families, and *ZEITLUPE* (*ZTL*) and *FLAVIN-BINDING, KELCH REPEAT, F-BOX* (*KFK1*) genes, all of which detect blue light (Ahmad et al. 1998; Briggs et al. 2001; Imaizumi et al. 2003; Kim et al. 2007); and lastly, the *UV RESISTANCE LOCUS 8* (*UVR8*) gene involved in detection of ultraviolet light (Christie et al. 2012). Similarly, a number of genes are involved in regulating the circadian clock, including: the *EARLY FLOWERING* gene family and *LUX-ARRHYTHMO* (*LUX*), which together form an evening complex (EC) that regulates the evening loop of the circadian clock and additionally have some influence on flowering time given that mutants show early flowering phenotypes (Nusinow et al. 2011); the *PSEUDO-RESPONSE REGULATOR* (*PRR*) gene family, which is involved in stabilising the expression of photoperiodic genes, such as *CONSTANS* (*CO*), that upregulate the floral integrator gene, *FT* (Hayama et al. 2017); and *GIGANTEA* (*GI*), which regulates circadian rhythms and is involved in upregulating *CO* and *FT* (Mizoguchi et al. 2005; Sawa and Kay 2011).

Relative to *M. truncatula*, the draft Tanjil reference genome (Hane et al. 2017) has revealed that narrow-leafed lupin contains an abundance of photoreceptor and circadian clock-related homologues (Table 9.5), including *PHYA, PHYB, CRY1, ZTL, URV8, LUX, ELF4, PRR59a,* and *GI*. Particularly interesting is that narrow-leafed lupin has multiple *PHYA* and *PHYB* homologues and an absence of *PHYE* representation. While the role of *PHYE* is still not completely understood in *Arabidopsis*, *phye* mutants have a subtle phenotype and it is likely that the gene is not an essential photoreceptor (Fankhauser 2001). The absence of *PHYE* in narrow-leafed lupin could suggest that this homologue also has a limited role, if any, within the Fabaceae. Contrastingly, *PHYA* appears to have a very important role in the photoperiod induction of flowering in long-day and short-day legumes, such as pea (Weller et al. 1997) and soybean (Liu et al. 2008), and there is evidence for functional variation in these genes being associated with modified *FT* homologue

Table 9.5 A list of prominent genes involved in the circadian clock, light detection, light signalling and photoperiodic response in dicotyledonous species indicated as being present or absent in narrow-leafed lupin (*Lupinus angustifolius* L.), adapted from Hane et al. (2017) to include narrow-leafed lupin scaffold and pseudochromosome coordinates

Functional role	Gene	Encoded protein	*Medicago truncatula* ID	*Lupinus angustifolius* ID	Narrow-leafed lupin scaffold coordinates	Pseudochromosome coordinates
Circadian clock	*ELF4*	MYB transcription factor	*3g070490*	*Lup016731*	Scaffold_3_1: 1361759–1362091	NLL-14: 1957756–1958088
	ELF4-like		*3g070490, 4g125590, 2g041310*	*Lup020202*	Scaffold_361: 420289–420636	NLL-02: 5389423–5389770
				Lup026070	Scaffold_56: 134624–134968	NLL-12: 16573466–16573810
	LHY		*7g118630*	*Lup021567*	Scaffold_40_1: 895730–900479	NLL-16: 18435585–18440334
	LUX		*4g064730*	*Lup004491*	Scaffold_13_1: 1864109–1864969	NLL-03: 1864109–1864969
				Lup029750	Scaffold_74_1: 167702–168571	NLL-10: 2350094–2350963
	ELF3a	Pseudo-response-regulator	*3g103970*	*Lup030650*	Scaffold_78_9: 766476–769585	NLL-03: 23827343–23830452
	ELF3b		*1g016920*	*Lup000361*	Scaffold_1_1096: 96100–101281	Not on pseudochromosome
	GI		*1g098160*	*Lup002034*	Scaffold_11_93: 741708–749955	NLL-20: 569302–577549
				Lup004730	Scaffold_13_17: 605276–615524	NLL-04: 26718452–26728700
				Lup025070	Scaffold_51_1: 790755–800911	NLL-13: 12513091–12523247
	PRR37a		*4g061360*	*Lup017167*	Scaffold_30_7: 892983–898198	NLL-13: 2361278–2366493
	PRR37b		*1g067110*	*Lup018664*	Scaffold_322: 352948–361243	NLL-20: 11464368–11472663
	PRR59a		*3g092780*	*Lup000501*	Scaffold_10_8: 725521–729620	NLL-15: 13463106–13467205
				Lup009602	Scaffold_19_75: 36327–40451	NLL-03: 3269049–3273173
	PRR59b/c		*7g118260, 8g024260*	*Lup002529*	Scaffold_113_48: 35416–39588	Not on pseudochromosome
				Lup008505	Scaffold_174_1: 54272–57584	NLL-11: 4914588–4917900
	TOC1a/b		*4g108880, 3g037390*	*Lup010052*	Scaffold_197: 397815–399379	NLL-15: 8784400–8785964
				Lup010053	Scaffold_197: 404149–407736	NLL-15: 8776043–8779630
				Lup026326	Scaffold_56: 1954572–1961135	NLL-12: 18393414–18399977

(continued)

Table 9.5 (continued)

Functional role	Gene	Encoded protein	*Medicago truncatula* ID	*Lupinus angustifolius* ID	Narrow-leafed lupin scaffold coordinates	Pseudochromosome coordinates
Light signalling	*PIF1*	bHLH transcription factor	*7g099540, 1g069155*	*Lup028329*	Scaffold_65_1: 1126795–1133916	NLL-05: 26066285–26073406
	PIF3		*1g084980, 7g111320*	*Lup015480*	Scaffold_28_48: 1070656–1074321	NLL-04: 22157166–22160831
	PIF4/5		*3g449770*	*Lup019016*	Scaffold_33: 2185854–2199589	NLL-02: 23580064–23593799
	PIF6		*7g110810*	*Lup005332*	Scaffold_137_1: 511630–513835	NLL-04: 11771040–11773245
	SPT		*5g017040*	*Lup001007*	Scaffold_101: 1198984–1202085	NLL-18: 15073223–15076324
				Lup006191	Scaffold_144_66: 213037–216040	NLL-17: 2282115–2285118
				Lup008225	Scaffold_17_1: 892304–894599	NLL-03: 4735616–4737911
	HY5	BZIP transcription factor	*3g436010*	*Lup008440*	Scaffold_172_80: 136068–138458	NLL-07: 17581908–17584298
	HYH		*7g057160*	*Lup028646*	Scaffold_68: 1269705–1272456	NLL-03: 9336280–9339031
	COP1	Ubiquitin ligase component	*5g085250*	*Lup000696*	Scaffold_100: 716900–720943	NLL-09: 20785163–20789206
	SPA1		*5g009530, 8g027985*	*Lup004875*	Scaffold_130: 960584–969591	NLL-18: 1576674–1585681
				Lup005516	Scaffold_139_1: 548770–554373	NLL-02: 15876233–15881836
				Lup032080	Scaffold_9_221: 657099–667282	NLL-06: 12546014–12556197
	SPA4		*8g091170, 2g084980, 2g085210*	*Lup023005*	Scaffold_440: 87379–91541	NLL-19: 1511011–1515173

(continued)

Table 9.5 (continued)

Functional role	Gene	Encoded protein	*Medicago truncatula* ID	*Lupinus angustifolius* ID	Narrow-leafed lupin scaffold coordinates	Pseudochromosome coordinates
Photoperiod response	*COLa*	B-box transcription factor	*7g018170*	*Lup001378*	Scaffold_105_5: 87432–91159	NLL-20: 20118710–20122437
				Lup011967	Scaffold_220: 853766–856299	NLL-08: 2372138–2374671
				Lup012039	Scaffold_223: 492343–494461	NLL-09: 737633–739751
	COLb/c		*1g013450, 3g015710*	*Lup009876*	Scaffold_190: 979124–980264	NLL-15: 19708384–19709524
				Lup030682	Scaffold_78_9: 1053290–1054411	NLL-03: 23542517–23543638
	COLd		*4g128930*	*Lup009465*	Scaffold_189: 580514–581696	NLL-01: 2648045–2649227
				Lup026154	Scaffold_56: 772645–773774	NLL-12: 17211487–17212616
	COLe/f		*3g082630, 5g069480*	*Lup022494*	Scaffold_43: 1228463–1231562	NLL-02: 1453723–1456822
				Lup025079	Scaffold_51_104: 9188–14031	Not on pseudochromosome
				Lup028850	Scaffold_7_13: 1213844–1216809	NLL-09: 1705973–1708938
				Lup032195	Scaffold_90_118: 468744–471175	NLL-02: 20922936–20925367
				Lup032539	Scaffold_94_15: 576654–579038	NLL-09: 19160935–19163319
	COLg		*7g108350*	*Lup002871*	Scaffold_117: 1266338–1268467	NLL-04: 15248849–15250978
				Lup006262	Scaffold_145_72: 72399–75049	NLL-16: 15596594–15599244
	COLh		*7g083540*	Not present		
	COLi		*8g104190*	*Lup011055*	Scaffold_21_1: 426608–428586	NLL-06: 20903202–20905180
				Lup029233	Scaffold_71_8: 7557–9533	Not on pseudochromosome
				Lup032792	Scaffold_96_2: 1291676–1293277	NLL-06: 10392104–10393705
	COLj		*2g088900*	*Lup006522*	Scaffold_15_15: 113295–115094	NLL-13: 3675219–3677018
	COLk		*1g110870*	*Lup016892*	Scaffold_3_382: 240333–242188	NLL-20: 3378242–3380097
				Lup018321	Scaffold_32_1: 183284–185175	NLL-08: 6799796–6801687
				Lup030499	Scaffold_77_11: 24873–26952	Not on pseudochromosome
	CDFa	DOF transcription factor	*3g435480*	*Lup015794*	Scaffold_286: 272039–274724	NLL-15: 9455917–9458602
				Lup018442	Scaffold_32_11: 600719–602903	NLL-01: 729678–731862
	CDFb		*7g010950, 6g012450*	*Lup004204*	Scaffold_129: 1193797–1197232	NLL-05: 1193797–1197232
				Lup011979	Scaffold_221_1: 381091–384222	NLL-20: 18759134–18762265
	CDFc		*4g082060, 5g014530*	*Lup006734*	Scaffold_150_38: 70471–73925	NLL-11: 8186631–8190085
				Lup008651	Scaffold_176: 954190–957225	NLL-09: 14283213–14286248
	CDFd		*8g044220, 7g086780*	*Lup001785*	Scaffold_109_4: 146989–148362	NLL-08: 19443251–19444624

(continued)

Table 9.5 (continued)

Functional role	Gene	Encoded protein	*Medicago truncatula* ID	*Lupinus angustifolius* ID	Narrow-leafed lupin scaffold coordinates	Pseudochromosome coordinates
Photoreceptor	*CRY1*	Blue light photoreceptor	*5g063920*	*Lup002802*	Scaffold_116_20: 406039–409586	NLL-02: 19994437–19997984
				Lup004976	Scaffold_133: 109119–112741	NLL-11: 3485437–3489059
				Lup012646	Scaffold_23: 2726555–2730196	NLL-14: 13521582–13525223
	CRY2		*1g043180, 1g076190*	*Lup008367*	Scaffold_171_1: 476879–480609	NLL-08: 2810441–2814171
	FKF1		*8g105590*	*Lup031570*	Scaffold_87_1: 372816–375476	NLL-17: 11817647–11820307
	PHOT1		*4g061610, 2g095980*	*Lup011301*	Scaffold_212: 467136–478146	NLL-06: 30542866–30553876
	PHOT2		*8g070530*	*Lup016543*	Scaffold_298: 195933–205721	NLL-18: 410270–420058
				Lup028173	Scaffold_64_13: 214212–227464	NLL-06: 22684932–22698184
	ZTL		*2g036510*	*Lup007315*	Scaffold_16_13: 12472–17780	Not on pseudochromosome
				Lup021384	Scaffold_4_25: 2419998–2426746	NLL-04: 1742781–1749529
	PHYA	Red/far-red light photoreceptor	*1g085160*	*Lup002691*	Scaffold_115_1: 596390–601232	NLL-08: 5897623–5902465
				Lup015476	Scaffold_28_48: 1016677–1021672	NLL-04: 22209815–22214810
				Lup017016	Scaffold_3_382: 1331546–1336613	NLL-20: 4469455–4474522
	PHYB		*2g034040*	*Lup014358*	Scaffold_26_21: 55113–59565	NLL-19: 14695569–14700021
				Lup027042	Scaffold_6_5: 489879–495700	NLL-16: 1716888–1722709
	PHYE		*2g049520*	Not present		
	UVR8	UV photoreceptor	*3g096780*	*Lup003585*	Scaffold_122_11: 97975–105142	NLL-15: 11986541–11993708
				Lup025974	Scaffold_55_94: 444193–462548	NLL-14: 9502628–9520983
				Lup031272	Scaffold_84: 216138–220976	NLL-01: 26004713–26009551

transcription profiles (Kong et al. 2010; Weller and Ortega 2015). It will be interesting in future to discover whether any of the *PHYA* homologues in narrow-leafed lupin have a similar capacity to moderate flowering time by regulation of any of the *FT* homologues.

The draft reference genome (Hane et al. 2017) additionally indicates that, relative to *M. truncatula*, narrow-leafed lupin also possesses a high number of *CO*-like homologues mediating photoperiod responsiveness, including 21 homologues from all three of the major *COL* family groups: Group I (*COLa-COLd*), Group II (*COLi* and *COLj*), and Group III (*COLe-COLh*, *COLj*). A total of 11 *CO*-like homologues are present in *M. truncatula*. While some *CO*-like genes are thought to be important in the photoperiodic responsiveness of the short-day legume, soybean (Wu et al. 2014), a similar functional role has not been found in the long-day legume, *M. truncatula* (Wong et al. 2014). Future investigation into the functionality of *CO*-like genes in narrow-leafed lupin will undoubtedly help us to improve our understanding of the role and evolution of this gene family in the Fabaceae.

9.3.2 Future Prospects of Improved Adaptive Genetic Variation Discovery with New and Improved Genetic and Genomic Resources

9.3.2.1 Genetic Maps, Genome Assemblies and Transcriptome Resources

Continual improvements to the reference genetic map and cultivar Tanjil reference genome assembly will greatly benefit studies of all agronomic and adaptive traits in narrow-leafed lupin, including flowering time. The ability to place scaffolds which are currently unmapped in the most recent genome assemblies (Hane et al. 2017; Zhou et al. 2018; Sect. 3.3) onto pseudochromosomes will allow new opportunities for genetic and association mapping. For instance, fewer genetic markers generated through genotyping by sequencing technologies, such as DArTseq, would be filtered out during the early stages of some mapping analyses as a result of lacking known mapping coordinates for the reference genome assembly. A greater abundance of genetic markers with a more even distribution throughout the genome would likely lead to improved candidate gene discovery, for example by increasing the chances of finding positive associations between genotypes and flowering time in GWAS studies, or increasing the resolution for genetic mapping of known loci for phenology.

The upcoming construction of a pan-genome assembly incorporating more than 40 diverse domesticated and wild narrow-leafed lupin accessions (Gagan Garg, Lars G. Kamphuis and Karam B. Singh unpublished data; Sect. 3.6.2) will similarly open up new opportunities to study flowering time control. One key advantage of a pan-genome assembly over a reference assembly generated from a single accession is the ability to assess large-scale polymorphisms and structural variations, including presence–absence variants (PAV; i.e. genomic sequences that are present in one individual but missing in another) and copy number variants (CNV; i.e. genomic sequences that are present in different numbers of copies between individuals) (Saxena et al. 2014). Such polymorphisms are gaining recognition for their importance in reflecting species genomic diversity (Scheben et al. 2016) and have been linked to phenotypic diversity for adaptive traits, including flowering time. For example, *Gdh7*, a known repressor of flowering under long-day conditions (Xue et al. 2018), was shown to be absent in two of 66 diverse rice (*Oryza sativa* and *O. rufipogon*) accessions (one landrace and one modern cultivar) within the recently developed pan-genome for rice species (Zhao et al. 2018). In addition to gene PAV or CNV, structural variations in regulatory elements can also be identified. For example, the usefulness of the upcoming narrow-leafed lupin pan-genome has already been demonstrated through the discovery of two new large (>1 Kb) deletions in the promoter of *LanFTc1*, a major flowering time gene, which reduces vernalisation responsiveness and results in early flowering time (Taylor et al. 2019). Lastly, pan-genomes also provide an opportunity for improved genotyping efficiency by enabling read mapping and variant calling in variable genome regions containing PAVs (Hurgobin and Edwards 2017). Again, this would serve to increase the number of genetic markers for mapping studies, thus potentially enhancing candidate gene discovery.

Lastly, current and future transcriptomes (Chap. 5) represent important complementary resources for flowering time candidate gene discovery and functional genomics. In particular, tissue and accession specific transcriptomes (Kamphuis et al. 2015) enable a greater capacity to dissect differential regulation and expression of flowering time genes, including those (if any) which may not be currently assembled or annotated in reference genome assemblies (Hane et al. 2017; Zhou et al. 2018). The development of future transcriptomes for plants grown under different environmental conditions would similarly be beneficial for teasing apart the biological role of candidate genes and their connection to other genes within the photoperiodic, vernalisation and autonomous pathways, as demonstrated in *Arabidopsis thaliana* (Torti et al. 2012).

9.3.2.2 TILLING and EcoTILLING Populations and Tools

A new resource being developed that may in future assist in dissecting the biological role(s) of candidate genes, plus aid the discovery of new polymorphisms in known flowering genes that result in novel phenotypic effects, is a Targeting Induced Local Lesion In Genomes (TILLING) population. TILLING is a reverse genetics approach whereby a large population containing thousands of random mutations throughout the genome is created by chemical mutagenesis, and amplification of the pooled DNA from multiple individuals is used to detect mismatches in heteroduplexes (i.e. the formation of double-stranded DNA) established between wild type and mutant genotypes for selected genes (Colbert et al. 2001; Till et al. 2006). These heteroduplex mismatches are revealed via plant endonucleases; enzymes, such as CEL I, that cleave DNA at the sites of mismatches in double-stranded DNA to produce two or more fragments. Following the detection of one or more heteroduplex mismatches, DNA from the individual(s) carrying the mutation(s) are used to sequence the gene of interest to identify the specific mutation(s) (Colbert et al. 2001; Till et al. 2006).

Such populations have proved to be useful in better understanding the role of flowering-related genes in major crop species. For example, in tetraploid wheat (*Triticum turgidum*), a TILLING population comprising 1,368 individuals was used to characterise *VRN1*; a vernalisation responsive gene that de-represses flowering (Chen and Dubcovsky 2012). A preliminary M_2 narrow-leafed lupin TILLING population has recently been developed in the cultivar Tanjil background through mutagenesis with ethyl-methanesulphonate (EMS) (Rhonda C. Foley and Karam B. Singh, unpublished data) and can now be used for high-throughput screening of SNP mutations, including G/C to A/T transversions typically produced by EMS (Colbert et al. 2001), in known flowering pathway genes. However, due to the dominant nature of the *Ku* allele in Tanjil, it may be difficult to discern the phenotypic effect of any mutations identified in this TILLING population.

An adaptation of TILLING that is capable of overcoming this potential issue is Ecotype TILLING (EcoTILLING). Here, naturally occurring SNP and INDEL mutations are screened for in genes of interest within natural populations of unrestricted size, origin and domestication status (Till et al. 2006). This approach has the benefit of surveying species-wide genetic diversity, provided that the population size is sufficiently large and representative of the species. EcoTILLING has been successfully used to identify new variants of vernalisation and photoperiod pathway genes that are significantly associated with flowering time and bolting in cereals, including hexaploid wheat (*T. aestivum*) (Chen et al. 2011) and rice (*Oryza sativa*) (Du et al. 2017), plus sugar beet (Frerichmann et al. 2013). With growing phenotypic datasets for large collections of wild and domestic narrow-leafed lupins from field-based trials (e.g. from the Australian Lupin Collection) and controlled environments under vernalising and non-vernalising conditions (e.g. from an upcoming GWAS study for narrow-leafed lupin (Candy Taylor, unpublished data)), EcoTILLING presents itself as an attractive option for future allele mining of flowering time variation for known genes, such as *LanFTc1*, and the validation of new candidate genes.

9.4 Using Genomic Information and Resources to Shape Adaptive Flowering Time Diversity in Domestic Narrow-Leafed Lupins

Pre-breeding efforts to identify novel, adaptive phenotypic diversity and to elucidate the genetic pathways for flowering time in narrow-leafed lupin generate valuable knowledge for breeding purposes. Below, we discuss how this knowledge can be applied to improve the efficiency of crosses and selections within breeding programmes using two current and emerging technologies, including marker-assisted selection (MAS) and genome editing.

9.4.1 Marker-Assisted Selection to Improve Breeding Efficiency

Identifying the genes involved in coordinating flowering time has significant implications for narrow-leafed lupin breeding as it presents an opportunity to improve the efficiency of plant selections and crossings using MAS. (For more information on MAS, please see Chap. 6). This is especially the case when the underlying causal polymorphisms for variations in gene function or expression which affect phenotypes can also be determined. For example, the recent discoveries of a series of INDEL polymorphisms in the promoter of *LanFTc1* that reduce or eliminate vernalisation requirement and which underlie the *Ku* and *Jul* loci (Nelson et al. 2017; Taylor et al. 2019) will enable perfect markers to be used within breeding programmes to select for these economically important loci. Molecular genetic markers may initially appear to be of limited value for flowering time, which is not only a relatively easy trait to score, particularly for the dominant *Ku* and *Jul* loci that are favoured in Australian and European narrow-leafed lupin breeding, but also a qualitative trait under the control of several loci. However, the importance of genetic markers becomes more apparent when the desirable variation is for a singular recessive allele that is easily masked by dominant loci, and is therefore more likely to be lost during the early stages of a breeding cycle. The PCR-based marker designed by Nelson et al. (2017) to assay INDELs in the promoter of *LanFTc1* will serve as a valuable tool to select for a rare and valuable variant identified in an Israeli accession (P22660) that facilitates a sought-after intermediate flowering time for long-season environments (Taylor et al. 2019). In addition, genetic markers for flowering time also provide certainty of genotype irrespective of environmental interactions during phenotyping, enabling single plant selections for homozygotes at an early stage of plant development, plus can be used to identify parents with valuable combinations of genotypes for one or more flowering time genes for crossing (Collard and Mackill 2008; Jung and Müller 2009). This again improves the efficiency of selection and crossing within breeding programmes. Continuation of the efforts to identify genes in the flowering time pathway will provide more opportunities and options to breeders for introducing and retaining valuable adaptive variation within breeding programmes, and should remain an important pre-breeding research objective.

9.4.2 Future Genome Editing for Rapid Germplasm Improvement

Candidate gene identification and functional genomics to decipher gene function and regulation also present opportunities to drastically improve breeding efficiency and increase adaptive genetic variation within elite domesticated germplasm through genome editing. Genome editing is a modern genetic modification (GM) approach that makes use of programmable endonuclease enzymes to induce site-specific double-stranded DNA breaks, which are subsequently repaired (Scheben et al. 2017). Although several genome editing technologies exist, including zinc finger nucleases (ZFNs) (Urnov et al. 2010) and transcription activator-like effector nucleases (TALENs) (Sun and Zhao 2013), the most widely known and adopted is the clustered regularly interspaced short palindromic repeat (CRISPR)/CRISPR-associated protein (Cas) system.

The CRISPR/Cas system originally evolved as an immune defence against viruses and plasmids in archaea and bacteria to silence foreign DNA (Gasiunas et al. 2012; Jinek et al. 2012). However, in recent years, it has been adapted to facilitate genome editing within higher eukaryotic organisms, including plants. Very simply, the adapted system relies on a synthetic single-guide RNA (sgRNA) containing a sequence (termed a spacer) of approximately 20 nucleotides to direct an associated Cas nuclease to targeted sites within the genome comprising complementary DNA sequences. Where the targeted sequences are adjacent to short protospacer adjacent motif (PAM) sequences, which are recognised and then bound by the Cas enzyme, a double-stranded break is introduced. These breaks are repaired by one of two endogenous repair mechanisms. The most dominant of these repair pathways is Non-Homologous End-Joining (NHEJ), in which the ends of the double-stranded break are ligated together in an efficient yet error-prone manner, frequently introducing small INDELs (Ran et al. 2013). This can result in a range of mutations, including frame-shifts or premature stop codons in gene coding sequences, therefore loss-of-function mutations. The second repair mechanism is Homology Directed Repair (HDR), which generally occurs only in dividing cells and is where an exogenous template is required for high-fidelity homologous recombination (Ran et al. 2013). While the 5′ and 3′ ends of the template must contain sequences homologous to those adjacent to the break (i.e. homology arms), the composition of the intermediate sequence is unrestricted, meaning that desirable deletions, insertions or point mutations can be specified. As the spacer in sgRNAs can easily and cheaply be synthesised to target any desired sequence (provided it is adjacent to a PAM), and a number of Cas nucleases originating from different prokaryotes can be utilised for different targeting specificity (Ran et al. 2013; Tang et al. 2017) and editing applications (Tang et al. 2017; Zetsche et al. 2017), CRISPR/Cas is an extremely versatile tool for genome editing.

In terms of application to breeding, CRISPR/Cas technology has tremendous potential to assist with rapid germplasm improvement and enhanced breeding efficiency. Unlike traditional GM approaches, whereby transgenes (i.e. genes from unrelated species) conferring desirable traits are inserted at indiscriminate locations within the genome, genome editing makes it possible to not only introduce transgenes or variable genes from the same species at specific positions but to also modify existing genetic sequences such that they mimic natural alleles or represent novel variation (Scheben et al. 2017). Additionally, CRISPR/Cas can be used to break the genetic linkage between loci for domestication or essential agronomic traits with adverse traits (Scheben et al. 2017), and the capacity to multiplex targets means that the homozygous fixation of several favourable genotypes (which is particularly useful for modifying quantitative traits, such as flowering time) can be achieved within a single generation (Cong et al. 2013). These features ultimately mean that new breeding lines and varieties can be generated in a much reduced time frame compared to conventional breeding pipelines, as the introduction of potentially deleterious alleles through crossing with certain germplasm (including wild types) is prevented and repeated backcrossing to re-establish elite domestic backgrounds is no longer necessary (Scheben et al. 2017). Furthermore, genome editing surpasses earlier GM approaches for crop improvement as transgene-free crops can be bred (Schaeffer and Nakata 2015). Most commonly, first-generation mutants (T_0) are made hemizygous (i.e. contain a single copy) for CRISPR/Cas constructs through stable transformation methods, such as *Agrobacterium tumefaciens*-mediated transformation. Selfing or crossing the T_0 mutants therefore allows for the selection of plants lacking the CRISPR/Cas transgene in subsequent generations (Schaeffer and Nakata 2015; Scheben et al. 2016). Alternatively, transient transformation methods in which the CRISPR/Cas constructs do not even integrate into the genome, such as preformed ribonucleoprotein complexes (Woo et al. 2015), ensure that the T_0 generation of mutants is transgene-free. The ability to produce transgene-free, improved crops through genome editing is highly beneficial in terms of potentially reshaping consumer and farmer concerns regarding GM improved crops and relaxing strict governmental regulations towards genome-edited crops in some jurisdictions (Scheben et al. 2017; Wolter and Puchta 2017).

The power of CRISPR/Cas to successfully engineer new flowering time phenotypes has recently been demonstrated in several crop species from diverse plant families. The first reported case was in tomato (*Solanum lycopersicum*), where mutations of the *SELF PRUNING 5G* (*SP5G*) floral repressor gene were created using CRISPR/Cas9 to firstly verify the role of SP5G within the photoperiod induction pathway, and secondly, to demonstrate the capacity to create day-neutral, early flowering and high-yielding cherry tomato mutants within a single generation (Soyk et al. 2017). Similarly, CRISPR/Cas9 was used in soybean to induce frameshift mutations in *GmFT2a*, a photoperiod responsive *FT* homologue responsible for floral induction under short-days, which were shown to delay flowering time in the T_1 and T_2 generations under long- and short-day conditions (Cai et al. 2018). Lastly, CRISPR/Cas9 has been used to delay flowering time in green bristlegrass (*Setaria viridis*) by creating loss-of-function mutants of the maize (*Zea mays*) *INDETERMINATE1* (*ID1*) homologue (Jaganathan et al. 2018).

Future use of the CRISP/Cas system and other genome editing technologies in narrow-leafed lupin could similarly open up new opportunities to manipulate flowering time for breeding purposes. One potential application is to replicate functionally desirable mutations identified through future TILLING or EcoTILLING studies, or alternatively the 1.2 Kb deletion in the *LanFTc1* promoter region of P22660 (Taylor et al. 2019), to immediately introduce novel phenotypes within an elite, domestic background. Additionally, genome editing can be used as another tool to engineer new alleles within known flowering genes, such as *LanFTc1*, on a trial and error basis in pre-breeding research. The continuation of candidate gene discovery within the flowering pathways of narrow-leafed lupin and functional genomic research to uncover the function and regulation of candidate genes will certainly increase the potential use of genome editing technology for improved crop adaption.

9.5 Conclusions

Phenology arguably represents the most influential trait for the adaptation of wild narrow-leafed lupin populations to varying habitats within the species' natural distribution around the Mediterranean and of domesticated cultivars to agricultural environments across the globe (see Chap. 2). Flowering time is a crucial aspect of phenological adaptation, and its regulation in response to environmental and endogenous signals ensures that plants flower at an appropriate time where abiotic resources and conditions are conducive to flower and grain production. Manipulation of this trait has been pivotal to the success of narrow-leafed lupin crop production in both Australian and Europe.

Initiatives to develop bi-parental mapping populations in breeding programmes have enabled genetic mapping, estimation of the phenotypic contribution to flowering time, and the identification of *LanFTc1*; the gene underlying *Ku* and *Jul*, the two main flowering time alleles conferring early, vernalisation-insensitive flowering in domesticated varieties of narrow-leafed lupin. However, little advances have yet been made to map other loci for flowering time and phenology-related characteristics (e.g. growth habit, early vigour, time from flowering to grain maturity), particularly in wild germplasm, and this should be a strong focus for future lupin crop adaptation research. The continual emergence of new genomic technologies and their decreasing costs will continue to assist the improvement of current and development of new genetic resources (e.g. the narrow-leafed lupin genetic map and reference pan-genome assembly) and techniques (e.g. genome editing). These resources and methodologies will be invaluable for further dissecting the genetic pathways affecting phenology, which is an important objective not only from a breeding perspective (e.g. for genetic marker development and the introduction of valuable genetic variation into a domesticated background), but also in terms of building knowledge of the genetic regulation of flowering time in *Lupinus* and the Papilionoideae subfamily of legumes.

References

Abe M et al (2005) FD, a bZIP protein mediating signals from the floral pathway integrator FT at the shoot apex. Science 309:1052

Adamczyk BJ, Lehti-Shiu MD, Fernandez DE (2007) The MADS domain factors AGL15 and AGL18 act redundantly as repressors of the floral transition in Arabidopsis. Plant J 50:1007–1019

Adrian J, Torti S, Turck F (2009) From decision to commitment: the molecular memory of flowering. Mol Plant 2:628–642

Ahmad M, Jarillo JA, Smirnova O, Cashmore AR (1998) Cryptochrome blue-light photoreceptors of Arabidopsis implicated in phototropism. Nature 392:720

Andrés F, Coupland G (2012) The genetic basis of flowering responses to seasonal cues. Nat Rev Genet 13:627

Azani N et al (2017) A new subfamily classification of the Leguminosae based on a taxonomically comprehensive phylogeny. Taxon 66:44–77

Berger J, Shrestha D, Ludwig C (2017) Reproductive strategies in Mediterranean legumes: trade-offs between phenology, seed size and vigor within and between wild and domesticated *Lupinus* species collected along aridity gradients. Front Plant Sci 8:548

Berger JD, Clements JC, Nelson MN, Kamphuis LG, Singh KB, Buirchell B (2013) The essential role of genetic resources in narrow-leafed lupin improvement. Crop Pasture Sci 64:361–373

Berry S, Dean C (2015) Environmental perception and epigenetic memory: mechanistic insight through *FLC*. Plant J 83:133–148

Bertioli DJ et al (2016) The genome sequences of *Arachis duranensis* and *Arachis ipaensis*, the diploid ancestors of cultivated peanut. Nat Genet 48:438

Blümel M, Dally N, Jund C (2015) Flowering time regulation in crops—what did we learn from Arabidopsis? Curr Opin Biotech 32:121–129

Boersma JG, Li C, Leśniewska K, Sivasithamparam K, Yang H (2008) Identification of quantitative trait loci (QTLs) influencing early vigour, height, flowering date, and seed size and their implications for breeding of narrow-leafed lupin (*Lupinus angustifolius* L.). Aust J Agric Res 59:527–535

Boersma JG, Pallotta M, Li C, Buirchell BJ, Sivasithamparam K, Yang H (2005) Construction of a genetic linkage map using MFLP and identification of molecular markers linked to domestication genes in narrow-leafed lupin (*Lupinus angustifolius* L.). Cell Mol Biol Lett 10:331–344

Bortesi L, Fischer R (2015) The CRISPR/Cas9 system for plant genome editing and beyond. Biotechnol Adv 33:41–52

Brachi B, Morris GP, Borevitz JO (2011) Genome-wide association studies in plants: the missing heritability is in the field. Genome Biol 12:232

Brien SJ, Cowling WA, Potter RH, O'Brien PA, Jones RACJ, Jones MGK (1999) A molecular marker for early maturity (*Ku*) and marker-assisted breeding of *Lupinus angustifolius*. In: van Santen E, Wink M, Weissmann S, Römer P (eds) Lupin, an ancient crop for the new millennium: proceedings of the 9th international lupin conference, Klink/Muritz, Germany, 20–24 June 1999, pp 115–117

Briggs WR et al (2001) The phototropin family of photoreceptors. Plant Cell 13:993

Cai Y et al (2018) CRISPR/Cas9-mediated targeted mutagenesis of *GmFT2a* delays flowering time in soya bean. Plant Biotechnol J 16:176–185

Cannon SB et al (2015) Multiple polyploidy events in the early radiation of nodulating and nonnodulating legumes. Mol Biol Evol 32:193–210

Cardoso D, Pennington RT, de Queiroz LP, Boatwright JS, Van Wyk BE, Wojciechowski MF, Lavin M (2013)

Reconstructing the deep-branching relationships of the papilionoid legumes. S Afr J Bot 89:58–75

Chardon F, Damerval C (2005) Phylogenomic analysis of the PEBP gene family in cereals. J Mol Evol 61:579–590

Chen A, Dubcovsky J (2012) Wheat TILLING mutants show that the vernalization gene *VRN1* down-regulates the flowering repressor *VRN2* in leaves but is not essential for flowering. PLoS Genet 8:e1003134

Chen C, Berger J, Fletcher A, Lawes R, Robertson M (2016a) Genotype × environment interactions for phenological adaptation in narrow-leafed lupin: a simulation study with a parameter optimized model. Field Crops Res 197:28–38

Chen Y, Shan F, Nelson MN, Siddique KH, Rengel Z (2016b) Root trait diversity, molecular marker diversity, and trait-marker associations in a core collection of *Lupinus angustifolius*. J Exp Bot 67:3683–3697

Chen C, Fletcher A, Lawes R, Berger J, Robertson M (2017) Modelling phenological and agronomic adaptation options for narrow-leafed lupins in the southern grainbelt of Western Australia. Eur J Agron 89:140–147

Chen L, Wang S-Q, Hu Y-G (2011) Detection of SNPs in the *VRN-A1* gene of common wheat (*Triticum aestivum* L.) by a modified Ecotilling method using agarose gel electrophoresis. Aust J Crop Sci 5:321–329

Christie JM et al (2012) Plant UVR8 photoreceptor senses UV-B by tryptophan-mediated disruption of cross-dimer salt bridges. Science 335:1492

Clements J, Cowling W (1994) Patterns of morphological diversity in relation to geographical origins of wild *Lupinus angustifolius* from the Aegean region. Genet Resour Crop Ev 41:109–122

Colbert T et al (2001) High-throughput screening for induced point mutations. Plant Physiol 126:480–484

Collard BC, Mackill DJ (2008) Marker-assisted selection: an approach for precision plant breeding in the twenty-first century. Philos Trans R Soc Lond B Biol Sci 363:557–572

Cong L et al (2013) Multiplex genome engineering using CRISPR/Cas systems. Science 1231143

Cowling WA (1999) Pedigrees and characteristics of narrow-leafed lupin cultivars released in Australia from 1967–1998. Agric W Aust Bull 4365:1–11

Croser JS et al (2016) Time to flowering of temperate pulses in vivo and generation turnover in vivo-in vitro of narrow-leafed lupin accelerated by low red to far-red ratio and high intensity in the far-red region. Plant Cell Tissue Organ 127:591–599

De Bodt S, Raes J, Van de Peer Y, Theißen G (2003) And then there were many: MADS goes genomic. Trends Plant Sci 8:475–483

Du A et al (2017) The DTH8-Hd1 module mediates day-length-dependent regulation of rice flowering. Mol Plant 10:948–961

Fankhauser C (2001) The phytochromes, a family of red/far-red absorbing photoreceptors. J Biol Chem 276:11453–11456

Frerichmann SLM, Kirchhoff M, Müller AE, Scheidig AJ, Jung C, Kopisch-Obuch FJ (2013) EcoTILLING in *Beta vulgaris* reveals polymorphisms in the *FLC*-like gene *BvFL1* that are associated with annuality and winter hardiness. BMC Plant Biol 13:52

Gao L-L, Hane JK, Kamphuis LG, Foley R, Shi B-J, Atkins CA, Singh KB (2011) Development of genomic resources for the narrow-leafed lupin (*Lupinus angustifolius*): construction of a bacterial artificial chromosome library and BAC-end sequencing. BMC Genomics 12:521

Gasiunas G, Barrangou R, Horvath P, Siksnys V (2012) Cas9-crRNA ribonucleoprotein complex mediates specific DNA cleavage for adaptive immunity in bacteria. Proc Natl Acad Sci USA 109:E2579–E2586

Gendall AR, Levy YY, Wilson A, Dean C (2001) The *VERNALIZATION 2* gene mediates the epigenetic regulation of vernalization in *Arabidopsis*. Cell 107:525–535

Gladstones JS, Crosbie GB (1979) Lupin wild types introduced into Western Australia to 1973. Department of Agriculture, Western Australia, Technical Bulletin No. 43

Gregis V et al (2013) Identification of pathways directly regulated by SHORT VEGETATIVE PHASE during vegetative and reproductive development in *Arabidopsis*. Genome Biol 14:R56–R56

Gu Q, Ferrandiz C, Yanofsky MF, Martienssen R (1998) The FRUITFULL MADS-box gene mediates cell differentiation during Arabidopsis fruit development. Development 125:1509

Hanano S, Goto K (2011) *Arabidopsis* TERMINAL FLOWER1 is involved in the regulation of flowering time and inflorescence development through transcriptional repression. Plant Cell 23:3172–3184

Hane JK et al (2017) A comprehensive draft genome sequence for lupin (*Lupinus angustifolius*), an emerging health food: insights into plant-microbe interactions and legume evolution. Plant Biotechnol J 15:318–330

Hayama R, Sarid-Krebs L, Richter R, Fernández V, Jang S, Coupland G (2017) PSEUDO RESPONSE REGULATORs stabilize CONSTANS protein to promote flowering in response to day length. EMBO J 36:904–918

Hecht V et al (2005) Conservation of Arabidopsis flowering genes in model legumes. Plant Physiol 137:1420–1434

Hecht V et al (2011) The pea *GIGAS* gene is a *FLOWERING LOCUS T* homolog necessary for graft-transmissible specification of flowering but not for responsiveness to photoperiod. Plant Cell 23:147–161

Hedman H, Källman T, Lagercrantz U (2009) Early evolution of the MFT-like gene family in plants. Plant Mol Biol 70:359–369

Helliwell CA, Wood CC, Robertson M, Peacock WJ, Dennis ES (2006) The Arabidopsis FLC protein interacts directly *in vivo* with *SOC1* and *FT* chromatin and is part of a high-molecular-weight protein complex. Plant J 46:183–192

Huang X, Han B (2014) Natural variations and genome-wide association studies in crop plants. Annu Rev Plant Biol 65:531–551

Hurgobin B, Edwards D (2017) SNP discovery using a pangenome: has the single reference approach become obsolete? Biology 6:21

Imaizumi T, Tran HG, Swartz TE, Briggs WR, Kay SA (2003) FKF1 is essential for photoperiodic-specific light signalling in *Arabidopsis*. Nature 426:302–306

Jaganathan D, Ramasamy K, Sellamuthu G, Jayabalan S, Venkataraman G (2018) CRISPR for crop improvement: an update review. Front Plant Sci 9:985

Jakoby M, Weisshaar B, Dröge-Laser W, Vicente-Carbajosa J, Tiedemann J, Kroj T, Parcy F (2002) bZIP transcription factors in *Arabidopsis*. Trends Plant Sci 7:106–111

Jaudal M, Yeoh CC, Zhang L, Stockum C, Mysore KS, Ratet P, Putterill J (2013) Retroelement insertions at the Medicago *FTa1* locus in *spring* mutants eliminate vernalisation but not long-day requirements for early flowering. Plant J 76:580–591

Jaudal M et al (2016) *MtVRN2* is a polycomb *VRN2-like* gene which represses the transition to flowering in the model legume *Medicago truncatula*. Plant J 86:145–160

Jinek M, Chylinski K, Fonfara I, Hauer M, Doudna JA, Charpentier E (2012) A programmable dual-RNA-guided DNA endonuclease in adaptive bacterial immunity. Science 337:816–821

Jung C-H, Wong CE, Singh MB, Bhalla PL (2012) Comparative genomic analysis of soybean flowering genes. PLoS ONE 7:e38250

Jung C, Müller AE (2009) Flowering time control and applications in plant breeding. Trends Plant Sci 14:563–573

Kamphuis LG, Hane JK, Nelson MN, Gao L, Atkins CA, Singh KB (2015) Transcriptome sequencing of different narrow-leafed lupin tissue types provides a comprehensive uni-gene assembly and extensive gene-based molecular markers. Plant Biotechnol J 13:14–25

Karlgren A, Gyllenstrand N, Källman T, Sundström JF, Moore D, Lascoux M, Lagercrantz U (2011) Evolution of the PEBP gene family in plants: functional diversification in seed plant evolution. Plant Physiol 156:1967–1977

Kasprzak A, Šafář J, Janda J, Doležel J, Wolko B, Naganowska B (2006) The bacterial artificial chromosome (BAC) library of the narrow-leafed lupin (*Lupinus angustifolius* L.). Cell Mol Biol Lett 11:396–407

Kim DH, Doyle MR, Sung S, Amasino RM (2009) Vernalization: winter and the timing of flowering in plants. Annu Rev Cell Dev Biol 25:277–299

Kim W-Y et al (2007) ZEITLUPE is a circadian photoreceptor stabilized by GIGANTEA in blue light. Nature 449:356

Klintenäs M, Pin PA, Benlloch R, Ingvarsson PK, Nilsson O (2012) Analysis of conifer *FLOWERING LOCUS T/TERMINAL FLOWER1*-like genes provides evidence for dramatic biochemical evolution in the angiosperm *FT* lineage. New Phytol 196:1260–1273

Kobayashi Y, Kaya H, Goto K, Iwabuchi M, Araki T (1999) A pair of related genes with antagonistic roles in mediating flowering signals. Science 286:1960

Kong F et al (2010) Two coordinately regulated homologs of *FLOWERING LOCUS T* are involved in the control of photoperiodic flowering in soybean. Plant Physiol 154:1220–1231

Korte A, Farlow A (2013) The advantages and limitations of trait analysis with GWAS: a review. Plant Methods 9:29

Kroc M, Koczyk G, Święcicki W, Kilian A, Nelson MN (2014) New evidence of ancestral polyploidy in the Genistoid legume *Lupinus angustifolius* L. (narrow-leafed lupin). Theor Appl Genet 127:1237–1249

Książkiewicz M et al (2017) A high-density consensus linkage map of white lupin highlights synteny with narrow-leafed lupin and provides markers tagging key agronomic traits. Sci Rep 7:15335

Książkiewicz M, Rychel S, Nelson M, Wyrwa K, Naganowska B, Wolko B (2016) Expansion of the phosphatidylethanolamine binding protein family in legumes: a case study of *Lupinus angustifolius* L. *FLOWERING LOCUS T* homologs, *LanFTc1* and *LanFTc2*. BMC Genomics 17:820

Laurie RE et al (2011) The Medicago *FLOWERING LOCUS T* homolog, *MtFTa1*, is a key regulator of flowering time. Plant Physiol 156:2207–2224

Lavin M, Herendeen PS, Wojciechowski MF (2005) Evolutionary rates analysis of Leguminosae implicates a rapid diversification of lineages during the tertiary. Syst Biol 54:575–594

Lee JH, Yoo SJ, Park SH, Hwang I, Lee JS, Ahn JH (2007) Role of *SVP* in the control of flowering time by ambient temperature in *Arabidopsis*. Genes Dev 21:397–402

Levy YY, Mesnage S, Mylne JS, Gendall AR, Dean C (2002) Multiple roles of *Arabidopsis VRN1* in vernalization and flowering time control. Science 297:243–246

Liljegren SJ, Ditta GS, Eshed Y, Savidge B, Bowman JL, Yanofsky MF (2000) SHATTERPROOF MADS-box genes control seed dispersal in Arabidopsis. Nature 404:766–770

Liu B, Kanazawa A, Matsumura H, Takahashi R, Harada K, Abe J (2008) Genetic redundancy in soybean photoresponses associated with duplication of the phytochrome A gene. Genetics 180:995–1007

Michaels SD, Amasino RM (1999) *FLOWERING LOCUS C* encodes a novel MADS domain protein that acts as a repressor of flowering. Plant Cell 11:949–956

Mikołajczyk J (1966) Genetic studies in *Lupinus angustifolius*. Part. III. Inheritance of the alkaloid content, seed hardness and length of the growing season in blue lupin. Genet Pol 7:181–196

Mizoguchi T et al (2005) Distinct roles of GIGANTEA in promoting flowering and regulating circadian rhythms in Arabidopsis. Plant Cell 17:2255–2270

Moon J et al (2003) The *SOC1* MADS-box gene integrates vernalization and gibberellin signals for flowering in *Arabidopsis*. Plant J 35:613–623

Mousavi-Derazmahalleh M et al (2018a) Exploring the genetic and adaptive diversity of a pan-Mediterranean crop wild relative: narrow-leafed lupin. Theor Appl Genet 131:887–901

Mousavi-Derazmahalleh M et al (2018b) The western Mediterranean region provided the founder population of domesticated narrow-leafed lupin. Theor Appl Genet 131:2543–2554

Nelson MN et al (2017) The loss of vernalization requirement essential to domestication in narrow-leafed lupin is associated with a deletion in the promoter and de-repressed expression of an *FT* homologue. New Phytol 213:220–232

Nelson MN et al (2010) Aligning a new reference genetic map of *Lupinus angustifolius* with the genome sequence of the model legume, *Lotus japonicus*. DNA Res 17:73–83

Nelson MN et al (2006) The first gene-based map of *Lupinus angustifolius* L.-location of domestication genes and conserved synteny with *Medicago truncatula*. Theor Appl Genet 113:225–238

Nusinow DA et al (2011) The ELF4–ELF3–LUX complex links the circadian clock to diurnal control of hypocotyl growth. Nature 475:398–402

Ogiso-Tanaka E et al (2013) Natural variation of the *RICE FLOWERING LOCUS T 1* contributes to flowering time divergence in rice. PLoS ONE 8:e75959

Pin PA, Benlloch R, Bonnet D, Wremerth-Weich E, Kraft T, Gielen JJL, Nilsson O (2010) An antagonistic pair of *FT* homologs mediates the control of flowering time in sugar beet. Science 330:1397–1400

Rahman MS, Gladstones JS (1972) Control of lupin initiation by vernalization, photoperiod and temperature under controlled envrionment. Aust J Exp Agric Anim Husb 12:638–645

Ran FA, Hsu PD, Wright J, Agarwala V, Scott DA, Zhang F (2013) Genome engineering using the CRISPR-Cas9 system. Nat Protoc 8:2281–2308

Ream TS, Woods DP, Amasino RM (2012) The molecular basis of vernalization in different plant groups. Cold Spring Harbor Symp Quant Biol 77:105–115

Rodriguez-Medina C, Atkins C, Mann A, Jordan M, Smith P (2011) Macromolecular composition of phloem exudate from white lupin (*Lupinus albus* L.). BMC Plant Biol 11:36

Sawa M, Kay SA (2011) GIGANTEA directly activates *Flowering Locus T* in *Arabidopsis thaliana*. Proc Natl Acad Sci USA 108:11698–11703

Saxena RK, Edwards D, Varshney RK (2014) Structural variations in plant genomes. Brief Funct Genomics 13:296–307

Schaeffer SM, Nakata PA (2015) CRISPR/Cas9-mediated genome editing and gene replacement in plants: transitioning from lab to field. Plant Sci 240:130–142

Scheben A, Wolter F, Batley J, Puchta H, Edwards D (2017) Towards CRISPR/Cas crops—bringing together genomics and genome editing. New Phytol 216:682–698

Scheben A, Yuan Y, Edwards D (2016) Advances in genomics for adapting crops to climate change. Current Plant Biol 6:2–10

Schmutz J et al (2010) Genome sequence of the palaeopolyploid soybean. Nature 463:178–183

Schönrock N, Bouveret R, Leroy O, Borghi L, Köhler C, Gruissem W, Hennig L (2006) Polycomb-group proteins repress the floral activator AGL19 in the FLC-independent vernalization pathway. Genes Dev 20:1667–1678

Schwartz C et al (2009) *Cis*-regulatory changes at *FLOWERING LOCUS T* mediate natural variation in flowering responses of *Arabidopsis thaliana*. Genetics 183:723–732

Searle I, Coupland G (2004) Induction of flowering by seasonal changes in photoperiod. EMBO J 23:1217–1222

Shore P, Sharrocks D (1995) The MADS-box family of transcription factors. Eur J Biochem 229:1–13

Song YH, Ito S, Imaizumi T (2013) Flowering time regulation: photoperiod- and temperature-sensing in leaves. Trends Plant Sci 18:575–583

Soyk S et al (2017) Variation in the flowering gene *SELF PRUNING 5G* promotes day-neutrality and early yield in tomato. Nat Genet 49:162–168

Sun N, Zhao H (2013) Transcription activator-like effector nucleases (TALENs): a highly efficient and versatile tool for genome editing. Biotechnol Bioeng 110:1811–1821

Sung S, Amasino RM (2004) Vernalization in *Arabidopsis thaliana* is mediated by the PHD finger protein VIN3. Nature 8:159–164

Sussmilch FC, Berbel A, Hecht V, Vander Schoor JK, Ferrándiz C, Madueño F, Weller JL (2015) Pea *VEGETATIVE2* is an *FD* homolog that is essential for flowering and compound inflorescence development. Plant Cell 27:1046–1060

Tang X et al (2017) A CRISPR-Cpf1 system for efficient genome editing and transcriptional repression in plants. Nat Plants 3:17018

Tao Z, Shen L, Liu C, Liu L, Yan Y, Yu H (2012) Genome-wide identification of SOC1 and SVP targets during the floral transition in Arabidopsis. Plant J 70:549–561

Tapia-López R et al (2008) An *AGAMOUS*-related MADS-box gene, *XAL1* (*AGL12*), regulates root meristem cell proliferation and flowering transition in Arabidopsis. Plant Physiol 146:1182–1192

Taylor CM et al (2019) INDEL variation in the regulatory region of the major flowering time gene *LanFTc1* is associated with vernalisaiton response and flowering time in narrow-leafed lupin (*Lupinus angustifolius* L.). Plant Cell Environ 42:174–187

Theissen G et al (2000) A short history of MADS-box genes in plants. Plant Mol Biol 42:115–149

Till BJ, Zerr T, Comai L, Henikoff S (2006) A protocol for TILLING and Ecotilling in plants and animals. Nat Protoc 1:2465–2477

Torti S et al (2012) Analysis of the *Arabidopsis* shoot meristem transcriptome during floral initiation identifies distinct regulatory patterns and a leucine-rich repeat protein that promotes flowering. Plant Cell 24:444–462

Turck F, Fornara F, Coupland G (2008) Regulation and identity of florigen: FLOWEIRNG LOCUS T moves center stage. Annu Rev Plant Biol 59:573–594

Urnov FD, Rebar EJ, Holmes MC, Zhang S, Gregory PD (2010) Genome editing with engineered zinc finger nucleases. Nat Rev Genet 11:636–646

Wang Z et al (2015) Functional evolution of phosphatidylethanolamine binding proteins in soybean and Arabidopsis. Plant Cell 27:323–336

Weller JL, Murfet IC, Reid JB (1997) Pea mutants with reduced sensitivity to far-red light define an important role for phytochrome A in day-length detection. Plant Physiol 114:1225

Weller JL, Ortega R (2015) Genetic control of flowering time in legumes. Front Plant Sci 6:207

Wickland DP, Hanzawa Y (2015) The *FLOWERING LOCUS T/TERMINAL FLOWER 1* gene family: functional evolution and molecular mechanisms. Mol Plant 8:983–997

Wigge PA, Kim MC, Jaeger KE, Busch W, Schmid M, Lohmann JU, Weigel D (2005) Integration of spatial and temporal information during floral induction in *Arabidopsis*. Science 309:1056–1059

Wojciechowski MF, Lavin M, Sanderson MJ (2004) A phylogeny of legumes (Leguminosae) based on analysis of the plastid *matK* gene resolves many well-supported subclades within the family. Am J Bot 91:1846–1862

Wolko B, Clements JC, Naganowska B, Nelson MN, Yang H (2011) Lupinus. In: Kole C (ed) Wild crop relatives: genomic and breeding resources. Springer, Heidelberg, pp 153–206

Wolter F, Puchta H (2017) Knocking out consumer concerns and regulator's rules: efficient use of CRISPR/Cas ribonucleoprotein complexes for genome editing in cereals. Genome Biol 18:43

Wong ACS et al (2014) Isolation and functional analysis of *CONSTANS-LIKE* genes suggests that a central role for CONSTANS in flowering time control is not evolutionarily conserved in *Medicago truncatula*. Front Plant Sci 5:486

Woo JW et al (2015) DNA-free genome editing in plants with preassembled CRISPR-Cas9 ribonucleoproteins. Nat Biotechnol 33:1162–1164

Wu F, Price BW, Haider W, Seufferheld G, Nelson R, Hanzawa Y (2014) Functional and evolutionary characterization of the *CONSTANS* gene family in short-day photoperiodic flowering in soybean. PLoS ONE 9:e85754

Xue W et al (2018) Natural variation in *Ghd7* is an important reuglator of heading data and yield potential in rice. Nat Genet 40:761–767

Yamashino T, Yamawaki S, Hagui E, Ueoka-Nakanishi H, Nakamichi N, Ito S, Mizuno T (2013) Clock-controlled and FLOWERING LOCUS T (FT)-dependent photoperiodic pathway in *Lotus japonicus* I: verification of the flowering-associated function of an FT homolog. Biosci Biotech Biochem 77:747–753

Yu L-H, Miao Z-Q, Qi G-F, Wu J, Cai X-T, Mao J-L, Xiang C-B (2014) MADS-box transcription factor AGL21 regulates lateral root development and responds to multiple external and physiological signals. Mol Plant 7:1653–1669

Zetsche B et al (2017) Multiplex gene editing by CRISPR-Cpf1 using a single crRNA array. Nat Biotechnol 35:31–34

Zhai H et al (2014) *GmFT4*, a homolog of *FLOWERING LOCUS T*, is positively regulated by *E1* and functions as a flowering repressor in soybean. PLoS ONE 9: e89030

Zhang H, Ransom C, Ludwig P, van Nocker S (2003) Genetic analysis of early flowering mutants in Arabidopsis defines a class of pleiotropic developmental regulator required for expression of the flowering-time switch Flowering Locus C. Genetics 164:347–358

Zhao Q et al (2018) Pan-genome analysis highlights the extent of genomic variation in cultivated and wild rice. Nat Genet 50:278–284

Zhou G et al (2018) Construction of an ultra-high density consensus genetic map, and enhancement of the physcial map from genome sequencing in *Lupinus angustifolius*. Theor Appl Genet 131:209–223

Genetic and Genomic Resources in White Lupin and the Application of Genomic Selection

10

P. Annicchiarico, N. Nazzicari and B. Ferrari

Abstract

Landraces represent extremely valuable and largely untapped genetic resources for white lupin improvement. The development of high-throughput, relatively low-cost genotyping techniques, such as genotyping-by-sequencing (GBS), has allowed to develop dense genetic maps and to explore the application of genomic selection to predict breeding values of inbred lines or germplasm accessions for complex polygenic traits. We provide an unprecedented assessment of genomic selection in lupins, by assessing the ability of two selection models (Ridge regression BLUP and Bayesian Lasso) to predict grain yield and other traits of 83 landraces from nine historical cropping regions and eight varieties of white lupin that were autumn-sown in Northern Italy. GBS was applied to 3–4 genotypes per landrace and two genotypes per variety, analyzing cultivar allele frequencies for 6,578 polymorphic SNP markers. The two selection models displayed similar predictive ability. Predictions proved highly accurate for grain yield, winter survival and onset of flowering, which displayed predictive abilities of 0.865, 0.852 and 0.838, respectively, based on cross-validation results. Moderately high predictive ability (0.626–0.495) emerged for pod fertility, individual seed weight, plant height, leaf size, and mainstem proportion of seeds and number of leaves. Genomic selection holds high promise for white lupin based on these results.

10.1 White Lupin Cultivation and Breeding Targets

White lupin (*Lupinus albus* L.) was seemingly domesticated in Ancient Greece from the locally occurring wild type var. *graecus*, spreading thereafter eastward, southward and westward (Gladstones 1998). It was a major food legume in the Roman Empire, and has a long history of cultivation in the Mediterranean basin, East Africa and the Atlantic islands of the Northern hemisphere (Buirchell and Cowling 1998; Kurlovich 2002).

The current white lupin cropping area exceeds the threshold of 10,000 ha in just a few countries (Chile, Australia, Morocco), also because of small investment in the crop improvement. An increasing interest in this crop, which is justified by its outstanding seed protein content and

P. Annicchiarico (✉) · N. Nazzicari · B. Ferrari
Council for Agricultural Research and Economics (CREA), Centre for Animal Production and Aquaculture, Lodi, Italy
e-mail: paolo.annicchiarico@crea.gov.it

N. Nazzicari
e-mail: nelson.nazzicari@crea.gov.it

B. Ferrari
e-mail: barbara.ferrari@crea.gov.it

K. B. Singh et al. (eds.), *The Lupin Genome*, Compendium of Plant Genomes,
https://doi.org/10.1007/978-3-030-21270-4_10

protein quality (Duranti et al. 2008) and its suitability as a component of healthy (Arnoldi et al. 2015) or vegetarian food products, is leading to new breeding efforts, particularly in Europe. White lupin displayed greater grain yield potential and protein production per unit area than other lupin species in various temperate-climate, moderately favourable environments, such as those of Southern Europe in Annicchiarico (2008) and of Chile in Mera and Alcalde (2019).

White lupin breeding efforts are mainly aimed to overcome the economically insufficient crop grain yield (Gresta et al. 2017), also by means of greater tolerance to abiotic and biotic stresses. Besides the grain yielding ability per se, other key breeding targets are represented by: (i) improved adaptation to calcareous soils (Kerley et al. 2001; Annicchiarico and Thami-Alami 2012); (ii) greater winter-hardiness, which is aimed to expand autumn sowing in cold-prone regions (Huyghe and Papineau 1990; Annicchiarico and Iannucci 2007); (iii) greater tolerance to drought (Rodrigues et al. 1995; Annicchiarico et al. 2018a); and (iv) greater resistance to anthracnose (*Colletotrichum lupini*), which is very important for cultivation in Central Europe and a few other regions (Jacob et al. 2017; Adhikari et al. 2013). Phenology as determined by vernalization requirement has crucial importance for plant adaptation to specific cropping regions (Huyghe 1997; Annicchiarico et al. 2010). Novel plant architectures that exploit dwarfing genes and/or a semi-determinate habit may increase the tolerance to lodging, harvest index and/or yield stability of the crop (Julier et al. 1993; Harzic et al. 1995; Annicchiarico et al. 2018b).

10.2 White Lupin Genetic Resources

Genetic resources for white lupin breeding are largely represented by cultivated landraces, because of the limited distribution of var. *graecus* (Gladstones 1998). They are limited to the primary gene pool, because of effective genetic barriers to interspecific hybridization (Atkins et al. 1998).

The main germplasm collections listed in Buirchell and Cowling (1998) hold over 6300 white lupin accession. Major regional landrace plant types were described by Buirchell and Cowling (1998) and Kurlovich (2002). The multi-environment evaluation of a world white lupin landrace collection including material from all historical growing regions revealed remarkable genotype × environment interaction across three contrasting growing regions of Europe (Annicchiarico et al. 2010). An ecological grouping of landrace germplasm based on the observed adaptive responses and major agronomic traits was defined by Annicchiarico et al. (2011), to integrate earlier descriptions of regional plant types and to support the incorporation of exotic genetic resources in breeding programs as a function of the targeted agro-climatic conditions.

Farm landraces represent extremely valuable and largely untapped genetic resources for white lupin improvement. The multi-environment comparison of landrace vs. elite variety germplasm highlighted the outstanding grain yielding ability of different landrace germplasm pools for specific agro-climatic conditions (Annicchiarico et al. 2010). The molecular characterization of the same material revealed a trend for variety germplasm to be fairly distinct from landrace material, suggesting that just a minor portion of the available genetic diversity has been exploited by breeders (Annicchiarico et al., unpublished data). The choice by breeders to exploit mainly sweet-seed variety germplasm as a genetic resource may have contributed to this diversity bottleneck.

10.3 White Lupin Genomic Resources

Until recently, white lupin genomic resources included a transcriptome assembly (O'Rourke et al. 2013; reviewed in detail in Sect. 5.2.6), two mapping populations with associated low-density linkage maps (Phan et al. 2007; Croxford et al.

2008; Vipin et al. 2013) and sequence-tagged site (STS) markers linked to low alkaloid content (*PauperM1*) and anthracnose resistance (*WANR1*, *WANR2* and *WANR3*) (Lin et al. 2009; Yang et al. 2010). In addition a white lupin reference genome assembly project is currently underway (Benjamin Péret, pers. comm.; Sect. 3.6.4). The main recombinant inbred line mapping population, developed from the cross Kiev Mutant × P27174, segregates for agronomic traits such as onset of flowering, alkaloid content and profile, plant height, pod shape, restricted branching and resistance to anthracnose and Phomopsis stem blight (*Diaporthe toxica*) (Phan et al. 2007). Most of these traits were localized on white lupin linkage maps (Phan et al. 2007; Vipin et al. 2013; Cowley et al. 2014; Raman et al. 2014) but quantitative trait loci (QTL) mapping efforts were hindered by low marker density.

The development of next-generation sequencing techniques has allowed to genotype large germplasm sets by thousands of single-nucleotide polymorphism (SNP) markers at a relatively low cost. While array-based procedures require prior knowledge of target sequences, genotyping-by-sequencing (GBS; Elshire et al. 2011) and RAD sequencing (Hohenlohe et al. 2010) skip sequence discovery and explore SNP polymorphism in DNA fragments cut by a restriction enzyme. Their efficient multiplexing procedure allows for lower genotyping costs than array-based techniques.

The study by Książkiewicz et al. (2017) confirmed the ability of GBS to generate thousands of polymorphic SNP markers also for white lupin. It produced a high-density linkage map, and mapped various QTL for vernalization responsiveness, resistance to anthracnose and Phomopsis, and seed alkaloid content. Incidentally, this study revealed no coincidence of these QTL with those previously mapped in narrow-leafed lupin (*L. angustifolius* L.), highlighting a limit for synteny-based approaches and the importance of QTL investigation in the target species. However, probably the greatest opportunity offered by novel high-throughput genotyping techniques is the application of genomic selection to predict breeding values of inbred lines or germplasm accessions for complex polygenic traits.

10.4 Potential Adoption of Genomic Selection in White Lupin

Genomic selection may enable breeders to predict breeding values of plant material for grain yield or other complex traits by means of a statistical model constructed from genome-wide marker information (Meuwissen et al. 2001). The genomic selection model is developed by the joint analysis of phenotyping and genotyping data of a germplasm sample well representing the target genetic base. Following its successful validation on independent material of the same genetic base, the model can be applied to large germplasm sets, thereby reducing largely the phenotypic evaluation effort (Heffner et al. 2009; Lorenz et al. 2011). Simulation and empirical studies, performed essentially on cereal crops, proved that genomic selection is superior to conventional marker-assisted selection based on limited marker numbers for prediction of breeding values for complex traits such as crop yield (Bernardo and Yu 2007; Heffner et al. 2011).

Genomic selection has represented a breakthrough for dairy cattle improvement (Wiggans et al. 2017). However, its requirement for high marker numbers delayed its application to plant breeding until recently, when new techniques have allowed to genotype large sets of genotypes by large SNP marker numbers at a relatively low cost. Pioneer genomic selection studies were encouraging for prediction of breeding values for yield in legumes. Predictive ability values (assessed as Pearson's correlation between genome-enabled modelled and observed data) were in the range of high to moderately high ($r > 0.40$) for soybean (Jarquín et al. 2014; Ma et al. 2016; Duhnen et al. 2017), pea (Annicchiarico et al. 2017a) and chick pea (Roorkiwal et al. 2016), and moderate ($r = 0.30$) in a second pea study (Annicchiarico et al. 2019a). Predictive ability values in the range 0.30–0.40 were found in alfalfa (Annicchiarico et al. 2015a), which can be viewed as quite favourable when considering the crop outbreeding system. Most importantly, various studies suggested greater efficiency of genome-enabled selection over phenotypic selection in terms of predicted genetic gain per unit time and/or unit cost in legume crops (Annicchiarico et al. 2015a, 2017a, 2017b, 2019a), as well as in various cereal crops (e.g., Heffner et al. 2010).

Genomic selection could be exploited not only for selection of inbred lines but also for identification of promising genetic resources in germplasm collections, as showed by results in Jarquín et al. (2016) for the USDA collection of soybean. Germplasm collections, including white lupin ones, have limited opportunities for thorough germplasm evaluation, owing to limited budgets. While the ecological grouping of landrace germplasm could provide a first criterion for the choice of exotic genetic resources in white lupin breeding programs (Annicchiarico et al. 2011), the development of genome-enabled models with good ability to predict key agronomic traits of germplasm accessions could provide a sharper criterion for selecting promising genetic resources, when considering that decreasing genotyping costs may soon allow for the genotyping of entire germplasm collections.

Because of its high perspective interest, this paper aims to provide a preliminary assessment of the ability of GBS marker-based genomic selection to predict grain yield and several major agronomic traits of white lupin germplasm accessions. Genomic predictions concerned a large subset of the world landrace collection and the elite varieties that were evaluated by Annicchiarico et al. (2010) in Northern Italy. The agronomic traits included, *inter alia*, winter survival and onset of flowering, whose adaptive importance was anticipated; pod fertility, which displayed special importance as a component of seed yield (Annicchiarico et al. 2010); the proportion of pod wall on pod biomass, which was associated with higher crop grain yield in various studies (Lagunes-Espinoza et al. 1999; Mera et al. 2006); and the number of mainstem leaves, which depends on vernalization and is associated with the number of first-order branches (Julier and Huyghe 1993). Trait predictions were assessed using two major statistical models, namely, Ridge regression BLUP (rrBLUP; Meuwissen et al. 2001; Searle et al. 2009) and Bayesian Lasso (BL; Park and Casella 2008).

10.5 Case Study of White Lupin Genomic Selection—Materials and Methods

We used phenotypic data for 83 landraces and eight French varieties of white lupin that were evaluated by Annicchiarico et al. (2010) in Lodi, Northern Italy (45°19′ N, 9°30′ E, 81 m a.s.l.). The landraces belonged to nine historical cropping regions, i.e. Egypt, Italy, Portugal, Spain, Turkey, Maghreb (Algeria and Morocco), East Africa (Ethiopia, Kenya and Sudan), West Asia (Syria, Lebanon, Israel and Jordan) and North Atlantic islands (Canaries and Madeira). Each region was represented by five to fourteen landraces. The material was evaluated in a rain-fed experiment sown on 14 October 2004, which was designed as a randomized complete block with three replications. Each plot was sown with 36 seeds, adopting a sowing density of 23.5 seeds/m^2. The test environment was characterized by 78 frost days and −9.0 °C absolute minimum temperature, and received 492 mm rainfall over the period October 2004 to June 2005.

Detailed information on the experiment management and the recorded traits was provided in Annicchiarico et al. (2010). Genomic predictions were investigated for dry grain yield recorded on a plot basis and 10 agronomic traits, namely: winter plant survival; onset of flowering (as days from 1 January to flowering of 50% of the plants); plant height at onset of flowering; number of leaves on the main stem; leaf size (measured on the first leaf below the primary reproductive branch); harvest index; proportion of seeds on the main stem; number of seeds per pod; individual seed dry weight; and proportion of pod wall on pod biomass (measured on pods of the main stem). We used phenotyping data for the total number of entries (91 landraces or varieties) for grain yield, winter plant survival and onset of flowering, and data for a subset of 69 entries (62 landraces and seven varieties) featuring high winter survival for the other traits (whose reliable assessment required the availability of a sizeable number of plants per plot). The data were transformed into best linear unbiased prediction (BLUP) values according to DeLacy et al. (1996) before using them for genomic model construction.

The molecular characterization was performed on three to four individual plants per landrace (given some degree of expected within-landrace diversity), and two plants per variety (given the expected genetic homogeneity of this material). In general, DNA isolation and GBS library construction and sequencing were performed as described by Annicchiarico et al. (2017a) with respect to DNA samples of pea, which implied the adoption of Elshire et al.'s (2011) GBS protocol based on *Ape*KI DNA digestion with modifications in the PCR. These modifications, adopted also for lupin, included the use of the KAPA *Taq* polymerase in place of the NEB *Taq* polymerase according to the results by Annicchiarico et al. (2017b), 25 nmoles of each primer instead of 5 nmoles, and 10 cycles of reaction instead of 18.

We used the UNEAK pipeline (Lu et al. 2013) that is part of the Tassel 5 software for SNP discovery and genotype calling. The raw reads (100 bp, single end reads) that began with the expected barcodes and cut site remnant were trimmed to 64 bp, grouping identical reads into one tag. Tags with 10 or more reads across all individuals were retained for pairwise alignment, which aimed to find tag pairs that differed by one bp. For each SNP marker, the read distribution of the paired tags in each genotype was used for SNP genotype calling, which was based on at least four aligned reads per genotype. We removed polymorphic markers with minor allele frequency below 2.5% or with missing rate over genotypes greater than 30%. Following Nazzicari et al. (2016), we estimated missing data using the K-nearest neighbors imputation algorithm ($K = 4$) coupled with the simple matching coefficient distance function as implemented in the R package Scrime version 1.3.5 (Schwender and Fritsch 2013).

Genome-enabled predictions were based on SNP marker allele frequencies of each genotype as estimated from the available genotype sample. Predictions were envisaged using rrBLUP (Meuwissen et al. 2001; Searle et al. 2009) and BL (Park and Casella 2008), as these models stood out for predictive ability among several models compared on different legume species (Annicchiarico et al. 2017b). The rrBLUP model, which assumes the effects of all loci to have a common variance, is well suited for traits that are influenced by a large number of minor genes. Bayesian models assume relatively few markers with large effects and allow markers to have different effects and variances (Wang et al. 2018). They assign prior densities to markers effects, thereby inducing different types of shrinkage.

Predictive abilities were assessed as correlations between modelled and observed data based on tenfold stratified cross validations with 20 repetitions (meaning 9 entries out of 91, or 7 out of 69, that were randomly excluded from modelling and used for validation in each cross-validation step). Predictions were assessed not only for the pooled set of landrace and variety germplasm but also for the landrace germplasm alone. Regression models, cross validations and predictive ability estimation were carried out using the R package GROAN version 1.2 (Nazzicari and Biscarini 2017).

10.6 Case Study of White Lupin Genomic Selection—Results

As reported in Annicchiarico et al. (2010), the genetic variation among tested accessions (landraces or varieties) was significant ($P < 0.01$) for all recorded traits, and when expressed as genetic coefficient of variation, proved relatively small ($CV_g \leq 6.2\%$) only for proportion of pod wall on pod biomass and harvest index (excluding onset of flowering, whose CV_g value was affected by the arbitrarily chosen measurement unit). Modest variation for harvest index was partly due to the tall plant type that featured all test genotypes except two elite varieties. The analysis of phenotypic data revealed several trait interrelationships, including those among higher grain yield, greater winter survival and later phenology in the test environment (Annicchiarico et al. 2010).

On average, GBS issued 1.66 M reads per genotype, and generated 6,578 polymorphic SNP markers that were exploitable for genome-enabled trait predictions after passing the different filtering stages. In general, predictive abilities estimated for the pooled landrace and variety germplasm were similar to those estimated for the landrace germplasm alone. Their value averaged across the eleven traits and two prediction models was 0.580 for the whole germplasm set and 0.568 for the landrace germplasm, a slight difference that paralleled the somewhat larger genotype sample size available for predictions of the former germplasm set. Given the absence of a variety germplasm effect on predictions, results were provided for the whole set of entries.

The two tested genomic selection models performed comparably. Their predictive ability averaged across the eleven traits was 0.581 for rrBLUP and 0.578 for BL. A slight advantage of rrBLUP over BL emerged for prediction of grain yield, onset of flowering and plant height, whereas BL was somewhat superior for predicting leaf size and harvest index (Figs. 10.1 and 10.2).

Genome-enabled predictions proved highly accurate for the three traits whose model construction could rely on larger genotype sample size, namely, grain yield, winter survival and onset of flowering, which displayed rrBLUP model-based predictive ability of 0.865, 0.852 and 0.838, respectively (Fig. 10.1). Most other traits, i.e. plant height, number of leaves on the main stem, leaf size, proportion of seeds on the main stem, number of seeds per pod and individual seed dry weight, exhibited moderately high predictive ability (ranging from 0.626 to 0.495 according to the rrBLUP model: Fig. 10.2). Only the two traits that featured narrower genetic variation according to CV_g values,

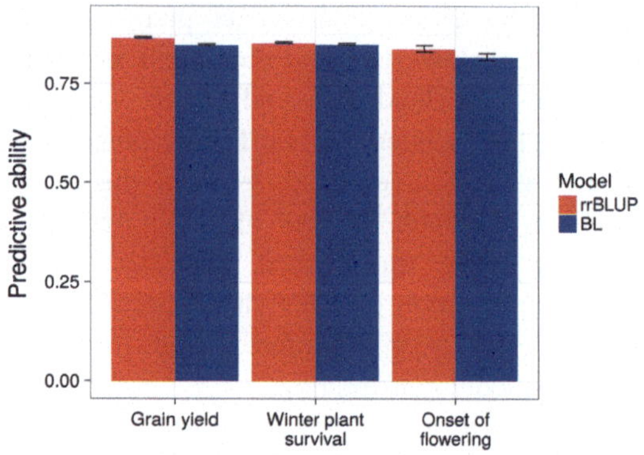

Fig. 10.1 Predictive ability of genomic selection using Bayesian Lasso (BL) or Ridge regression BLUP (rrBLUP) models for grain yield, winter plant survival and onset of flowering of 83 landraces and eight varieties of white lupin, as correlation between modelled data based on 6,578 SNP markers and observed data averaged across 20 independent tenfold stratified cross validations (vertical bars represent $P = 0.95$ confidence intervals)

namely, proportion of pod wall on pod biomass and harvest index, exhibited modest predictive ability (Fig. 10.2).

10.7 Perspective and Conclusion

Based on results from various studies summarized in Heffner et al. (2009), Bayesian models such as BL are expected to be especially valuable in the presence of limited phenotyping records and large sets of marker data, as it was generally the case in this study. A slight advantage of BL over rrBLUP emerged for two traits with smaller genotype sample (leaf size and harvest index), while a slight advantage of rrBLUP was found for two traits with larger genotype sample (grain yield and onset of flowering). However, these models performed comparably over the 11 lupin traits of the current assessment. In earlier studies on legume species, rrBLUP (or a similar model such as GBLUP) outperformed BL for predicting pea agronomic traits in Burstin et al. (2015) and alfalfa forage yield in Annicchiarico et al. (2015a), whereas the two models performed comparably for pea grain yield in Annicchiarico et al. (2017a, 2019a) and various alfalfa forage quality traits in Biazzi et al. (2017). On the

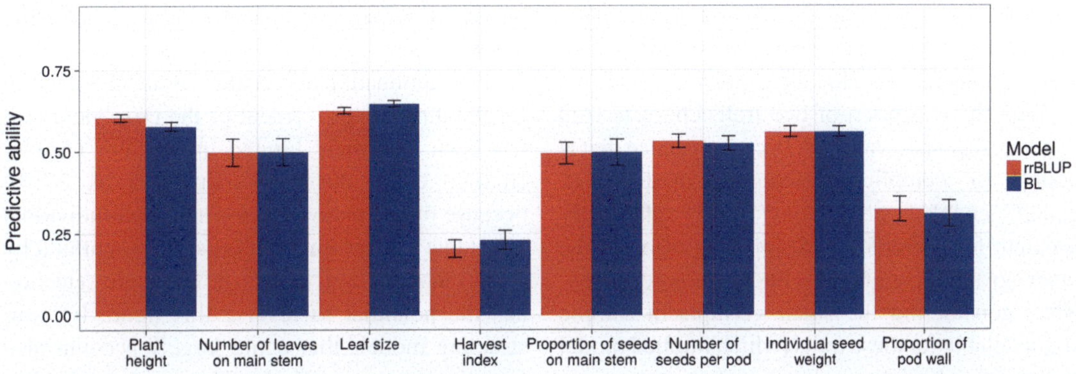

Fig. 10.2 Predictive ability of genomic selection using Bayesian Lasso (BL) or Ridge regression BLUP (rrBLUP) models for plant height at onset of flowering, number of leaves on the main stem, leaf size, harvest index, proportion of seeds on the main stem, number of seeds per pod, individual seed dry weight and proportion of pod wall on pod biomass of 62 landraces and seven varieties of white lupin, as correlation between modelled data based on 6578 SNP markers and observed data averaged across 20 independent tenfold stratified cross validations (vertical bars represent $P = 0.95$ confidence intervals)

whole, the similar predictive ability of rrBLUP and BL supports the testing of both models in future white lupin genomic selection studies.

The genome-enabled predictive ability that we observed for grain yield was beyond expectations, when considering earlier results for crop yield of other inbred grain legume crops and two currently unfavourable circumstances. The first circumstance was the genotype sample size, which was suboptimal (Viana et al. 2017) and smaller than in the cited genomic selection studies. The second circumstance was the target germplasm type, which included mainly landrace populations (rather than pure lines), whose SNP allele frequency values could not be estimated precisely by the available genotype sample of three to four genotypes per population (also in consideration of the sizeable within-population diversity that featured these landraces: Annicchiarico et al., unpublished data). However, predictive accuracies tended to be biased upward by the current adoption of intra-experiment cross validations rather than cross-environment validations (Lorenz et al. 2011), where the latter can also account for genotype × environment interactions within a target region.

A parallel study on white lupin genomic selection based on a larger set of entries and more than one test environment confirmed the high predictive ability of white lupin grain yield even across climatically-different Italian environments or contrasting drought-stress levels, showing a predictive accuracy in the range 0.48-0.64 across such contrasting environments (Annicchiarico et al. 2019b).

The accurate genomic prediction of onset of flowering is encouraging, given the key importance of this trait for adaptation to different agro-climatic regions of white lupin cultivars (Annicchiarico and Carroni 2009) and genetic resources (Annicchiarico et al. 2010, 2011). The development of accurate predictions was favoured by the lower genetic complexity of this trait relative to crop yield. Early onset of flowering was controlled by two complementary dominant genes in Adhikari et al. (2011), while showing a polygenic control with respect to vernalization requirement in Książkiewicz et al. (2017).

The current high predictive ability of breeding values for winter survival has remarkable practical importance, to widen the adaptation and economic sustainability of white lupin in temperate regions by cultivation of autumn-sown, winter-hardy germplasm (Huyghe 1997). Winter survival depends on a few substantially independent plant characteristics, such as delayed floral initiation (as determined mainly by the vernalization requirement), intrinsic frost tolerance of the hardened plant and, to a lesser extent, larger seed size (via the thicker root parenchyma of the larger seedling) (Huyghe and Papineau 1990; Papineau and Huyghe 2004).

With the exception of two traits characterized by modest genetic variation (a circumstance that is bound to complicate genomic predictions), genome-enabled predictions were sufficiently accurate for germplasm exploitation also for the other agronomic traits. This finding is noteworthy, when considering the small germplasm sample size available for these traits (which implied model construction based on only 62 phenotypes).

While producing encouraging results for the perspective genomic selection of white lupin inbred lines in breeding programs, the specific germplasm type used in this study emphasizes the use of genome-enabled selection to identify promising genetic resources within large landrace germplasm collections.

The high potential interest of genomic selection highlighted by this study for white lupin is likely to apply as well to another major lupin species such as narrow-leafed lupin (*L. angustifolius* L.). The genome of this species displayed high sequence collinearity with that of white lupin (Książkiewicz et al. 2017), although possessing a larger genome (Naganowska et al. 2003). The narrow-leafed lupin genome sequencing effort revealed a large proportion of repetitive DNA (57%; Hane et al. 2017). This finding reinforces the adoption in GBS for lupins of a methylation sensitive restriction enzyme such as *Ape*KI, which tends to avoid highly repetitive DNA regions, thereby helping targeting restriction sites that are relatively random and evenly distributed along the genome in gene-rich regions.

Genomic selection studies may be associated with genome-wide association studies (e.g. Annicchiarico et al. 2017a), to gain information on the genetic control of target traits and possibly explore, particularly for less complex traits, the application of marker-assisted selection. On the other hand, genomic selection may be devised also for oligogenic traits, e.g. resistance to anthracnose (Annicchiarico et al., unpublished data). This may be particularly convenient when breeding simultaneously for several traits by trait-specific genomic selection models whose predicted values are combined into a selection index.

The application of marker-assisted selection for the direct improvement of the crop yield trait has been extremely limited in legumes (Annicchiarico et al. 2015b; Varshney et al. 2015), because of the many genes with small individual effect that control this trait [as shown in alfalfa by Annicchiarico et al. (2015a)]. Genomic selection has the potential to reverse this trend, by constructing models that could largely account also for minor genetic effects. This study and additional results in Annicchiarico et al. (2019b) provide evidence that this avenue can be feasible and successful for improving the grain yield and other major agronomic traits of lupin crops.

Acknowledgements This work was part of the project 'Legumes for the agriculture of tomorrow (LEGATO)', which received funding from EU's 7th Framework Programme under Grant Agreement No. 613551.

References

Adhikari K, Buirchell B, Yan G, Sweetingham M (2011) Two complementary dominant genes control flowering time in albus lupin (*Lupinus albus* L.). Plant Breed 130:496–499

Adhikari KN, Thomas G, Diepeveen D, Trethowan R (2013) Overcoming the barriers of combining early flowering and anthracnose resistance in white lupin (*Lupinus albus* L.) for the Northern Agricultural Region of Western Australia. Crop Pasture Sci 64:914–921

Annicchiarico P (2008) Adaptation of cool-season grain legume species across climatically-contrasting environments of southern Europe. Agron J 100:1647–1654

Annicchiarico P, Carroni AM (2009) Diversity of white and narrow-leafed lupin genotype adaptive response across south-European environments and implications for selection. Euphytica 166:71–81

Annicchiarico P, Iannucci A (2007) Winter survival of pea, faba bean and white lupin cultivars across contrasting Italian locations and sowing times, and implications for selection. J Agric Sci 145:611–622

Annicchiarico P, Thami-Alami I (2012) Enhancing white lupin (*Lupinus albus* L.) adaptation to calcareous soils through lime-tolerant plant germplasm and *Bradyrhizobium* strains. Plant Soil 350:131–144

Annicchiarico P, Harzic N, Carroni AM (2010) Adaptation, diversity, and exploitation of global white lupin (*Lupinus albus* L.) landrace genetic resources. Field Crops Res 119:114–124

Annicchiarico P, Harzic N, Huyghe C, Carroni AM (2011) Ecological classification of white lupin landrace genetic resources. Euphytica 180:17–25

Annicchiarico P, Nazzicari N, Li X, Wei Y, Pecetti L, Brummer EC (2015a) Accuracy of genomic selection for alfalfa biomass yield in different reference populations. BMC Genomics 16:1020

Annicchiarico P, Barrett B, Brummer EC, Julier B, Marshall AH (2015b) Achievements and challenges in improving temperate perennial forage legumes. Crit Rev Plant Sci 34:327–380

Annicchiarico P, Nazzicari N, Pecetti L, Romani M, Ferrari B, Wei Y, Brummer EC (2017a) GBS-based genomic selection for pea grain yield under severe terminal drought. Plant Genome 10:10.3835

Annicchiarico P, Nazzicari N, Wei Y, Pecetti L, Brummer EC (2017b) Genotyping-by-sequencing and its exploitation for forage and cool-season grain legume breeding. Front Plant Sci 8:679

Annicchiarico P, Romani M, Pecetti L (2018a) White lupin variation for adaptation to severe drought stress. Plant Breed 137:782–789

Annicchiarico P, Carroni AM, Manunza P, Huyghe C, Pecetti L (2018b) Grain yield and morphology of dwarf vs tall white lupin in Mediterranean environments. In: Brazauskas G, Statkevičiūtė G, Jonavičienė K (eds) Breeding grasses and protein crops in the era of genomics. Springer, Dordrecht, The Netherlands, pp 113–117

Annicchiarico P, Nazzicari N, Pecetti L, Romani M, Russi L (2019a) Pea genomic selection for Italian environments. BMC Genomics 20:603

Annicchiarico P, Nazzicari N, Ferrari B, Harzic N, Carroni AM, Romani M, Pecetti L (2019b) Genomic prediction of grain yield in contrasting environments for white lupin genetic resources. Mol Breed 39:142

Arnoldi A, Boschin G, Zanoni C, Lammi C (2015) The health benefits of sweet lupin seed flours and isolated proteins. J Funct Foods 18:550–563

Atkins CA, Smith PMC, Gupta S, Jones MGK, Caligari PDS (1998) Genetics, cytology and biotechnology. In: Gladstones JS, Atkins C, Hamblin J (eds) Lupins as crop plants: biology, production and utilization. CABI, Wallingford, UK, pp 67–92

Bernardo R, Yu J (2007) Prospects for genomewide selection for quantitative traits in maize. Crop Sci 47:1082–1090

Biazzi E, Nazzicari N, Pecetti L, Brummer EC, Palmonari A, Tava A, Annicchiarico P (2017) Genome-wide association mapping and genomic selection for alfalfa (*Medicago sativa*) forage quality traits. PLoS ONE 12:e0169234

Buirchell BJ, Cowling WA (1998) Genetic resources in lupins. In: Gladstones JS, Atkins C, Hamblin J (eds) Lupins as crop plants: biology, production and utilization. CABI, Wallingford, UK, pp 41–66

Burstin J, Salloignon P, Chabert-Martinello M, Magnin-Robert J-B, Siol M, Jacquin F, Chauveau A, Pont C, Aubert G, Delaitre C, Truntzer C, Duc G (2015) Genetic diversity and trait genomic prediction in a pea diversity panel. BMC Genomics 16:105

Cowley R, Luckett DJ, Ash GJ, Harper JD, Vipin CA, Raman H, Ellwood S (2014) Identification of QTLs associated with resistance to Phomopsis pod blight (*Diaporthe toxica*) in *Lupinus albus*. Breed Sci 64:83–89

Croxford AE, Rogers T, Caligari PDS, Wilkinson MJ (2008) High-resolution melt analysis to identify and map sequence-tagged site anchor points onto linkage maps: a white lupin (*Lupinus albus*) map as an exemplar. New Phytol 180:594–607

DeLacy IH, Basford KE, Cooper M, Bull JK, McLaren CG (1996) Analysis of multi-environment trials—an historical perspective. In: Cooper M, Hammer GL (eds) Plant adaptation and crop improvement. CABI, Wallingford, UK, pp 39–124

Duhnen A, Gras A, Teyssèdre S, Romestant M, Claustres B, Daydé J, Manfin B (2017) Genomic selection for yield and seed protein content in soybean: a study of breeding program data and assessment of prediction accuracy. Crop Sci 57:1325–1337

Duranti M, Consonni A, Magni C, Sessa F, Scarafoni A (2008) The major proteins of lupin seed: characterization and molecular properties for use as functional and nutraceutical ingredients. Trends Food Sci Technol 19:624–633

Elshire RJ, Glaubitz JC, Sun Q, Poland JA, Kawamoto K, Buckler ES, Mitchell SE (2011) A robust, simple genotyping-by-sequencing (GBS) approach for high diversity species. PLoS ONE 6:e19379

Gladstones JS (1998) Distribution, origin, taxonomy, history and importance. In: Gladstones JS, Atkins C, Hamblin J (eds) Lupins as crop plants: biology, production and utilization. CABI, Wallingford, UK, pp 1–39

Gresta F, Wink M, Prins U, Abberton M, Capraro J, Scarafoni A, Hill G (2017) Lupins in European cropping systems. In: Murphy-Bokern D, Stoddard FL, Watson CA (eds) Legumes in cropping systems. CABI, Wallingford, UK, pp 88–108

Hane JK, Ming Y, Kamphuis LG, Nelson MN, Garg G, Atkins CA, Bayer PE, Bravo A, Bringans S, Cannon S, Edwards D, Foley R, Gao L, Harrison MJ, Huang W, Hurgobin B, Li S, Liu C, McGrath A, Morahan G, Murray J, Weller J, Jian J, Singh KB (2017) A comprehensive draft genome sequence for lupin (*Lupinus angustifolius*), an emerging health food: insights into plant–microbe interactions and legume evolution. Plant Biotechnol J 15:318–330

Harzic N, Huyghe C, Papineau J (1995) Dry matter accumulation and seed yield of dwarf autumn-sown white lupin (*Lupinus albus* L.). Can J Plant Sci 75:549–555.

Heffner EL, Sorrells ME, Jannink J-L (2009) Genomic selection for crop improvement. Crop Sci 49:1–12

Heffner EL, Lorenz AJ, Jannink JL, Sorrells ME (2010) Plant breeding with genomic selection: gain per unit time and cost. Crop Sci 50:1681–1690

Heffner EL, Jannink J-L, Iwata H, Souza E, Sorrells ME (2011) Genomic selection accuracy for grain quality

traits in biparental wheat populations. Crop Sci 51:2597–2606
Hohenlohe PA, Bassham S, Etter PD, Stiffler N, Johnson EA, Cresko WA (2010) Population genomics of parallel adaptation in threespine stickleback using sequenced RAD tags. PLoS Genet 6:e1000862
Huyghe C (1997) White lupin (*Lupinus albus* L.). Field Crops Res 53:147–160
Huyghe C, Papineau J (1990) Winter development of autumn-sown white lupin: agronomic and breeding consequences. Agronomie 10:709–716
Jacob I, Feuerstein U, Heinz M, Schott M, Urbatzka P (2017) Evaluation of new breeding lines of white lupin with improved resistance to anthracnose. Euphytica 213:236
Jarquín D, Kocak K, Posadas L, Hyma K, Jedlicka J, Graef G, Lorenz A (2014) Genotyping by sequencing for genomic prediction in a soybean breeding population. BMC Genomics 15:740
Jarquín D, Specht J, Lorenz A (2016) Prospects of genomic prediction in the USDA soybean germplasm collection: historical data creates robust models for enhancing selection of accessions. G3 (Bethesda) 6:2329–2341
Julier B, Huyghe C (1993) Description and model of the architecture of four genotypes of determinate autumn-sown white lupin (*Lupinus albus* l.) as influenced by location, sowing date and density. Ann Bot 72:493–501
Julier B, Huyghe C, Papineau J, Milford GFJ, Day JM, Billot C, Mangin P (1993) Seed yield and yield stability of determinate and indeterminate autumn-sown white lupins (*Lupinus albus*) grown at different locations in France and the UK. J Agric Sci 121:177–186
Kerley SJ, Shield IF, Huyghe C (2001) Specific and genotypic variation in the nutrient content of lupin species in soils of neutral and alkaline pH. Aust J Agric Res 52:93–102
Książkiewicz M, Nazzicari N, Yang H, Nelson N, Renshaw D, Rychel S, Ferrari B, Carelli M, Tomaszewska M, Stawiński S, Naganowska B, Wolko B, Annicchiarico P (2017) A high-density consensus linkage map of white lupin highlights synteny with narrow-leafed lupin and provides markers tagging key agronomic traits. Sci Rep 7:15335
Kurlovich BS (2002) Lupins (geography, classification, genetic resources and breeding). Publishing House Intan, St. Petersburg, Russia
Lagunes-Espinoza L, Huyghe C, Papineau J, Pacault D (1999) Effect of genotype and environment on pod wall proportion in white lupin: consequences to seed yield. Aust J Agric Res 50:575–582
Lin R, Renshaw D, Luckett D, Clements J, Yan G, Adhikari K, Yang H (2009) Development of a sequence-specific PCR marker linked to the gene "pauper" conferring low-alkaloids in white lupin (*Lupinus albus* L.) for marker assisted selection. Mol Breed 23:153–161
Lorenz AJ, Chao S, Asoro FG, Heffner EL, Hayashi T, Iwata H, Smith KP, Sorrells ME, Jannink J-L (2011) Genomic selection in plant breeding: knowledge and prospects. Adv Agron 110:77–123
Lu F, Lipka AE, Glaubitz J, Elshire R, Cherney JH, Casler MD, Buckler ES, Costich DE (2013) Switchgrass genomic diversity, ploidy, and evolution: novel insights from a network-based SNP discovery protocol. PLoS Genet 9:e1003215
Ma Y, Reif JC, Jiang Y, Wen Z, Wang D, Liu Z, Guo Y, Wei S, Wang S, Yang C, Wang H, Yang C, Lu W, Xu R, Zhou, Wang R, Sun Z, Chen H, Zhang W, Wu J, Hu G, Liu C, Luan X, Fu Y, Guo T, Han T, Zhang M, Sun B, Zhang L, Chen X, Han D, Yan H, Li W (2016) Potential of marker selection to increase prediction accuracy of genomic selection in soybean (*Glycine max* L.). Mol Breed 36:113
Mera M, Alcalde JM (2019) *Lupinus albus* is the species that achieves greatest grain and protein yields in chile. In: Developing lupin crop into a modern and sustainable food and feed source. Fundación PROINPA, Cochabamba, Bolivia, p 44
Mera M, Beltran L, Miranda H, Rouanet JL (2006) Strong heritability across years and sites for pod wall proportion and specific weight in *Lupinus albus* and genotypic correlation with other pod and seed attributes. Plant Breed 125:161–166
Meuwissen THE, Hayes BJ, Goddard ME (2001) Prediction of total genetic value using genome-wide dense marker maps. Genetics 157:1819–1829
Naganowska B, Wolko B, Śliwińska E, Kaczmarek Z (2003) Nuclear DNA content variation and species relationships in the genus *Lupinus* (Fabaceae). Ann Bot 92:349–355
Nazzicari N, Biscarini F (2017) GROAN: genomic regression workbench (version 1.2.0). https://cran.r-project.org/package=GROAN. Accessed 19 July 2019
Nazzicari N, Biscarini F, Cozzi P, Brummer EC, Annicchiarico P (2016) Marker imputation efficiency for genotyping-by-sequencing data in rice (*Oryza sativa*) and alfalfa (*Medicago sativa*). Mol Breed 16:69
O'Rourke JA, Yang SS, Miller SS, Bucciarelli B, Liu J, Rydeen A, Gronwald JW (2013) An RNA-Seq transcriptome analysis of orthophosphate-deficient white lupin reveals novel insights into phosphorus acclimation in plants. Plant Physiol 161:705–724
Papineau J, Huyghe C (2004) Le Lupin Doux Protéagineux. Editions France Agricole, Paris
Park T, Casella G (2008) The Bayesian Lasso. J Am Stat Assoc 103:681–686
Phan HTT, Ellwood SR, Adhikari K, Nelson MN, Oliver RP (2007) The first genetic and comparative map of white lupin (*Lupinus albus* L.): identification of QTLs for anthracnose resistance and flowering time, and a locus for alkaloid content. DNA Res 14:59–70
Raman R, Vipin C, Luckett DJ, Cowley RB, Ash GJ, Harper JD, Raman, H (2014) Localisation of loci involved in resistance to *Diaporthe toxica* and

Pleiochaeta setosa in white lupin (*Lupinus albus* L.). Open J Genet 4:210

Rodrigues ML, Pacheco CMA, Chaves MM (1995) Soil-plant water relations, root distribution and biomass partitioning in *Lupinus albus* L. under drought conditions. J Exp Bot 46:947–956

Roorkiwal M, Rathore A, Das RR, Singh MK, Jain A, Srinivasan S, Gaur PM, Chellapilla B, Tripathi S, Li Y, Hickey JM, Lorenz A, Sutton T, Crossa J, Jannink J-L, Varshney RK (2016) Genome-enabled prediction models for yield related traits in chickpea. Front Plant Sci 7:1666

Schwender H, Fritsch A (2013) Scrime: analysis of high-dimensional categorical data such as SNP data (version 1.3.5). https://cran.r-project.org/web/packages/scrime/index.html. Accessed 19 July 2019

Searle SR, Casella G, McCulloch CE (2009) Variance components. Wiley, New York, USA

Varshney RK, Kudapa H, Pazhamala L, Chitikineni A, Thudi M, Bohra A, Gaur PM, Janila P, Fikre A, Kimurto P, Ellis N (2015) Translational genomics in agriculture: some examples in grain legumes. Crit Rev Plant Sci 34:169–194

Viana JMS, Piepho H-P, Silva FF (2017) Quantitative genetics theory for genomic selection and efficiency of genotypic value prediction in open-pollinated populations. Sci Agric 74:41–50

Vipin CA, Luckett DJ, Harper JD, Ash GJ, Kilian A, Ellwood SR, Raman H (2013) Construction of integrated linkage map of a recombinant inbred line population of white lupin (*Lupinus albus* L.). Breed Sci 63:292–300

Wang X, Xu Y, Hu Z, Xu C (2018) Genomic selection methods for crop improvement: current status and prospects. Crop J 6:330–340

Wiggans GR, Cole JB, Hubbard SM, Sonstegard TS (2017) Genomic selection in dairy cattle: the USDA experience. Annu Rev Anim Biosci 5:309–327

Yang H, Lin R, Renshaw D, Li C, Adhikari K, Thomas G, Yan G (2010) Development of sequence-specific PCR markers associated with a polygenic controlled trait for marker-assisted selection using a modified selective genotyping strategy: a case study on anthracnose disease resistance in white lupin (*Lupinus albus* L.). Mol Breed 25:239–249

Genomics of Yellow Lupin (*Lupinus luteus* L.)

11

Muhammad Munir Iqbal, William Erskine,
Jens D. Berger, Joshua A. Udall and Matthew N. Nelson

Abstract

Yellow lupin (*Lupinus luteus* L.) is a minor annual legume crop valued for its productivity in highly infertile, acidic soils and for its very high protein seeds. Yellow lupin belongs to the 'Old World' group of lupin species and is closely related to narrow-leafed lupin. Yellow lupin shares similar climatic adaptation to narrow-leafed lupin over which it offers some additional advantages such as greater water-logging tolerance and disease resistance. Despite its promise, yellow lupin is grown only as a niche crop in Australia, Europe and South America, and has attracted very limited breeding attention to date. Major constraints to the wider uptake of yellow lupin as a crop include lack of diversity in the domesticated gene pool and a historic focus on adaptation to a limited range of environments. Current varieties are also sensitive to some abiotic stresses (notably drought, extreme temperatures, salinity and alkalinity) and to sap-sucking insects such as aphids. Good genetic resources are available for yellow lupin including extensive seed collections that capture much of the species-wide diversity and three recombinant inbred line populations. Until recently, yellow lupin has lagged behind its well-resourced sister species narrow-leafed lupin in terms of genomic resources but is now catching up. Transcriptomic datasets have been used to generate molecular markers and to investigate the causes of flower and pod abortion. The first genetic map for yellow lupin was recently released, which is being used to investigate phenology, domestication traits and productivity under water-limiting conditions. Transgenesis methods have been developed for yellow lupin, a key enabling technology for future genome editing activities. Efforts are underway to develop a high-quality reference genome sequence for yellow lupin. These developing resources will help researchers acquire knowledge and

M. M. Iqbal
School of Agriculture and Environment,
The University of Western Australia, Perth, Australia
e-mail: munirqbl@gmail.com

M. M. Iqbal · W. Erskine · M. N. Nelson (✉)
UWA Institute of Agriculture, The University of Western Australia, Perth, Australia
e-mail: matthew.nelson@csiro.au

W. Erskine
e-mail: william.erskine@uwa.edu.au

M. M. Iqbal · W. Erskine
Centre for Plant Breeding and Genetics,
The University of Western Australia, Perth, Australia

J. D. Berger · M. N. Nelson
CSIRO Agriculture & Food, Perth, Australia
e-mail: jens.berger@csiro.au

J. A. Udall
USDA-ARS Southern Plains Agricultural Research Center, College Station, TX, USA
e-mail: jaudall1@gmail.com

M. N. Nelson
Royal Botanic Gardens, Kew, Ardingly, UK

K. B. Singh et al. (eds.), *The Lupin Genome*, Compendium of Plant Genomes,
https://doi.org/10.1007/978-3-030-21270-4_11

molecular tools to equip lupin breeders to overcome the restraints on broader adoption of this promising legume crop.

11.1 Background

Yellow lupin (*Lupinus luteus* L.) belongs to the Genistoid clade of the Fabaceae, one of the largest flowering plant families comprising approximately 765 genera and 19,500 species (Naganowska et al. 2003; Legume Phylogeny Working Group 2017). Yellow lupin is a minor annual pulse crop valued for its adaptation to highly infertile, acidic soils and for its high protein seeds. It is also valued by horticulturalists for its attractive yellow, sweetly scented flowers. Yellow lupin originated in the Mediterranean region, although its precise geographical origin is unclear due to the widespread traditional cultivation of wild and semi-wild types (Gladstones 1998).

Yellow lupin is part of the 'Old World' group of 12–13 lupin species distributed around the Mediterranean region and North Africa. They are divided into smooth-seeded and rough-seeded groups (Naganowska et al. 2003). Yellow lupin is very closely related to *L. hispanicus*, which shares the same chromosome number ($2n = 52$) and similar genome size ($2C = 2.44$ and 2.15 pg, respectively) (Naganowska et al. 2003; Fig. 11.1). More recently, phylogenetic analysis revealed that narrow-leafed lupin is closely related to yellow lupin despite the reduced chromosome number of narrow-leafed lupin ($2n = 40$) (Drummond et al. 2012; Fig. 11.1). Given that the most recent common ancestor of Old-World lupins is estimated to be 4.6–12.5 million years ago (MYA), yellow lupin is expected to share the same triplicated genome structure that is evident in narrow-leafed lupin, which is believed to have arisen by a whole-genome triplication event(s) 24.6 MYA (Hane et al. 2017; Kroc et al. 2014; Drummond et al. 2012).

A major step in the development of yellow lupin as a crop was its introduction for green manure and animal feed to the acid sands of the Baltic region in the eighteenth century, far outside its natural distribution (Hondelmann 1984). In this environment, the focus was on early crop phenology in order to obtain mature ripe seeds, as part of the lupin's new role as a grain crop in the cool wet environments characteristic of the end of summer. A range of trials based on time to sowing, response to vernalisation, photoperiod and temperature were conducted in order to understand crop phenology (Hackbarth 1955; Troll 1940a, b). These trials suggested that yellow lupin should be sown early enough in the cool spring to fulfil the crop's vernalisation requirements in order to initiate flowering, but that there was a trade-off between meeting the vernalisation requirement and avoiding frost damage due to early sowing (Hackbarth 1951). The next major step was the selection of low alkaloid, 'sweet' types in the 1920s in Germany made possible by the development of simple assays to detect alkaloids (based on reaction of alkaloids to iodine-mercury-potassium iodide solution; Hondelmann 1984). This paved the way for a broader range of uses in livestock and poultry feed and in the human diet.

Yellow lupins were brought to Australia in the 1950s and were selected for adaptation traits—

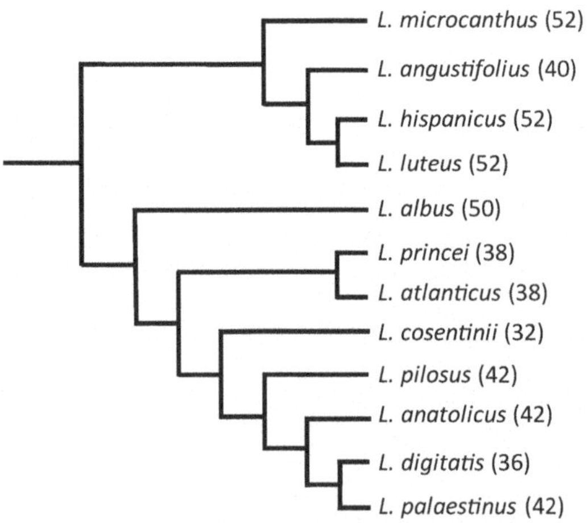

Fig. 11.1 Simplified and unscaled phylogenetic tree of Old-World *Lupinus* species, redrawn from the more comprehensive consensus tree of Drummond et al. (2012). Diploid chromosome numbers are shown in parentheses and taken from Naganowska et al. (2003)

especially drought escape—for the short-season target environments of South-Western Australia (Iqbal 2019). During the next four decades, research focus was also given to the domestication traits such as selection of low alkaloid genotypes and weakening or removal of vernalisation response in order to achieve an early flowering genotype. The first yellow lupin variety in Australia, Wodjil, was developed from a selection of Polish cultivar Teo in 1997 (French et al. 2001). The main cultivation areas today are in northern Europe, Belarus, Ukraine and recently in Western Australia. However, production is much lower than for narrow-leafed lupin (*Lupinus angustifolius* L.) and white lupin (*Lupinus albus* L.) (Iqbal et al. 2019).

Yellow lupin shares similar climatic adaptation to narrow-leafed lupin, where it offers some additional advantages such as greater tolerance to aluminium toxicity, high efficiency of accessing soil phosphorus in depleted soils, greater root-rot resistance and immunity to cucumber mosaic virus (Clements et al. 2009; Lambers et al. 2013). Yellow lupin seed also has higher protein content than narrow-leafed lupin seed (38% compared with 30% for narrow-leafed lupin) and excellent protein quality (among the highest cysteine and methionine contents of any grain legume), which makes it an excellent option for aquaculture feed (Gladstones 1970; Glencross et al. 2008). Despite this promise, yellow lupin remains a niche crop in Northern Europe, Australia and South America due to a range of constraints (see Sect. 11.4). However, yellow lupin's fortunes may be turning with recent advances in genetic and genomic resources that aim to overcome these constraints.

11.2 Genetic Resources for Yellow Lupin

Understanding the genetic basis of traits responsible for a crop's survival and sustainable production in diverse growth environments has been, and continues to be, a key task for crop improvement (Stinchcombe and Hoekstra 2008). A prerequisite for genetic analysis is the availability of good genetic resources. According to Wolko et al. (2011), there were 3,732 accessions of yellow lupin held in 21 ex situ seedbanks, constituting around 10% of all lupin seed collections. The Vavilov collection in St. Petersburg, Russia houses the largest number of *L. luteus* accessions (751) followed by the Australian Lupin Collection in Perth, Australia (463 accessions; currently moving to the Australian Grains Genebank in Horsham, Australia) and Poznań Plant Breeders station at Wiatrowo, Poland (421 accessions) (Wolko et al. 2011). These accessions include released cultivars, breeding lines, experimental populations and wild accessions. Wild germplasm is expected to show the greatest variation in adaptation and domestication traits (Berger et al. 2008, 2012). These collections form a valuable resource for a range of applications in future including understanding species-wide genetic diversity, deducing historic population migrations and bottlenecks, and for gene discovery through genome-wide association studies as used effectively in narrow-leafed lupin (Mousavi-Derazmahalleh et al. 2018a, b).

In addition to these diverse genetic resources, three recombinant inbred line (RIL) populations of yellow lupin have been developed to study agriculturally significant traits (Table 11.1). The most diverse population was developed at the Department of Primary Industries and Regional Development (DPIRD; Perth, Australia) from a cross between Australian cultivar Wodjil and P28213, a wild accession from a high rainfall region of the Azores. This population segregates for a range of adaptation and domestication traits, and was used to develop the first genetic map for yellow lupin (Iqbal et al. 2019). A second RIL population from a narrower cross of two breeding lines was developed at DPIRD with the intention of mapping CMV resistance. A third RIL population was developed in Chile from a cross between two cultivars with contrasting flowering time, height and architecture traits. So far, no results have been published from analysis of the second and third populations.

Table 11.1 Recombinant inbred line populations developed to study agriculturally relevant traits

Female parent	Male parent	Segregating traits	References
Wodjil (cultivar)	P28213 (wild, Azores)	Phenology, pod dehiscence, hard/soft seed, alkaloid content, flower colour	Iqbal et al. (2019)
96D001-18-12 (breeding line)	99D002-1 (breeding line)	CMV resistance	Berger et al. (2013)
Jantar (cultivar)	Pajbce (cultivar)	Flowering time, plant height, plant architecture	Berger et al. (2013)

11.3 Genomic Resources for Yellow Lupin

Yellow lupin has lagged behind in the development of genomic resources compared to its more widely grown sister species narrow-leafed lupin (see Chap. 3). The first transcriptomic dataset for yellow lupin was developed as a resource for marker discovery by Parra-González et al. (2012). They used 454 pyrosequencing to generate 71,655 contigs representing distinct transcript forms from young leaves, buds and flowers. They identified 2,572 contigs containing simple sequence repeat motifs, of which 222 were identified as polymorphic. They then used 50 polymorphic markers to sample genetic diversity in a panel of domesticated and wild germplasm (for more detail see Sect. 5.2.2). More recently, indel markers were developed by the same group (Osorio et al. 2018a).

Building on the seed protein catalogue produced by Ogura et al. (2014), seed transcriptomes were subsequently developed using Illumina HiSeq sequencing by Foley et al. (2015) to explore seed storage proteins across lupin crop species. Seed storage proteins in lupins comprise four classes of conglutin: α, β, γ and δ, each with multiple copies (16 distinct copies have been identified). They observed that >60% of conglutin transcripts comprised the δ-class in mature yellow lupin seeds, around threefold higher proportion than in narrow-leafed lupin (where β-conglutins constitute the most abundant class of transcripts). They found evidence for post-translational modifications that serve to diversify protein structure in lupin species.

Glazinska et al. (2017) generated flower and bud transcriptomes using HiSeq sequencing to study spontaneous flower and bud abortion, a phenomenon that reduces seed numbers and presumably total grain yield in yellow lupin crops (see also Sect. 5.2.8). Comparing abscising organs to control organs, they identified 1,343 (flowers), 2,933 (flower pedicels) and 1,491 (pods) differentially expressed genes (DEGs). A large portion of these DEGs were involved in hormone signalling and cell wall functioning. These DEGs provide targets for modified gene expression either in natural populations or through genetic modification to reduce the frequency of pod abortion.

The plastid genome of yellow lupin was reported by Martin et al. (2014), the first for any genistoid legume. Analysing the 155,894 bp sequence, they found a novel 36 kb inversion that may have occurred early in the divergence of genistoids from the rest of the Papillionoid legume subfamily.

Transgenic lines of yellow lupin were produced with different selectable markers to unlock the potential of yellow lupin crop improvement (Li et al. 2000). This is an important enabling technology for implementing CRISPR-Cas and other gene editing in the targeted improvement of traits in plant breeding (Bortesi and Fischer 2015; Lemmon et al. 2018).

The reference genome of narrow-leafed lupin cultivar Tanjil (Hane et al. 2017) has provided a solid foundation for genomic research in narrow-leafed lupin and to a lesser extent in other lupin species (see Chap. 3). No such reference genome is yet available for yellow lupin and remains a research priority for yellow lupin.

Fortunately, the development of a yellow lupin reference genome is currently underway led by Joshua Udall (unpublished data). The alignment of lupin genomes will shed light on the evolution of lupin genomes with their wide variation in chromosome numbers (Table 11.1). This will sharpen our currently blurry view gained from genetic and cytogenetic comparative investigations of lupin chromosome evolution (Nelson et al. 2006, 2010; Kroc et al. 2014; Phan et al. 2007; Susek et al. 2016, 2019). See Sects. 5 and 8 for more complete description of cytogenetic and genetic comparative studies, respectively. These genomic resources, together with proteomic and metabolomic resources (Ogura et al. 2013, 2014; Osorio et al. 2018b; Piornos et al. 2015), provide valuable tools for overcoming key constraints on yellow lupin cropping.

11.4 Using Genomics to Address Constraints on Yellow Lupin Cultivation

Several factors hinder the wider uptake of yellow lupin as a crop, which represent key targets for genomic analyses. One major constraint to the more widespread cultivation of yellow lupin is its sensitivity to abiotic stresses including terminal drought, extreme temperatures, salinity and alkalinity, which are frequently encountered in Mediterranean-type dryland cropping environments (Chaves et al. 2003). Plant stress responses are complex in nature, difficult to phenotype and likely controlled by many genes, making genetic improvement difficult and expensive (Stoddard et al. 2006). However, advances have been made in other crop stress systems through the development of accurate, high-throughput phenotyping methods in combination with advanced genomic analyses (e.g. in chickpea; Varshney et al. 2011).

Such efforts in yellow lupin are in their early stages but a recent advance was reported in the first genetic dissection of drought response in yellow lupin (Iqbal et al. 2019). They used genotyping-by-sequencing (GBS; Baird et al. 2008) and DArTseq™ (Sansaloni et al. 2011) methods to genotype the Wodjil x P28213 RIL population, enabling the generation of the first genetic map for yellow lupin comprising 2,458 markers distributed across 40 linkage groups (Iqbal et al. 2019). This excess number of linkage groups (26 were expected corresponding to the haploid chromosome number of yellow lupin), indicates that the map remains incomplete despite the large number of markers. Irregularities in segregating marker data seemed to point to the RIL population having been subject to a degree of cross-pollination to domesticated types during the single seed descent process. If proven to be the case, this is a concern as the Wodjil x P28213 RIL population is the obvious choice as a reference population among the lupin research community due to its greater diversity than the other available RIL populations (Table 11.1).

Yellow lupins are also subject to some biotic stresses, most notably aphid predation in ultra-low alkaloid cultivars such as Wodjil (Adhikari et al. 2012; Berlandier and Sweetingham 2003). Cowpea aphid behaviour monitoring of aphid-susceptible and aphid-resistant cultivars suggested that there may be phloem-based deterrence in resistant cultivars (Zehnder et al. 2001). However, further analysis using cowpea, bluegreen and green peach aphids showed that cultivar Teo gained its resistance from suppressing growth and survival of aphids rather than deterrence (Edwards et al. 2003). Evidence was found for weak anti-feeding factors against pea aphid located in non-phloem tissue in yellow lupin cultivar Perkoz (Kordan et al. 2018). In addition to stress caused by aphid predation per se, damage by viruses transmitted by aphids can be extremely damaging but resistance is available within the *L. luteus* gene pool (Robertson and Coyne 2009; Jones and Latham 1996). To our knowledge, genomic approaches (e.g. transcriptomics or genome-wide association studies) have not been applied to study the nature of resistance to biotic stresses in yellow lupin and remains a priority for future genomic research.

The domesticated breeding pool of narrow-leafed lupin is extremely shallow due to a series of population bottlenecks, starting with a handful of founder lines from the Iberian

Peninsula, extreme selection for domestication traits and then relative isolation of breeding programmes (Berger et al. 2012; Mousavi-Derazmahalleh et al. 2018a, b; Cowling et al. 1998; Cowling et al. 2009). The recent domestication of yellow lupin closely parallels that of narrow-leafed lupin (see Chap. 8), and adaptive diversity is more restricted in the domesticated gene pool of yellow lupin compared to the wild gene pool (Berger & Ludwig 2014). This is likely reflected also in reduced genome-wide diversity although this remains an open question in the absence of detailed diversity analyses. The only published molecular diversity analysis to date used 50 simple sequence repeat markers to investigate genetic diversity in a set of 64 mainly domesticated accessions, which appeared to support broader diversity among the wild accessions than among the domesticated accessions (Parra-González et al. 2012). However, a much more comprehensive analysis is required using wider sampling among wild accessions and more in-depth genotyping.

The genetic basis of domestication traits was investigated in the Wodjil x P28213 RIL population (Iqbal 2019). As was found in narrow-leafed lupin (Nelson et al. 2006), Iqbal found that vernalisation responsiveness in flowering, alkaloid content, flower colour and seed colour were all controlled by single genes that could be positioned on the genetic map. Segregation ratios pointed to two-gene control for seed dehiscence, plant growth habit and seed permeability, however these could not be mapped to the genetic map (Iqbal 2019). Yellow lupin can follow the example of narrow-leafed lupin where the first domestication/adaptation gene was identified and used to find new adaptive variation from wild germplasm (Nelson et al. 2017; Taylor et al. 2019).

Another important constraint in yellow lupin's genetic improvement is the historic focus on adaptation to a limited climatic range, again paralleling narrow-leafed lupin (Berger et al. 2012). Thus far, lupin breeders have focused on developing early cultivars with highly temperature-responsive phenologies that escape drought in their warm, short-season target environments in Australia or cold, wet autumn weather in northern Europe (Iqbal 2019), as described in detail for narrow-leafed lupin in Sect. 2.2.2. This resulted in cultivars with no adaptive mechanism other than appropriate phenology (Berger et al. 2008). The genetic control of time to flowering and time to maturity was recently dissected in the RIL population derived from a cross between Wodjil (short season, little or no vernalisation requirement) and P28213 (long season, strong vernalisation requirement) (Iqbal et al. 2019). This identified regions of the genome controlling these phenology traits, which can now be used to develop simple molecular marker assays to assist phenological diversification of breeding pools.

11.5 Conclusions

Yellow lupin holds much unrealized potential as a grain legume due to its high-quality seed and its unique adaptation to sandy, acidic soils. Recent genomic developments are starting to unlock that potential through tapping into the phenological and genetic diversity present in wild and domesticated gene pools, understanding agronomic traits such as pod abortion and drought response mechanisms as well as providing molecular marker tools for accelerating the breeding of resilient and diverse cultivars.

References

Adhikari KN, Edwards OR, Wang S, Ridsdill-Smith TJ, Buirchell B (2012) The role of alkaloids in conferring aphid resistance in yellow lupin (*Lupinus luteus* L.). Crop Pasture Sci 63:444–451

Baird N, Etter P, Atwood T, Currey M, Shiver A, Lewis Z, Selker E, Cresko W, Johnson E (2008) Rapid SNP discovery and genetic mapping using sequenced RAD markers. PLoS ONE 3:e3376

Berger JD, Adhikari K, Wilkinson D, Buirchel B, Sweetingham M (2008) Ecogeography of the Old World lupins. 1. Ecotypic variation in yellow lupin (*Lupinus luteus* L.). Aust J Agric Res 59:691–701

Berger JD, Buirchell B, Luckett D, Nelson M (2012) Domestication bottlenecks limit genetic diversity and constrain adaptation in narrow-leafed lupin (*Lupinus angustifolius* L.). Theor Appl Genet 124:637–652

Berger JD, John C, Nelson M, Kamphuis L, Singh K, Buirchell B (2013) The essential role of genetic resources in narrow-leafed lupin improvement. Crop Pasture Sci 64:361–373

Berger JD, Ludwig C (2014) Contrasting adaptive strategies to terminal drought-stress gradients in Mediterranean legumes: phenology, productivity, and water relations in wild and domesticated *Lupinus luteus* L. J Exp Bot 65:6219–6229

Berlandier F, Sweetingham M (2003) Aphid feeding damage causes large losses in susceptible lupin cultivars. Aust J Exp Agric 43:1357–1362

Bortesi L, Fischer R (2015) The CRISPR/Cas9 system for plant genome editing and beyond. Biotechnol Adv 33:41–52

Chaves M, Maroco J, Pereira J (2003) Understanding plant responses to drought; from genes to the whole plant. Funct Plant Biol 30:239–264

Clements J, Chong L, Quealy J, Prilyuk L, Yang H, Francis G, Buirchell B (2009) Interspecific hybrids between *Lupinus angustifolius* and *L. luteus*–an avenue to increase the value of narrow-leafed lupin in Australia. SABRAO J Breed Genet 41

Cowling W, Huyghe C, Swiecicki W, Gladstones J, Atkins C, Hamblin J (1998) Lupin breeding In: Gladstones JS, Atkins CA, Hamblin J (eds) Lupins as crop plants: biology, production and utilization. CAB International, Wallingford, UK

Cowling WA, Buirchell BJ, Falk DE (2009) A model for introducing novel genetic diversity from wild relatives into elite crop populations. Crop Pasture Sci 60:1009–1015

Drummond C, Eastwood R, Miotto S, Hughes C (2012) Multiple continental radiations and correlates of diversification in *Lupinus* (Leguminosae): testing for key innovation with incomplete taxon sampling. Syst Biol 61:443–460

Edwards O, Ridsdill-Smith T, Berlandier F (2003) Aphids do not avoid resistance in Australian lupin (*Lupinus angustifolius*, *L. luteus*) varieties. Bull Entomol Res 93:403–411

Foley RC, Jimenez-Lopez JC, Kamphuis LG, Hane JK, Melser S, Singh KB (2015) Analysis of conglutin seed storage proteins across lupin species using transcriptomic, protein and comparative genomic approaches. BMC Plant Biol 15:106

French R, Sweetingham M, Shea G (2001) A comparison of the adaptation of yellow lupin (*Lupinus luteus* L.) and narrow-leafed lupin (*L. angustifolius* L.) to acid sandplain soils in low rainfall agricultural areas of Western Australia. Crop Pasture Sci 52:945–954

Gladstones JS (1970) Lupins as crop plants. Field Crop Abstr 23:123–148

Gladstones J (1998) Distribution, origin, taxonomy, history and importance. In: Gladstones JS, Atkins CA, Hamblin J (eds) Lupins as crop plants: biology, production and utilization. CAB International, Wallingford, UK

Glazinska P, Wojciechowski W, Kulasek M, Glinkowski W, Marciniak K, Klajn N, Kesy J, Kopcewicz J (2017) *De novo* transcriptome profiling of flowers, flower pedicels and pods of *Lupinus luteus* (yellow lupine) reveals complex expression changes during organ abscission. Front Plant Sci 8:641

Glencross B, Palta J, Berger J (2008) Harvesting the benefits of lupin meals in aquaculture feeds. In: Lupins for health and wealth. Proceedings of the 12th international lupin conference, Fremantle, Western Australia. International Lupin Association, 14–18 Sept 2008

Hackbarth J (1951) Beobachtungen über den Entwicklungsrhythmus bei *Lupinus luteus*. Z für Pflanzenzücht 30:198–209

Hackbarth J (1955) Die oekologischen Anspruече der Lupinenarten. I. Anbau zur Koernergewinnung. Z für Pflanzenzücht 35:149–178

Hane J, Ming Y, Kamphuis L, Nelson M, Garg G, Atkins C, Bayer P, Bravo A, Bringans S, Cannon S et al (2017) A comprehensive draft genome sequence for lupin (*Lupinus angustifolius*), an emerging health food: insights into plant-microbe interactions and legume evolution. Plant Biotechnol J 15:318–330

Hondelmann W (1984) The lupin—ancient and modern crop plant. Theor Appl Genet 68:1–9

Iqbal MM (2019) Genetics of adaptation of yellow lupin to dryland conditions. School of Agricultural and Environment. The University of Western Australia, Perth

Iqbal M, Huynh M, Udall JA, Kilian A, Adhikari K, Berger J, Erskine W, Nelson MN (2019) The first genetic map for yellow lupin enables genetic dissection of adaptation traits in an orphan grain legume crop. BMC Genomics 20:68

Jones RAC, Latham LJ (1996) Natural resistance to cucumber mosaic virus in lupin species. Ann Appl Biol 129:523–542

Kordan B, Stec K, Słomiński P, Laszczak-Dawid A, Wróblewska-Kurdyk A, Gabryś B (2018) Antixenosis potential in pulses against the pea aphid (Hemiptera: Aphididae). J Econ Entomol 112:465–474

Kroc M, Koczyk G, Swiecicki W, Kilian A, Nelson M (2014) New evidence of ancestral polyploidy in the Genistoid legume *Lupinus angustifolius* L. (narrow-leafed lupin). Theor Appl Genet 127:1237–1249

Lambers H, Clements J, Nelson M (2013) How a phosphorus-acquisition strategy based on carboxylate exudation powers the success and agronomic potential of lupines (*Lupinus*, Fabaceae). Am J Bot 100:263–288

Legume Phylogeny Working Group (2017) A new subfamily classification of the Leguminosae based on a taxonomically comprehensive phylogeny. Taxon 66:44–77

Lemmon ZH, Reem NT, Dalrymple J, Soyk S, Swartwood KE, Rodriguez-Leal D, Van Eck J, Lippman ZB

(2018) Rapid improvement of domestication traits in an orphan crop by genome editing. Nat Plants 4:766–770

Li H, Wylie S, Jones M (2000) Transgenic yellow lupin (*Lupinus luteus*). Plant Cell Rep 19:634–637

Martin GE, Rousseau-Gueutin M, Cordonnier S, Lima O, Michon-Coudouel S, Naquin D, de Carvalho JF, Aïnouche M, Salmon A, Aïnouche A (2014) The first complete chloroplast genome of the Genistoid legume *Lupinus luteus*: evidence for a novel major lineage-specific rearrangement and new insights regarding plastome evolution in the legume family. Ann Bot 113:1197–1210

Mousavi-Derazmahalleh M, Bayer PE, Nevado B, Hurgobin B, Filatov D, Kilian A, Kamphuis LG, Singh KB, Berger JD, Hane JK, Edwards D, Erskine W, Nelson MN (2018a) Exploring the genetic and adaptive diversity of a pan-Mediterranean crop wild relative: narrow-leafed lupin. Theor Appl Genet 131:887–901

Mousavi-Derazmahalleh M, Nevado B, Bayer PE, Filatov D, Hane JK, Edwards D, Erskine W, Nelson MN (2018b) The western Mediterranean region provided the founder population of domesticated narrow-leafed lupin. Theor Appl Genet 131:2543–2554

Naganowska B, Wolko B, Sliwinska E, Kaczmarek Z (2003) Nuclear DNA content variation and species relationships in the genus *Lupinus* (Fabaceae). Ann Bot 92:349–355

Nelson MN, Książkiewicz M, Rychel S, Besharat N, Taylor C, Wyrwa K, Jost R, Erskine W, Cowling W, Berger J, Batley J (2017) The loss of vernalization requirement in narrow-leafed lupin is associated with a deletion in the promoter and de-repressed expression of a Flowering Locus T (FT) homologue. New Phytol 213:220–232

Nelson M, Phan H, Ellwood S, Moolhuijzen P, Hane J, Williams A, O'Lone C, Fosu-Nyarko J, Scobie M, Cakir M, Jones M, Bellgard M, Ksiazkiewicz M, Wolko B, Barker S, Oliver R, Cowling W (2006) The first gene-based map of *Lupinus angustifolius* L.—location of domestication genes and conserved synteny with *Medicago truncatula*. Theor Appl Genet 113:225–238

Nelson MN, Moolhuijzen P, Boersma J, Chudy M, Lesniewska K, Bellgard M, Oliver R, Swiecicki W, Wolko B, Cowling W (2010) Aligning a new reference genetic map of *Lupinus angustifolius* with the genome sequence of the model legume, *Lotus japonicus*. DNA Res 17:73–83

Ogura T, Hernández A, Aizawa T, Ogihara J, Sunairi M, Alcaino J, Salvo-Garrido H, Maureira-Butler IJ (2013) Identification of a low digestibility δ-Conglutin in yellow lupin (*Lupinus luteus* L.) seed meal for atlantic salmon (*Salmo salar* L.) by coupling 2D-PAGE and mass spectrometry. PloS One 8:e80369

Ogura T, Ogihara J, Sunairi M, Takeishi H, Aizawa T, Olivos-Trujillo MR, Maureira-Butler IJ, Salvo-Garrido HE (2014) Proteomic characterization of seeds from yellow lupin (*Lupinus luteus* L.). Proteomics 14:1543–1546

Osorio CE, Udall JA, Salvo-Garrido H, Maureira-Butler IJ (2018a) Development and characterization of InDel markers for *Lupinus luteus* L. (Fabaceae) and cross-species amplification in other lupin species. Electron J Biotechnol 31:44–47

Osorio CE, Amiard VS, Aravena-Calvo J, Udall JA, Doyle JJ, Maureira-Butler IJ (2018b) Chromatographic fingerprinting of *Lupinus luteus* L. (Leguminosae) main secondary metabolites: a case of domestication affecting crop variability. Genet Resour Crop Evol 65:1281–1291

Parra-González L, Aravena-Abarzúa G, Navarro-Navarro C, Udall J, Maughan J, Peterson L, Salvo-Garrido H, Maureira-Butler I (2012) Yellow lupin (*Lupinus luteus* L.) transcriptome sequencing: molecular marker development and comparative studies. BMC Genomics 13:1

Phan H, Ellwood S, Adhikari K, Nelson M, Oliver R (2007) The first genetic and comparative map of white lupin (*Lupinus albus* L.): identification of QTLs for anthracnose resistance and flowering time, and a locus for alkaloid content. DNA Res 14:59–70

Piornos JA, Burgos-Díaz C, Ogura T, Morales E, Rubilar M, Maureira-Butler I, Salvo-Garrido H (2015) Functional and physicochemical properties of a protein isolate from AluProt-CGNA: a novel protein-rich lupin variety (*Lupinus luteus*). Food Res Int 76:719–724

Robertson NL, Coyne CJ (2009) First report of bean yellow mosaic virus from diseased *Lupinus luteus* in Eastern Washington. Plant Dis 93:319–319

Sansaloni C, Petroli C, Jaccoud D, Carling J, Detering F, Grattapaglia D, Kilian A (2011) Diversity Arrays Technology (DArT) and next-generation sequencing combined: genome-wide, high throughput, highly informative genotyping for molecular breeding of *Eucalyptus*. BMC Proc 5:P54

Stinchcombe J, Hoekstra H (2008) Combining population genomics and quantitative genetics: finding the genes underlying ecologically important traits. Heredity 100:158–170

Stoddard F, Balko C, Erskine W, Khan H, Link W, Sarker A (2006) Screening techniques and sources of resistance to abiotic stresses in cool-season food legumes. Euphytica 147:167–186

Susek K, Bielski WK, Hasterok R, Naganowska B, Wolko B (2016) A first glimpse of wild lupin karyotype variation as revealed by comparative cytogenetic mapping. Front Plant Sci 7:1152

Susek K, Bielski W, Czyz KB, Hasterok R, Jackson SA, Wolko B, Naganowska B (2019) Impact of chromosomal rearrangements on the interpretation of lupin karyotype evolution. Genes 10:259

Taylor CM, Kamphuis LG, Zhang W, Garg G, Berger JD, Mousavi-Derazmahalleh M, Bayer PE, Edwards D, Singh KB, Cowling WA (2019) INDEL variation in the regulatory region of the major flowering time gene

LanFTc1 is associated with vernalization response and flowering time in narrow-leafed lupin (*Lupinus angustifolius* L.). Plant Cell Environ 42:174–187

Troll H (1940a) Saatzeitversuche mit Zucht-und Landsorten sowie Wildformen von *L. luteus* und *L. angustifolius*. Pflanzenbau 16:403–430

Troll H (1940b) Vegetationsbeobachtungen an Lupinen in verschiedenen geographischen Breiten. Der Zücht 12:129–139

Varshney R, Lekha P, Junichi K, Pooran M, Krishnamurthy L, Dave H (2011) Genomics and physiological approaches for root trait breeding to improve drought tolerance in chickpea (*Cicer arietinum* L.). In: Root genomics. Springer, Berlin, Germany

Wolko B, Clements JC, Naganowska B, Nelson MN, Yang H (2011) *Lupinus*. In: Kole C (ed) Wild crop relatives: genomic and breeding resources. Springer, Heidelberg, Germany

Zehnder GW, Nichols AJ, Edwards OR, Ridsdill-Smith TJ (2001) Electronically monitored cowpea aphid feeding behavior on resistant and susceptible lupins. Entomol Exp Appl 98:259–269

The Repetitive Content in Lupin Genomes

12

Abdelkader Aïnouche, Aurore Paris, Delphine Giraud, Jean Keller, Pauline Raimondeau, Frédéric Mahé, Pavel Neuman, Petr Novak, Jiri Macas, Malika Aïnouche, Armel Salmon and Guillaume E. Martin

Abstract

In this chapter, we present the first detailed evaluation of the repetitive compartment in *Lupinus* genomes. Low-depth next-generation sequencing (NGS) genomic resources from four closely related smooth-seeded Mediterranean lupin species (*L. albus*, *L. angustifolius*, *L. luteus*, and *L. micranthus*), exhibiting remarkable differences in genome size and chromosome number have been investigated. The repetitive compartment is composed of a wide diversity of repeats and represents 23–51% of the genomes. This compartment is essentially comprised of transposable elements (43–85%), mainly represented by *copia* and *gypsy* LTR retrotransposon families. Among the latter, some prominent families (*Tekay*, *Athila*, *Maximus-SIRE*) significantly contribute to genome size differences among species and in shaping different species-specific repeat profiles, regardless of their chromosome numbers. Also particular

A. Aïnouche (✉) · A. Paris · D. Giraud · J. Keller · P. Raimondeau · M. Aïnouche · A. Salmon
UMR CNRS 6553 ECOBIO, Université de Rennes 1, 35042 Rennes Cedex, France
e-mail: kader.ainouche@univ-rennes1.fr

A. Paris
e-mail: paris.aurore.11@gmail.com

D. Giraud
e-mail: delphine.giraud@univ-rennes1.fr

J. Keller
e-mail: jean.keller@lrsv.ups-tlse.fr

P. Raimondeau
e-mail: paulinerai@hotmail.fr

M. Aïnouche
e-mail: malika.ainouche@univ-rennes1.fr

A. Salmon
e-mail: armel.salmon@univ-rennes1.fr

J. Keller
Laboratoire de Recherche en Sciences Végétales, Université de Toulouse, CNRS, UPS, 24 chemin de Borde Rouge, BP42617, 31326 Auzeville, Castanet-Tolosan, France

F. Mahé
CIRAD (Centre de coopération Internationale en Recherche Agronomique pour le Développement), Campus international de Baillarguet, 34398 Montpellier Cedex 5, France
e-mail: frederic.mahe@cirad.fr

P. Neuman · P. Novak · J. Macas
Laboratory of Molecular Cytogenetics, Biology Centre of the Czech Academy of Sciences, Institute of Plant Molecular Biology, Branisovska 31, 37005 Ceske Budejovice, Czech Republic
e-mail: neumann@umbr.cas.cz

P. Novak
e-mail: petr@umbr.cas.cz

J. Macas
e-mail: macas@umbr.cas.cz

G. E. Martin
CIRAD (Centre de coopération Internationale en Recherche Agronomique pour le Développement), UMR AGAP, 34398 Montpellier, France
e-mail: guillaume.martin@cirad.fr

G. E. Martin
AGAP, Univ Montpellier, CIRAD, INRA, Montpellier SupAgro, Montpellier, France

K. B. Singh et al. (eds.), *The Lupin Genome*, Compendium of Plant Genomes,
https://doi.org/10.1007/978-3-030-21270-4_12

lineages of these elements have been differentially and recently amplified within species, such as in *L. luteus*, *L. albus*, and *L. angustifolius*. Moreover, this study highlighted the diversity of tandem repeats in lupin genomes, with minisatellites and satellites mostly being species-specific, whereas microsatellites (SSRs) are ubiquitously distributed. Strikingly, *L. angustifolius* exhibited a tremendous amount of tandem repeats in its genome (26%), including a noteworthy accumulation of one particular hexamer SSR (15.24% of the genome), which demonstrate that also tandem repeats may greatly contribute to genome obesity and dynamics in lupins. Therefore, differential lineage-specific amplifications of retrotransposons and tandem repeats occurred among lupins. Accordingly, this strongly suggests that different processes and mechanisms regulating amplification, proliferation, and clearance of repeats have differentially operated within the same genus among closely related Mediterranean species over the last ~10–12 Myr. Further extension of such evaluation to various representatives of the lupins diversity and outgroups will provide a better overview of the repetitive compartment and its evolutionary dynamics in the genus. Additionally, the genomic resources generated by this work represent a valuable basis to start building a repeats database specifically dedicated to best understand the genomic landscape, repeats distribution, and localization in lupins. This will facilitate further investigations on the functional and evolutionary impact of repeats on genes of interest, such as those responsive for important agronomical, adaptive, and defense features.

12.1 Introduction

Genome size variation, with a magnitude order of 200,000, is one of the most remarkable biological features in Eukaryotes (Bennetzen and Wang 2014; Biscotti et al. 2015). As this variation is not correlated with the morphological or physiological complexity of organisms it has been termed the "C-value paradox" (Thomas 1971) and later the "C-value enigma" (Gregory 2005). In angiosperms, Genome Size (GS) ranges from 60 Mb (in *Genlisea aurea*) to 150 Gb (in *Paris japonica*), which corresponds to approximately a 2,400-fold variation (Greilhuber et al. 2006; Leitch and Leitch 2008; Vu et al. 2015; Pellicer et al. 2018). Moreover, GS variation occurs at various taxonomic levels, including among closely related species within genera (Greilhuber et al. 2006; Hu et al. 2011) or even among various accessions within species, such as in *Hordeum spontaneum* (Kalendar et al. 2000) or in *Arabidopsis thaliana* (Schmuths 2004). Apart from whole genome duplication, triplication, or polyploidy (Soltis et al. 2009; Renny-Byfield and Wendel 2014), it is now obvious that repetitive sequences may account for a large proportion in the plant genomes, regardless of the number of protein coding genes, the ploidy level or the past paleopolyploid history (Bennetzen 2002, 2005; Wendel et al. 2016). While, the repetitive sequences were previously regarded as "junk," "parasitic," or "selfish" DNA (Doolittle and Sapienza 1980; Orgel et al. 1980; Lönnig and Saedler 1997), nowadays they are not only considered as a determinant fraction involved in GS variation (expansion/contraction) but also that they play a major role in their evolutionary dynamics and are crucial for living organisms (Biémont and Vieira 2006; Oliver et al. 2013; Wendel et al. 2016; Hosaka and Kakutani 2018; Pellicer et al. 2018). Two types of repetitive DNA sequences proliferate in the genomes: tandem repeats (or satellites sensu lato) and interspersed repeats (or transposable elements).

12.1.1 Tandem Repeats

Tandem Repeats (TR) consist of basic nucleotide units (or monomers) that are repeated head-to-tail to form TR arrays. According to the size of the repeated unit the tandem repeats are classified as: microsatellites or Simple Sequence Repeats (SSR) with motives shorter than 10–12 bp, minisatellites with motives between 12 and ~60 bp, and satellites with longer monomers (>60–100 bp or even several kilobases). Increase (or decrease) of the number of repeated units in microsatellites, for instance, generally results from a "slipped-strand mispairing" mechanism due to a polymerase shift during DNA replication (Levinson and Gutman 1987) or unequal cross-overs (Petes 1980). Satellite DNA can represent up to half of the genome in some eukaryotes (Satović et al. 2018). Microsatellites (SSRs) are ubiquitous in genomes and are widely used as genetic markers for genotyping (Parra-González et al. 2012; Raman et al. 2014; Kamphuis et al. 2015; Atnaf et al. 2017). The other larger arrays of TR, minisatellites, satelittes, including highly repetitive gene families such as nuclear ribosomal DNA, are helpful for chromosome fingerprinting. They are usually associated to centromeric, peri-centromeric and telomeric regions and seem to have a significant functional regulatory role (Streelman and Kocher 2002; Li 2004; Lower et al. 2018), but they yet remain poorly investigated and were the subject of only a few comparative genomics studies (Shi et al. 2013; Ruiz-Ruano et al. 2016; Usai et al. 2017).

12.1.2 Transposable Elements

Transposable Elements (TEs) (or jumping genetic elements of McClintock 1948) are very diversly interspersed repetitive DNA sequences that are able to duplicate themselves and to insert their copies at different positions in the genome via a transposition mechanism (Kumar and Bennetzen 1999; Bennetzen 2002). Following the classification of Wicker et al. (2007), TEs are divided into two main classes, according to the type of intermediate (DNA or RNA) used in their transposition mode. Class I elements, or retrotransposons, follow a transposition mode using an RNA intermediate called "copy/paste", which may dramatically increase their copy number in genomes (Vicient et al. 1999; Bennetzen 2002, 2005; Piegu et al. 2006). Five orders are distinguished within this class: Long Terminal Repeats (LTRs), Dictyostelium intermediate repeat sequence (DIRSs), Penelope-like elements (PLEs), Long Interspersed Nuclear Elements (LINEs), and Small Interspersed Nuclear Element (SINEs). Within each order, elements are clustered into superfamilies based on the structure of their protein and non-coding domains. The Class II elements (or DNA transposons) transpose via a DNA intermediate in a mode called "cut / paste," which results in their excision from their genomic location and their insertion elsewhere in the genome. Two subclasses are recognized: subclass 1 mainly corresponds to TIR elements, which are characterized by their Terminal Inverted Repeats (TIR) at their extremities; and subclass 2 which corresponds to Helitron and Maverick elements (Wicker et al. 2007).

In plants, amplification and accumulation of Class I elements represent the major source of GS increase. For example, LTR retrotransposons may reach between $\sim$70 and 76% of the genomes in maize, bread wheat, and barley (Mayer et al. 2011; Oliver et al. 2013; Wicker et al. 2018). TEs amplification can be activated by various environmental (biotic and abiotic) and genomic (e.g., hybridizations) stresses during the evolutionary history of organisms (Kalendar et al. 2000; Liu and Wendel 2000; Jiang et al. 2003; Grandbastien et al. 2005; Wessler 2006). In turn, different regulatory mechanisms are triggered at the cellular and molecular levels to control their proliferation and counteract genome expansion via epigenetic mechanisms (small RNA, DNA methylation, histone modification and removal) (Bennetzen 2005; Hawkins et al. 2006, 2009; Slotkin and Martienssen 2007; Lisch 2009; Yaakov and Kashkush 2012; Axtell 2013; Castel and Martienssen 2013). The repeated waves of TE amplification and regulatory mechanisms thus have a deep impact on the host genomes. They may drive structural genomic rearrangements and generate genetic diversity which accompanies the adaptation and diversification of species in their environments (Bennett 2005; Morgante et al. 2005; Chénais et al. 2012). Following their insertion into or near genes, they may modify expression and function of various genes which may induce variable phenotypic changes (Jiang et al. 2003; Kashkush et al. 2003; Lisch 2013). Also, there is evidence that they contribute to the formation of new genes and represent an important source of evolutionary novelties (Biémont and Vieira 2006; Oliver et al. 2013; Lynch et al. 2015).

12.1.3 Advancing the Discovery of Repetitive Sequences Using Next-Generation Sequencing Technology

Regarding their importance, investigations on the repetitive sequences greatly benefited from the advances of high-throughput sequencing technologies. Several strategies and bioinformatics programs have been developed for the detection and identification of repeated elements in fully sequenced genomes of model organisms (Quesneville et al. 2005; Lerat 2010; Flutre et al. 2011; Treangen and Salzberg 2011; Wajid and Serpedin 2012). However, assembly, annotation, and precise location of massive similar repeated short-reads, representing regions which underwent various processes of recombination/deletion, are challenging and generally results in incompletely assembled genomes with large gap-spaces and potentially chimerical structures (Jiang et al. 2004; Sequencing Project 2005). Combination of short-reads technologies (Illumina HiSeq) with long-reads sequencing ones (Pacific BioSiences and Oxford Nanopore) will yield higher quality genomes to accurately assemble and circumscribe repeated structures. Other programs have been designed to directly evaluate the repetitive content from raw unassembled short-read sequences generated from various high-throughput DNA sequencing technology platforms, such as for example: RepeatExplorer (Novák et al. 2010, 2013), Transposome (Staton and Burke 2015), REPdenovo (Chu et al. 2016). Such programs use various tools which allow detection, quantitative estimation, reconstruction, and annotation of repetitive elements in NGS data. They are based on all-to-all read sequence similarities, graph-based clustering methods, and repeats identification using complementary Blast methods and search of conserved specific TEs protein coding domains against reference databases. These toolkits demonstrated their efficiency for evaluating the repetitive compartment from a reduced sample of low-pass genome sequence data (even less than 1% genome coverage) in various plant taxa (Macas et al. 2007; Hřibová et al. 2010; Novák et al. 2010, 2013; Renny-Byfield et al. 2011; Piednoël et al. 2013; Staton and Burke 2015; Vu et al. 2015; Wu et al. 2019). They not only allow rapid investigation of the repetitive compartment in many non-model genomes but also may provide crucial information to assist assembly and annotation of complex genomes.

12.1.4 Lupinus: A System of Interest to Evaluate the Dynamics of the Genomic Repetitive Compartment

In this context, the Genistoid legume *Lupinus* (Fabaceae) is a system of particular interest to explore the evolutionary dynamics of repetitive sequences and their impact on the evolution of host genomes. Indeed, *Lupinus* is a large genus which is composed of hundreds of species adapted to very diverse ecological conditions which diversified during the last ~16 Myr (mean age of the stem node of the genus according to Hughes and Eastwood 2006) in two major regions of the World: about 250–300 species in the New World and around 12–13 in the Old World (Gladstones et al. 1998; Ainouche et al. 2004; Eastwood et al. 2008). Among the latter, the smooth-seeded lupins, which are mainly circum-Mediterranean are distinguished from the rough-seeded lupins (or *Scabrispermae*), which are predominantly North African. Previous studies have shown that lupins exhibit a remarkable variation of their chromosome number ($2n = 32$ to 52) and their genome size ($2C = 1$ to ~ 2.6 Gb), including between closely related taxa, regardless of their chromosome number (Gladstones et al. 1998; Naganowska 2003;

Naganowska et al. 2005; Conterato and Schifino-Wittmann 2006; Mahé 2009). A first PCR-based screening of the repeated compartment revealed a significant diversity of LTR retrotransposons in lupin genomes (Mahé 2009). Because of their beneficial properties for agriculture, human health, and nutrition (Gladstones et al. 1998; Cabello-Hurtado et al. 2016), and for their novel status as model plants for studying symbiosis, proteoid roots, and Pi uptake (O'Rourke et al. 2013; Keller et al. 2018), the smooth-seeded Mediterranean lupins, which include three crops (*L. albus*, *L. luteus*, and *L. angustifolius*), are under increasing attention. Several transcriptomes and genomic resources have been generated (see Sects. 4 and 6) (Parra-González et al. 2012; O'Rourke et al. 2013; Kamphuis et al. 2015; Keller et al. 2018) and a first draft genome has been recently released (Hane et al. 2017; this book), providing the raw material to best understand structure, evolution, and functional potential of the lupin genome, including its repetitive compartment.

Therefore, as a first step to develop our knowledge on this enigmatic genomic compartment, we report results from: (i) a preliminary survey of the diversity of LTR retotransposons (*Ty1-copia* and *Ty3-gypsy-like* elements) in *Lupinus* and allied Genistoid taxa, based upon analysis of their Reverse Transcriptase sequences (RTs) (Flavell et al. 1992; Alix and Heslop-Harrison 2004; Mahé 2009); and (ii) a detailed evaluation of the repetitive compartment of four smooth-seeded Mediterranean lupin taxa, from the analysis of low-depth NGS genomic resources, using different programs to identify and estimate the repetitive sequences (Benson 1999; Novák et al. 2010, 2013, 2017).

12.2 Exploring Retrotransposons Diversity in Genomes of Lupins and Allied Genistoids

Ty1-copia-like and *Ty3-gypsy-like* superfamilies of class I retrotransposons are ubiquitous in eukaryote genomes and most often involved in Genome Size (GS) variation. A preliminary investigation of their diversity, was conducted in 44 accessions belonging to 27 lupin taxa (16 from the Old World and 11 from the New World) and 8 other Genistoid representatives; Table 12.1). This was carried out through analysis of their constitutive Reverse Transcriptase sequences (RTs) (Mahé 2009). Accordingly,

Table 12.1 List of accessions from *Lupinus* and other Genistoid taxa surveyed for retrotransposons diversity. The origin, geographic distribution, and accession reference are indicated for each sample. OW = Old World species; NW = New World species; Med = Mediterranean; Afr = African; NA = North American; SA = South American

Taxon	2*n*	Origin/Distribution	Sample source and reference number
L. affinis	48	Oregon/NW, West NA	USDA/504315/N20
L. albus	50	Algeria/OW, Med	INAE-DZ/M20
L. anatolicus	42	Turkey/OW, Afr	AKA/K32
L. angustifolius subsp. reticulatus	40	France/OW, Med	AKA/T25
L. angustifolius subsp. angustifolius	40	Algeria/OW, Med	AKA-M1/T24
L. atlanticus	38	Morocco/OW, Afr	USDA/384612-FM83/T1
	38	Morocco/OW, Afr	INRA-SAPF/T11
	38	Morocco/OW, Afr	USDA/384613-FM87/T2
L. bracteolaris	32–34	Brazil/NW, South-East SA	USDA/404349/S80
L. concinnus	?	USA/NW	N19
L. cosentinii	32	?/OW, Med	INRAL-FR/T15
L. diffusus	?	Florida/NW	K35

(continued)

Table 12.1 (continued)

Taxon	2n	Origin/Distribution	Sample source and reference number
L. digitatus	36	Egypt/OW, Afr-Med	WADA-PI26877/T4
L. elegans	48	Mexico/NW, West SA	USDA/185099/S33
L. hirsutissimus	?	USA/NW	AKA/N85
L. hispanicus subsp. bicolor	52	Spain/OW, Med	USDA/PI 384554/T23
L. hispanicus subsp. hispanicus	52	Portugal /OW, Med	USDA/384555/T22
L. luteus	52	Algeria/OW, Med	AKA/M5
	52	Algeria/OW, Med	AKA/T20
	52	Algeria/OW, Med	AKA/T21
L. mariae-josephae	52?	Spain/OW, Med	H. Pascual/MJ1
L. micranthus	52	Algeria/OW, Med	AKA/T19
	52	Algeria/OW, Med	T 28
L. mutabilis	48	Perou/NW, West SA	INAE-DZ/S35/MU23
L. nanus	48	USA/NW	N42
L. palaestinus	42	Near-East/OW, Afr-Med	INRA-FR/T14
L. paraguariensis	36	Brazil/NW, East SA	BRA-02828/BZ1
L. pilosus	42	Algeria/OW, Afr-Med	INAE-DZ/T6
	42	Algeria/OW, Afr-Med	INAE-DZ/T9
	42	North-Africa/OW, Afr-Med	USDA/W6 PI 11995/T13
L. pilosus tassilicus	?	Lybia/OW, Afr	AKA/A641
L. polyphyllus	48	USA/NW, NA	USDA/504404/T26
L. princei	38	Kenya/OW, Afr	WADA P 23021/T0
	38	Kenya/OW, Afr	RP Chyulu 1800/T16
	38	Kenya/OW, Afr	RP Chyulu 1915/T17
L. texensis	36	USA/NW, South NA	USDA/577291/N45
Anarthrophyllum cumingii	?	?/NW, South SA	AKA/201
Argyrolobium uniflorum	?	OW	AKA/G25
Chamaecytisus mollis	?	OW	AKA/C84
Crotalaria podocarpa	?	OW	AKA/K50
Cytisus heterochrous	?	OW	AKA/G8
Genista tinctoria	?	OW	AKA/G56
Thermopsis rhombifolia	?	NW	AKA/G46
Ulex parviflorus	?	Spain/OW, Med	AKA/G24

conserved coding RT domains were amplified, cloned, and sequenced from genomic DNA samples using universal primers (Flavell et al. 1992) following the procedure described by Alix and Heslop-Harrison (2004). After (i) removing the low-quality sequences from the hundreds of amplicons generated, (ii) verifying their homology with known RTs from public databases (via Blastn, Blastx, and RepeatMasker; https://www.ncbi.nlm.nih.gov and https://www.repeatmasker.org/), and (iii) size-filtering, a total of 367 retrotransposon-*like* RT sequences were selected for further analysis. Among them 260 amplicons ranged in size from 248 to 295 bp, with pairwise identity varying from 38.6 to 100% for *copia* elements (GenBank accession numbers GU189754 to GU190013); and 107 ranged from 366 to 564 bp with a pairwise identity of 32.4–100% for *gypsy* elements (accession numbers GU190014 to GU190133).

Within this set of amplicons, 305 were from the lupin species (211 RT-copia, 89 RT-gypsy and five unidentified) and 62 were from the eight Genistoid representatives (including 40 RT-copia, 17 RT-gypsy, and five unannotated). Altogether, these 367 DNA sequences were aligned with MAFFT (Katoh and Standley 2013). The sequence data matrix was then subjected to a Maximum likelihood (ML) phylogenetic analysis, using the best-fitted evolutionary model (GTR + R7: General Time Reversible model, rates Gamma distributed) identified with ModelFinder (Kalyaanamoorthy et al. 2017) as implemented in IQ-TREE v1.5.5 (Nguyen et al. 2015). The robustness of the nodes was estimated with 10,000 ultrafast bootstrap replicates (Hoang et al. 2017). The phylogenetic tree resulting from this analysis is shown in Fig. 12.1, where each terminal branch representing RT-*copia* or RT-*gypsy* amplicons is colored according to its taxonomic and geographic origin (see Fig. legends) and its annotation assignment indicated by different colors in the outer circle.

A remarkable diversity of retrotransposons was detected within genomes of both lupin and Genistoid species. Random amplified RT sequences using universal primers allowed a clear segregation of the two retrotransposon

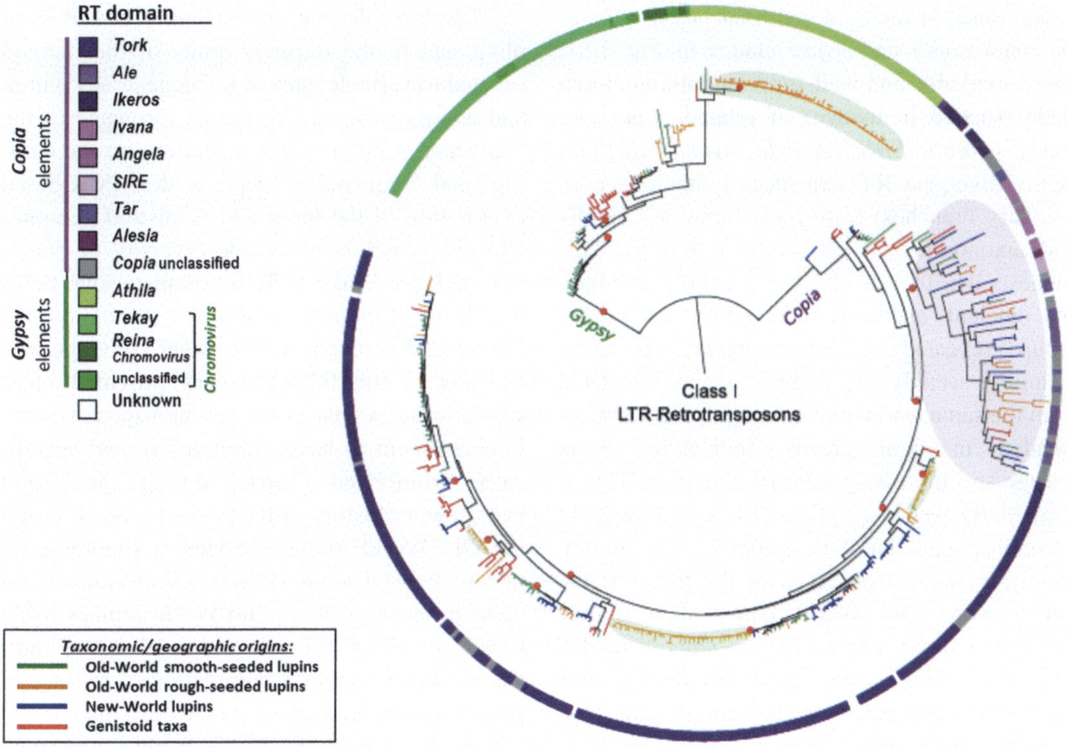

Fig. 12.1 Maximum likelihood phylogenetic tree of 367 *copia* and gypsy RT fragments amplified from 44 accessions belonging to 27 Old World and New World lupin taxa (305 sequences) and from 8 other Genistoid representatives (62 sequences). Each terminal branch is colored according to its taxonomical/geographical group of origin and its *copia* or *gypsy* annotation assignment indicated by different colors in the outer circle (see legends in the figure). Red dots on the tree indicate remarkable well supported nodes (by bootstrap estimate). Some remarkable groups of *copia* or *gypsy* amplicons likely representing ubiquitous elements of ancient origin (shaded in mauve) or recent lineage-specific amplifications (shaded in light green) are also indicated

superfamilies, with a higher amplification success for the *copia* ones. Within each of these superfamilies, several families have been identified: mainly *Tork copia*-like elements and at a lower scale other *copia* families (*Ale*, *Ikeros*, *Ivana*, *Angela*, *SIRE*, *Tar* and *Alesia*). Whereas *gypsy*-like elements were mainly represented by *Tekay* and *Athila* elements, and a few Reina and *Tat/Ogre* ones. Most of them showed significant levels of RT-identity with those from other Fabaceae available in databases for *Cicer* (52.5–88.4%), *Vigna* (51–86%), *Vicia*, or Soybean, suggesting that these elements were most likely inherited from a common Papilionoid ancestor or even from earlier origin.

Although the number of amplicons is low for some taxa, the main retrotransposon families detected appear to be ubiquitous throughout the lupin and Genistoid genomes, as illustrated by the presence of multicolored branches in each of the main *copia* and *gypsy* clades in Fig. 12.1. One remarkable and well-supported multicolored clade (shaded in mauve) of related *copia* elements (*Alesia*, *SIRE*, *Angela*, *Ivana*) includes highly divergent RT sequences (with long evolutionary branches) from most lupin and Genistoid taxonomic and geographical groups. This suggests that they likely derive from a common and ancient ancestor. The other noteworthy groups revealed by the phylogeny are those including weakly divergent sequences isolated from the same taxonomic or geographical lineage (shaded in light green), indicating recent lineage-specific transpositional activities. This is particularly well exemplified by several specific retrotransposon families (groups with monochrome branches) observed in the tree for: the rough-seeded Old World lupins with orange branches (*Athila* and *Tork*); the smooth-seeded Old World lupins with green branches (*Tekay* and *Tork*); the New World lupins with blue branches (*Tork* groups); and in the Genistoids with red branches (a *Tekay* group). Other homogeneous lineage-specific groups are composed of more divergent RT sequences likely deriving from earlier transposition events, such as for instance in the Genistoids (red branches) which show specific lines of *Athila* and *Tork* elements.

Within the collection of conserved RT domains generated from the lupins, five were amplified from a sample of *L. angustifolius* subsp. *angustifolius* (originating from North Africa), three RT-*gypsy* and two RT-*copia* clones. These clones have been used as queries in a rapid screening of the current reference NLL genome (of *L. angustifolius* cultivar. Tanjil; (Hane et al. 2017)) to estimate a potential number of PCR-based amplified products that could be expected from this genome. Interestingly, no less than 997 and 1209 non-redundant hits were found for the *gypsy* and the *copia* elements, respectively, using the easy Blast search tool (with evalue threshold: 1.0e−5) implemented in the Lupin Genome Portal https://www.lupinexpress.org/ (Priyam et al. 2015).

Therefore, despite an inevitably biased sampling due to the intrinsic limits of the method (related to variable rates of RT degeneracy within and among genomes, to the performance of the "universal primers," and depending on the cloning and sequencing depth), the PCR-based exploration of the lupin and Genistoid genomes allowed detection of a wide diversity of *copia*-like and *gypsy*-like LTR retrotransposons families. Most of them are ubiquitous throughout the lupins and Genistoids. Moreover, phylogenetic analysis of the RT sequences provided clues which suggest that some retrotransposons subfamilies seem to have differentially and specifically proliferated (bursts) during the recent evolutionary history of the genus in the New and the Old World lupins. Besides, a Fluorescence in situ hybridization (FISH) test performed on metaphase root cells of Old World lupines using *copia* ant *gypsy* RT-probes (Mahé 2009) indicated a much higher accumulation of retrotransposons in the large genome of the Mediterranean species *L. luteus* (2367 Mb/2C) than in the small genome of *L. micranthus* (1147 Mb/2C). Thus, altogether these results emphasized the need to more accurately identify and evaluate the diversity and relative abundance of transposable elements.

12.3 NGS-Based Evaluation of the Repetitive Compartment in Lupin Genomes

As highlighted, lupins are characterized by a noteworthy genome size variation (GS = 2C = nuclear DNA amount per cell) ranging from 1.05 to 2.6 Gb, regardless of their various chromosome numbers (varying from 2n = 32, 34, 36, 38, 40, 42, 48, 50 to 52). This is observable even between taxa having the same chromosome number (such as *L. luteus* which has more than twice the size of that of *L. micranthus*), as well as regardless of their Old World or New World origins and of their phylogenetic relationships.

Therefore, in order to deepen our understanding of the lupin genome dynamics, four lupin accessions with small and large genomes, belonging to different Mediterranean Old World smooth-seeded species (Table 12.2; Fig. 12.2), were subjected to a comparative NGS-based analysis of their genomic repetitive compartment: *L. albus* (2n = 50; 2C = 1.13 Gb), *L. angustifolius* (2n = 40; 2C = 1.85 Gb), *L. luteus* (2n = 52; 2C = 2.37 Gb), and *L. micranthus* (2n = 52; 2C = 1.15 Gb). For this purpose, a sequence data set of 1,200,000 paired-end 100 bp reads (120 Mb) per accession, extracted from resources generated by low-depth genomic Illumina HiSeq sequencing, and representing 5–10% of each genome (Table 12.2), was analyzed with RepeatExplorer (Novák et al. 2010, 2013). Following analysis of a combined data set (including 4,800,000 reads, each labeled according to its species origin), 293,635 clusters were obtained. Among the 744 clusters containing more than 48 reads (the largest having 265,540 reads), 176 were annotated as transposable elements and 207 as simple sequence repeats; the remaining clusters corresponded to organelle or to unclassified sequences.

12.3.1 Composition of the Repetitive Compartment in Lupin Genomes

Identification and distribution of the main elements of the repetitive compartment have been determined in the four targeted genomes. As summarized in Table 12.3 and illustrated in Fig. 12.2, the repetitive compartment (including transposable elements and tandem repeat satellites; excluding nuclear ribosomal DNA sequences or nrDNA) represents a large part of the genomes and varies from 23.27% in the small genome of *L. micranthus* to 50.36% in the largest genome of *L. luteus*, regardless of their same chromosome number (2n = 52). While *L. albus* shares a close chromosomes number (2n = 50) and a similar genome size with *L. micranthus*, it contains a much larger proportion of repeats (41.10%). In turn, the accession of *Lupinus angustifolius* analyzed here has a lower chromosome number (2n = 40), a relatively large genome (2C = 1.85 Gb), and exhibits a high repeats proportion (49,63%), which is underestimated compared to the 54% reported for

Table 12.2 Origins and characteristics of the genomic resources of four *Lupinus* species used in this study

Species	Accession code	Origins	2*n*	2C DNA amount[a] (in pg)	Genome Size[b] (2C in Mb)	Total length of reads sequenced (Gb)	Genome coverage (x folds)	% of genome analyzed with RE[c]
L. albus	M20	Egypt	50	1.16	1134.48	1.2	1.06	10.57%
L. angustifolius	IPG2	Morocco	40	1.89	1848.42	15.4	8.2	6.5%
L. luteus	M6	Algeria	52	2.42	2366.76	1.15	0.49	5.07%
L. micranthus	B12	Algeria	52	1.07	1147	1.13	1.06	10.46%

[a]According to Naganowska (2003) and Mahé (2009)
[b]Using 1 pg (picogramme) = 978 Mb according to Dolezel et al. (2003)
[c]Repeat Explorer (Novák et al. 2010, 2013)

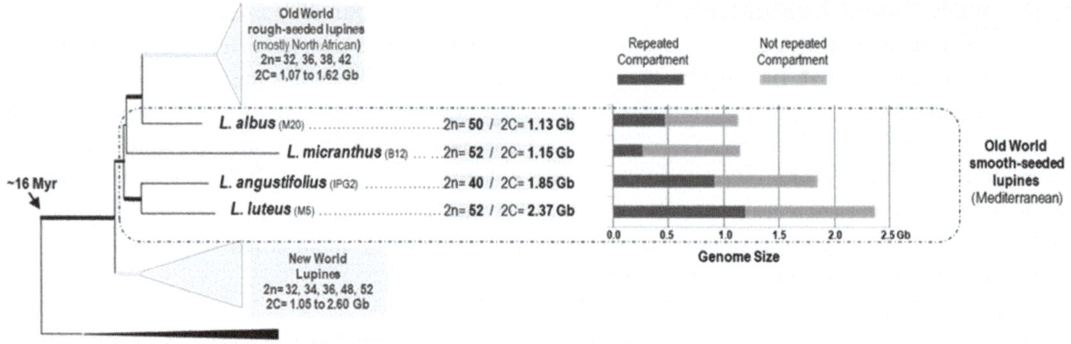

Fig. 12.2 Condensed phylogenetic tree of the lupins (on the left) redrawn from (Mahé et al. 2011), showing the position and relationships of the four Mediterranean smooth-seeded species subjected to a RepeatExplorer analysis of their repetitive genomic compartment. The Old World rough-seeded lupins and the New World lupin clades are presented. The mean age of the lupin stem node is indicated (according to Hughes and Eastwood 2006). Chromosome numbers (2*n*) and genome size (2C in Gb) of the taxa are given on the right of the figure, together with a histogram showing the genomic proportion of the repeated compartment in the four Mediterranean smooth-seeded species analyzed

Table 12.3 Proportions of the main DNA repeat categories (as % of the genome) in four Old World lupins. Repeats are classified is according to RepeatExplorer annotation

Genomic proportion of the different DNA repeat categories				
Repeats annotation	*L. albus*	*L. angustifolius*	*L. luteus*	*L. micranthus*
LTR retrotransposons (*copia* + *gypsy*)	**33.27**	**20.44**	**41.10**	**13.43**
LINE	0.07	0.04	0.03	0.08
DNA transposons	1.52	1.05	1.24	0.88
Satellites sensu lato	3.37	**25.97**	5.29	5.74
Unknown	2.94	2.13	2.69	3.15
nrDNA/45S	1.95	2.91	1.53	1.67
Repetitive Compartment (nrDNA excluded)	**41.18**	**49.63**	**50.36**	**23.27**

the NLL cultivar Tanjil sequenced genome (Hane et al. 2017).

In *L. albus*, *L. luteus*, and *L. micranthus*, the repetitive compartment is mainly composed of transposable elements, which are essentially represented by variable proportions of LTR retrotransposons (33,27%, 41,10%, and 13,43% of the genome, respectively), whereas LINEs and DNA transposons are present at less than 2% in each genome (Table 12.3; Fig. 12.3). Apart from the indeterminate repeats (around 2–3%), satellites (tandem repeats) are present at a low proportion in the three genomes, ranging from 3.37% in *L. albus*, to ~5–6% in *L. luteus*, and *L. micranthus*. While similar repeat categories were

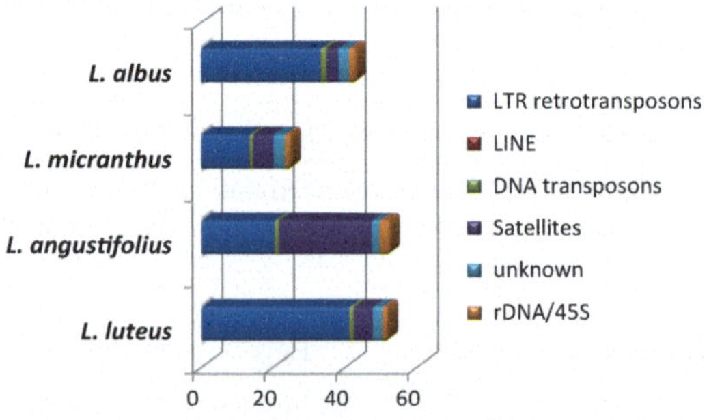

Fig. 12.3 Histogram showing the genomic proportions of the main DNA repeat categories (as % of the genome) in four Old World lupins. Repeats are classified according to RepeatExplorer annotation outputs (Novák et al. 2010)

detected in *L. angustifolius*, and it exhibited a noteworthy different pattern, with a repetitive compartment made up of a little more than half by satellites (around 26% of the genome). The remaining part is mainly composed of LTR retrotransposons (approximately 21% of WG). Although, the RepeatExplorer-based proportion of LTR retrotransposons was lower than that estimated from the Narrow-Leafed Lupin (NLL) sequenced genome (~28%), the above observations already demonstrate that not only TEs but also satellites may account for high proportions in lupin genomes where they may reach substantial amounts, ranging from ~38 Mb and ~65 Mb in the small genomes (of *L. albus* and *L. micranthus*, respectively) to ~126 Mb and ~481 Mb in the larger genomes of *L. luteus* and *L. angustifolius*, respectively. Otherwise, the nrDNA varies from 1.5 to 2% in the genomes of *L. albus*, *L. luteus*, and *L. micranthus*, while it displays a significant increase in *L. angustifolius* (~3%), which suggests the occurrence of different nrDNA evolutionary patterns among the Mediterranean lupins (Wolko and Weeden 1989; Kroc et al. 2014).

12.3.2 Identification and Distribution of LTR Retrotransposons in the Lupin Genomes

A more accurate analysis of the repetitive compartment shows that the *copia* and *gypsy* superfamilies of LTR retrotransposons are well represented in all species, at various proportions and different relative ratio (Table 12.4; Fig. 12.4). *Copia* elements constitute 5.48% (62.9 Mb) to 11.73% (216.8 Mb) of the genomes, with the highest proportions in large genomes (216.8 Mb for *L. angustifolius* and 231.5 Mb for *L. luteus*). In turn, the *gypsy* elements exhibited a wider range, from 3.73% (42.8 Mb) in the small genome of *L. micranthus* to 31.31% (741 Mb) in the largest genome of *L. luteus*, with however no correlation with GS regarding the substantial proportion of 20.31% (230.4 Mb) in *L. albus* (with a small GS) as compared to that of *L. angustifolius* (8.7%; 160.8 Mb) which has a larger GS. Accordingly, this observation reveals two distribution patterns of the LTR retrotransposon superfamilies. The first one is characterized by a *gypsy/copia* ratio lower than 1, where *copia* elements are ~1.3–1.5 times

Table 12.4 Proportions of the LTR retrotransposon *copia* and *gypsy* families (as % of the genome) detected in four Old World lupins (annotated according to the nomenclature of Wicker et al. 2007)

TE Superfamily & Family	*L. albus*	*L. luteus*	*L. micranthus*	*L. angustifolius*
Copia—AleI/Retrofit	0.09	0.03	0.07	0.03
Copia—AleII	0.05	0.38	0.17	0.09
Copia—Angela	1.13	1.41	0.92	1.49
Copia—Ivana/Oryco	0.21	0.42	0.09	0.22
Copia—TAR	0.44	0.50	0.37	0.54
Copia—Tork	0.77	0.96	0.70	1.30
Copia—Maximus/SIRE	5.06	6.06	3.15	8.06
Subtotal *Copia* (% of the genome)	**7.75**	**9.78**	**5.48**	**11.73**
Gypsy—Athila	1.09	16.12	0.97	3.30
Gypsy—Ogre/Tat	0.65	4.58	1.47	1.27
Gypsy—Chromovirus	18.57	10.62	1.29	4.14
Subtotal *gypsy* (% of the genome)	**20.31**	**31.31**	**3.73**	**8.70**
***Gypsy/Copia* ratio**	2.62	3.2	0.68	0.74

more abundant than the *gypsy* ones, such as in *L. micranthus* and *L. angustifolius*. The second pattern is defined by a *gypsy/copia* ratio much higher than 1, where *gypsy* elements clearly represent the prominent part of the LTR elements and are 2.6 and 3.2 times more abundant than the *copia* ones in *L. albus* and *L. luteus*, respectively.

A thorough annotation revealed that each of the LTR superfamilies is characterized by a fairly homogeneous and similar profile of *copia* and *gypsy* TE families in the lupin genomes surveyed, regardless of their variable proportions. Indeed, seven different *copia* (*Alel/Retrofit*, *AlelI*, *Angela*, *Ivana/Oryco*, *TAR*, *Tork*, and *Maximus/SIRE*) and three *gypsy* (*Athila*, *Ogre/Tat*, *Chromovirus*) families were identified in all species (Table 12.4; Fig. 12.4). The *Maximus/SIRE* family is the best represented in the *copia* superfamily with 3 to 8.06% of the genomes (in *L. micranthus* and *L. angustifolius*, respectively), followed at a lower level by the *Angela* (0.92 and 1.49%) and *Tork* (0.7 and

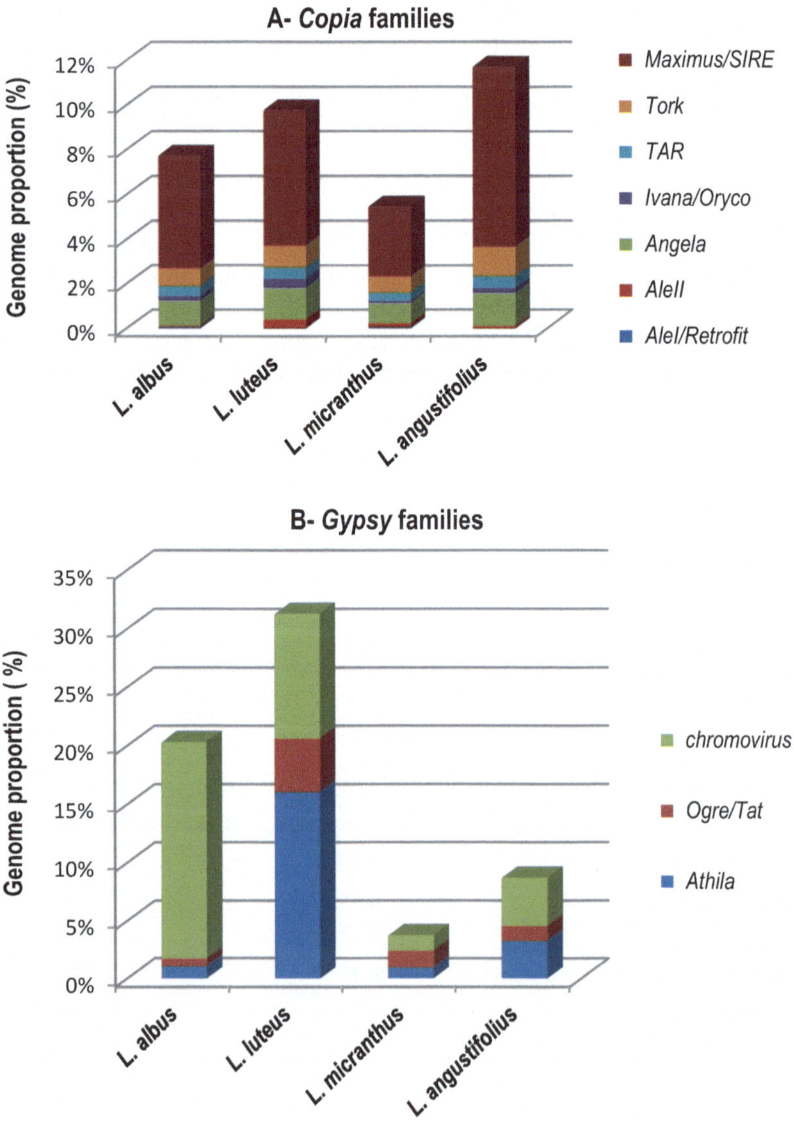

Fig. 12.4 Histogram showing the genomic proportions of the LTR retrotransposon *copia* (A) and *gypsy* (B) families (as % of the genome) detected in four Old World lupines. Retrotransposons are annotated according to the nomenclature of Wicker et al. 2007

1.3%) families. Together, the latter three families represent 86 and 92% of the *copia* elements of each genome (respectively), while the remaining families (*TAR*, *Tork*, *AleI*, *Allel/Retrofit*, *Ivana/Oryco*) are poorly represented, each at less than 0.6% of the nuclear genome. In the gypsy superfamily, the *Athila* family alone represents 16.2% (~382 Mb, i.e., half of the repetitive compartment) of the large *L. luteus* genome, whereas the *Chromovirus* family makes up 10.62 (~252 Mb) and 18.57% (210 Mb) of the genomes of *L. luteus* and *L. albus* (a small genome). The *Ogre/Tat* family is much less represented throughout the lupin genomes (less than 5%), with however a substantial amount (4.58%, i.e., ~109 Mb) in the large *L. luteus* genome. It is interesting to notice here: (i) that the amplification of *Athila* and *Chromovirus* elements played a decisive role in genome size increase in *L. luteus* (together representing 26.74% of the genome) compared to its counterpart in *L. micranthus* (2.26%), which has the same chromosome number and a smaller genome; (ii) that the latter elements were either only moderately amplified (or amplified then partly deleted via removal mechanisms; Devos 2002), such as in the other large genome of *L. angustifolius* (7.44%); but also, (iii) that *gypsy* elements may significantly proliferate in the small genomes, such as Chromovirus (18.57%) in *L. albus*.

12.3.3 Phylogenetic Analysis on LTR Retrotransposons RT Domains

In order to refine the annotation of LTR retrotransposons and to get insights into their diversity and dynamics in the Mediterranean lupine genomes, phylogenetic analyses were performed on RT (reverse transcriptase) domains extracted from clusters of reads generated by the RepeatExplorer analyses.

For each species, reads of each cluster (annotated as *copia* or *gypsy*) were assembled independently with Mira4 (Chevreux et al. 1999), and the consensus sequences obtained were submitted to BLASTx v. 2.6.0+ (Altschul et al. 1990; Camacho et al. 2009) against a public database of RT nucleotide sequences (Repbase v. 23.08; Bao et al. 2015). Sequences translated in protein which showed homology with RT domains, and that have at least 130 amino acids in length, were kept for further analyses. RT sequences from six angiosperms species (*Glycine max*, *Medicago truncatula*, *Arabidopsis thaliana*, *Zea mays*, *Oryza sativa*, *Triticum monococcum)* were selected in Repbase and added to the data set. Only potentially functional sequences without stop codon were retained. Each of the *gypsy* or *copia* RT protein sequences were aligned separately using Clustal Omega (Sievers et al. 2014). Informative blocks in multiple alignments were selected with the Gblocks package (Castresana 2000) prior to perform phylogenetic analyses with IQ-TREE (Nguyen et al. 2015). The LG+R6 and the LG+R7 protein evolution models were respectively retained (via ModelFinder; (Kalyaanamoorthy et al. 2017)) for phylogenetic reconstruction of *gypsy* and *copia* trees using the maximum likelihood method. The robustness of branches was estimated after 10,000 Ultrafast Bootstraps (Hoang et al. 2017). Annotation of *copia* and *gypsy* elements was determined according to the classification of Wicker et al. 2007.

The *copia* tree was built with 71 lupin sequences (43 from *L. angustifolius*, 2 from *L. micranthus*, 13 from *L. albus*, and 13 from *L. luteus*) and 244 sequences from other taxa (Fig. 12.5). Interestingly, all the most conserved RT sequences detected in *L. albus*, *L. luteus*, and *L. micranthus*, and about half of those detected *in L. angustifolius*, belong to the *Maximus/SIRE* family, which agrees with the prominence of this *copia* family in the lupin genomes. Moreover, this suggests that these elements, displaying well-conserved RT domains, most likely result from recent amplification events experienced by each species, as this seems corroborated by some specific groups of poorly divergent sequences with short branches (indicated in Fig. 12.5). All the other remaining conserved RT sequences represent diverse *Angela*, *TAR*, *Tork*, and *Ale* elements detected in *L. angustifolius*, which indicates that it is the only Mediterranean lupine species containing conserved copies of these *copia* families that are potentially able to proliferate. In particular, a distinct monophyletic group of *Angela* RTs suggests

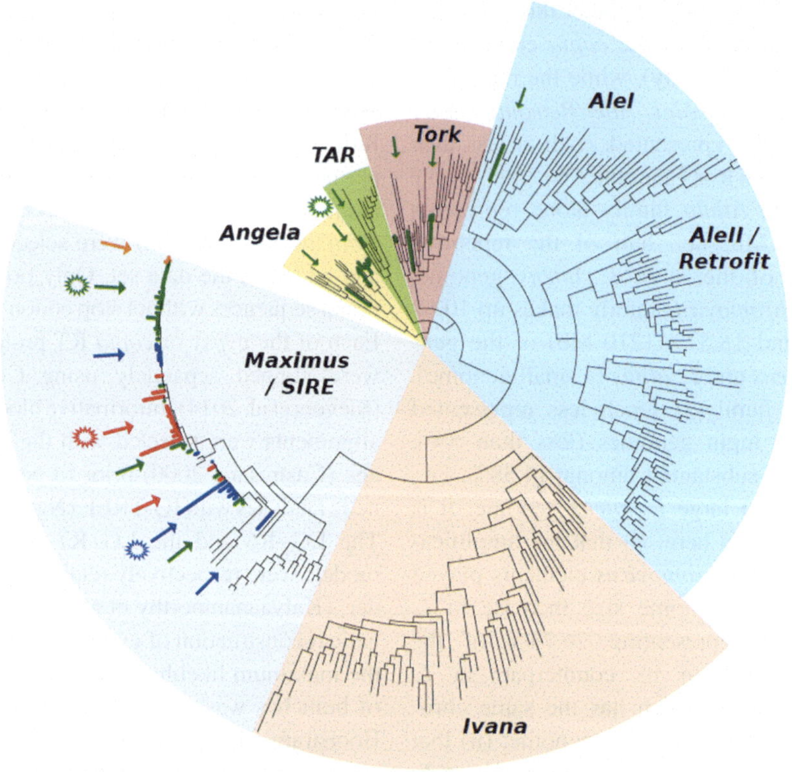

Fig. 12.5 Phylogenetic analysis of lupin LTR retrotransposon *copia* elements based on amino-acid sequences of their conserved RT domains, using the maximum likelihood method. The tree was built with 71 lupin sequences (43 from *L. angustifolius*, 2 from *L. micranthus*, 13 from *L. albus*, and 13 from *L. luteus*) and 244 reference sequences from databases. Annotation of *copia* families (colored clades named in black and bold) were determined following the classification of Wicker et al. (2007). Each terminal branch is colored according to its species origin: green for *L. angustifolius*, orange for *L. micranthus*, blue for *L. albus*, red for *L. luteus*, and black for reference taxa. Radiated/irregular circles likely represent recent species-specific amplification of particular *copia* lines

a lineage-specific amplification of one *Angela* line during the recent evolutionary history of this species. Although, *AleII/Retrofit* and *Ivana/Oryco* elements were detected in all lupins, indicating their common and ancient origin, no conserved RTs were found, which suggests that these poorly represented elements have undergone degenerative processes that tend toward their elimination from the genomes.

The *gypsy* tree was constructed with 72 lupine sequences (7 from *L. angustifolius*, 1 from *L. micranthus*, 29 from *L. albus*, and 35 from *L. luteus*) and 236 reference sequences from a set of plant genomes (Fig. 12.6). All the most conserved RT sequences detected belong to the three *gypsy* families identified via RepeatExplorer, *Chromovirus*, *Athila*, and *Ogre/Tat* (Table 12.4). The distribution of the conserved RTs among species appears correlated with the relative proportions of the *gypsy* families in the genomes. Conserved RTs of *Athila* elements were mostly found in *L. luteus* (13) and few in *L. angustifolius* (3) and *L. albus* (1). Few conserved RTs (1–3) of the *Ogre/Tat* elements were detected in lupins (with none in *L. albus*). With regard to *Chromovirus* elements, conserved RTs were mostly extracted from *L. luteus* and *L. albus*, the richest genomes in *gypsy* elements, and only three from *L. angustifolius*. Among the wide range of known *Chromovirus* elements, the phylogeny allowed to refine the classification of the lupin ones into two subfamilies, most of them

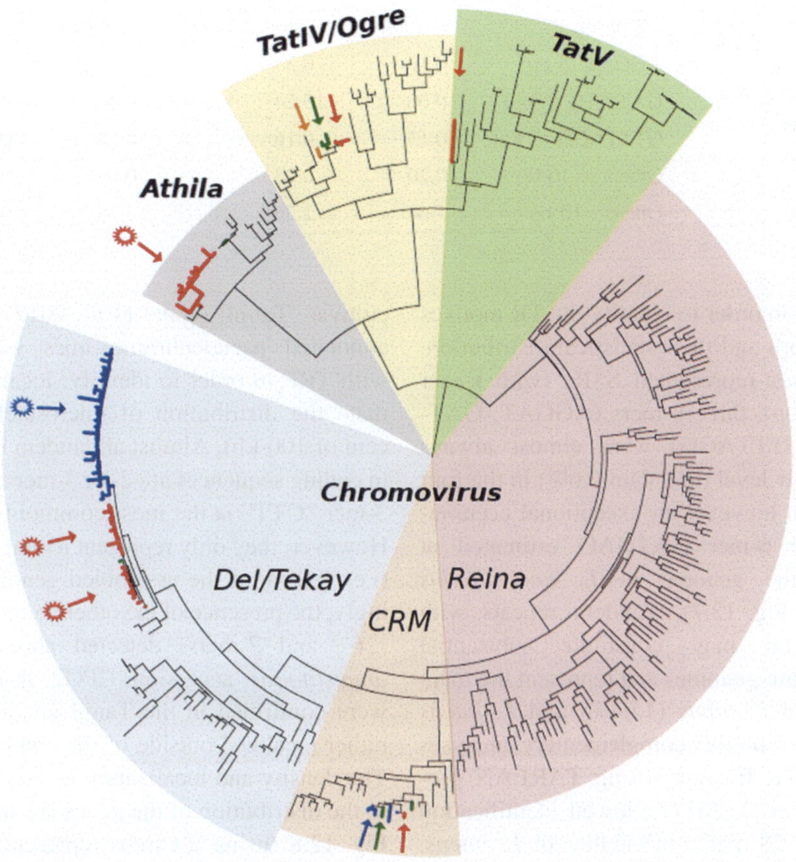

Fig. 12.6 Phylogenetic analysis of lupin LTR retrotransposon *gypsy* elements based on amino-acid sequences of their conserved RT domains, using the maximum likelihood method. The tree was built with 72 lupin sequences (7 from *L. angustifolius*, 1 from *L. micranthus*, 29 from *L. albus*, and 35 from *L. luteus*) and 236 reference sequences from databases. Annotation of *gypsy* families (colored clades named in black and bold) were determined following the classification of Wicker et al. (2007). Each terminal branch is colored according to its species origin: green for *L. angustifolius*, orange for *L. micranthus*, blue for *L. albus*, red for *L. luteus*, and black for reference taxa. Radiated/irregular circles likely represent recent species-specific amplification of particular *gypsy* lines

as *Tekay* homologs and the few others as *CRM* homologs (following RepBase annotation). Interestingly, the *gypsy* phylogeny reveals that *L. luteus* and *L. albus* most likely experienced recently independent and specific proliferation of *gypsy* elements, as this is illustrated by noteworthy monophyletic and monochromatic groups of poorly divergent RTs (with short branches) of *Tekay* and *Athila* retrotransposon lineages in Fig. 12.6. The other conserved RTs are minority lineages of *gypsy* elements represented in the Mediterranean lupin genomes that seem, however, yet potentially functional and able to proliferate.

12.3.4 Diversity and Abundance of Tandem Repeats in Lupin Genomes

As shown from the above RepeatExplorer-based analysis, the proportion of tandem repeats (excluding nrDNA) in the Mediterranean lupin genomes, varies from 3.37% in *L. albus* to a tremendous value of 26% in *L. angustifolius* (Table 12.3). In the latter species, TRs were even revealed more abundant than TEs. For each species, the reads contained in the clusters annotated as TRs were together analyzed using the TRF program v.4.09 (Tandem Repeat Finder;

Table 12.5 Proportion of the main types of tandem repeats (as % of the genome) detected in four Old World lupins

TR motifs	*L. albus*	*L. angustifolius*	*L. luteus*	*L. micranthus*
AGGAT (5 bp)	0.16	0.33	0.37	0.60
GATGAG (6 bp)	0.00	15.24	0.25	0.57
GTTTAGG (7 bp)	0.13	0.16	0.06	0.20
Others ≤ 10 bp	0.20	0.93	0.06	0.19
Others >10 bp	1.64	2.26	2.78	0.53

Benson 1999) in order to identify the TR motives (k-mers <50 bp) and their statistical distribution. Among the best-represented SSRs (with k-mer motives <10 bp), three k-mers (AGGAT, GATGAG, and GTTTAGG) were almost always present at a low level (less than 0.6%) in the four genomes, with however an exceptional accumulation of the 6-mer GATGAG estimated at 15.24% of the genome in *L. angustifolius* (Table 12.5; Fig. 12.7). Tandem repeats with k-mers >10 bp may constitute substantial amounts in lupin genomes and represent the main TR fraction in *L. albus* (1.64%) and *L. luteus* (2.78%). Interestingly, complementary analyses of the latter TR fraction (using TAREAN program; Novák et al. 2017) allowed identification of one major 28-mer minisatellite in *L. luteus*, one major 170-mer satellite and one 38-mer minisatellite in *L. albus*, as well as two 165-mer and 629-mer satellites in *L. micranthus*.

Moreover, taking advantage of the availability of a reference genome (*L. angustifolius* NLL cultivar. Tanjil; (Hane et al. 2017)), the twenty annotated pseudochromosomes were screened with TRF in order to identify, localize, and estimate the distribution of microsatellites (as per cent of 100-kb). Almost all tandem repeats found in coding sequences are 2- or 3-mers, of which the 3-mer "CTT" is the most commonly distributed. However, they only represent a total of 24,000 bp (i.e., 0.03% of the assembled genome). Interestingly, the presence of the other abundant SSRs (5-, 6-, and 7-mers) detected above in our *L. angustifolius* accession (IPG2 from Morocco) were confirmed in the Tanjil genome, but were rather localized outside of the coding sequences. The density and localization of the SSRs relative to the distribution of the genes are summarized in Fig. 12.8 (using a Circos representation; (Krzywinski et al. 2009)). The SSRs are distributed in all the genome and did not exhibit any chromosome specificity. The 6-mer SSR(GATGAG)^n previously identified in the IPG2 accession is confirmed as the major SSR in the NLL genome

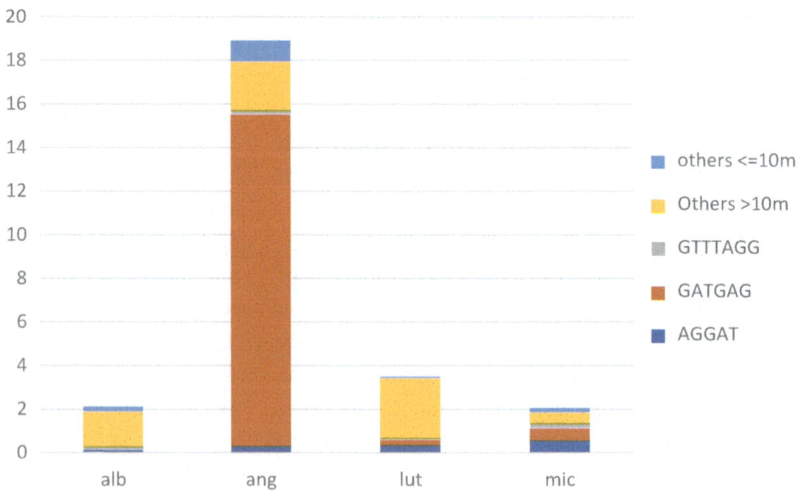

Fig. 12.7 Histogram representing the diversity and proportion of the simple sequence repeats (as % of the genome) detected in four Old World lupins using Tandem Repeat Finder program (Benson 1999; Lim et al. 2013): from *L. albus*(alb), *L. angustifolius* (ang), *L. luteus* (lut), and *L. micranthus* (mic)

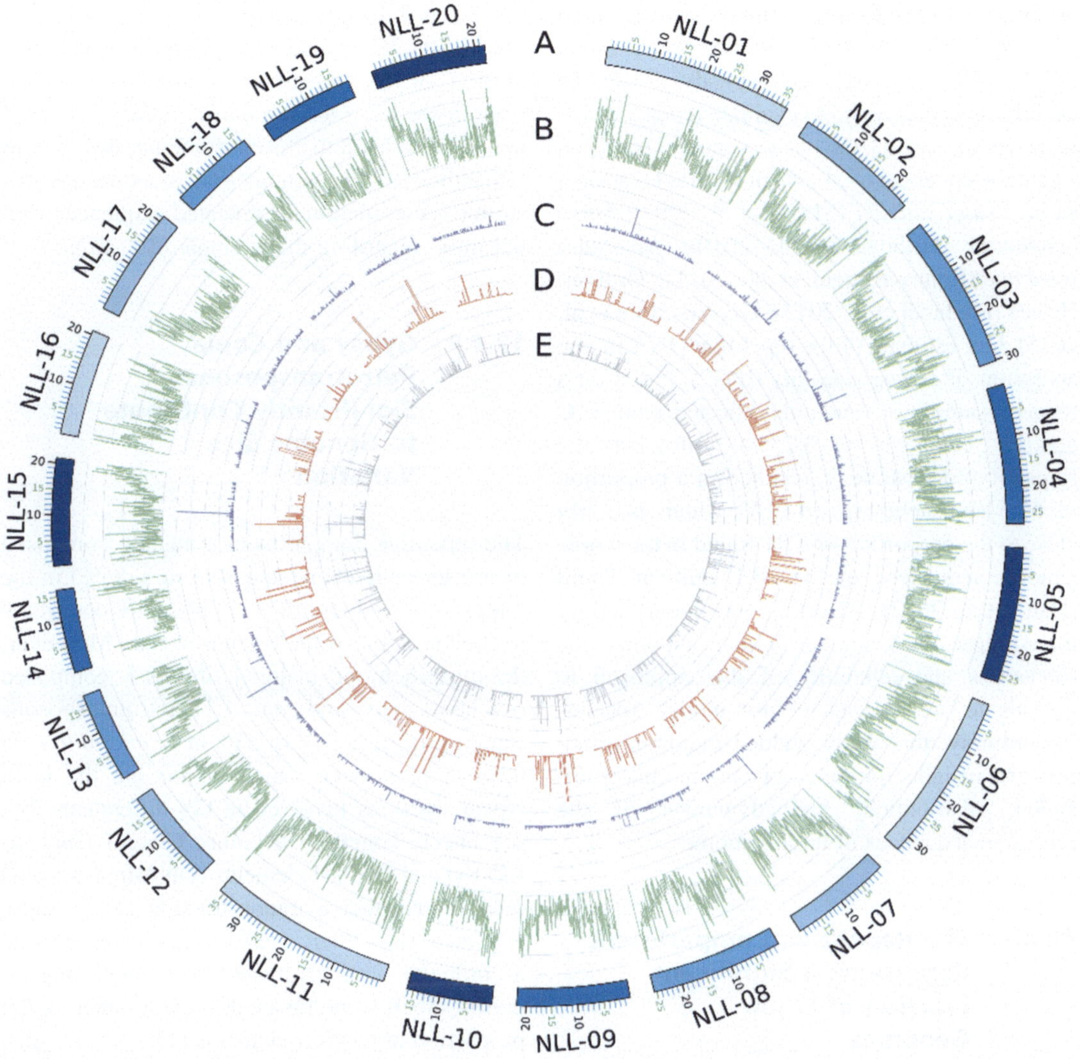

Fig. 12.8 Microsatellites distribution along pseudochromosomes of the first WG lupin sequenced, *L. angustifolius* (NLL var. Tanjil). The five consecutive circles from the outside to the inside of the figure represent: A, the 20 chromosomes (named NLL-01 to NLL-20) (in blue); B, Genes distribution and proportions per 100 kb (in green); C, proportions per 100 Kb of the 3-mer SSRs $(CTT)^n$ (×10 to be readable; in mauve); D, proportions per 100 Kb of 6-mers SSRs (in orange); and E, proportions per 100 Kb of all microsatellites (in gray). The higher the peaks are the higher is the proportion of genes or SSRs

cv. Tanjil, with pics of density mainly distributed in gene-poor regions. A thorough survey reveals that 1,143 genes include SSRs with the 3-mer (CTT) repeated at least four times. For example, a microsatellite with 65 perfect tandemly repeated (CTT) monomers was found in a putatively functional gene encoding a cytosolic oligopeptidase (ID: 109349122).

12.4 Repetitive Compartment of Lupin Genomes

In this chapter, we present the first detailed evaluation of the repetitive compartment in genomes of four smooth-seeded Mediterranean lupin taxa, based on the analysis of low-depth NGS genomic resources, using various bioinformatics programs to identify and estimate the repetitive sequences (Benson 1999; Novák et al. 2010, 2017, 2013). This approach already proved its usefulness to detect and to evaluate repeats which represent at least 0.01% of the genome, based on a genome coverage of >0.5% in several taxa, such as in: *Pisum sativum* (Macas et al. 2007); *Musa acuminata* (Hřibová et al. 2010); *Nicotiana tabacum* (Renny-Byfield et al. 2011); Orobanchacea (Piednoël et al. 2013); *Genlisea* (Vu et al. 2015). Our estimate of the repetitive DNA in one accession of *L. angustifolius* (IPG2), based on a reduced sample of randomly selected reads (1C genome coverage = 3.25%) following the RepeatExplorer strategy, resulted in a proportion of 52.54% (including nrDNA) which is fairly close to the proportion of 57% found in the whole genome sequenced of the NLL cultivar Tanjil (Hane et al. 2017). Also this was underlined by the studies cited above, which supports the robustness and reliability of this approach to investigate and compare non-model species. Accordingly, this study yielded major information and insights on the composition, characterization, distribution, and dynamics of the repetitive sequences in lupin genomes.

12.4.1 The Repetitive Compartment Represents a Significant Fraction of Lupin Genomes

As frequently observed in other angiosperms (Bennetzen 2000, 2005; Piegu et al. 2006; Hu et al. 2011; Bennetzen and Wang 2014; Vu et al. 2015; Wendel et al. 2016), the repetitive compartment represents a large proportion of the genomes (23–51%, excluding nrDNA) in the Mediterranean smooth-seeded lupins. The highest proportions were found in the largest genomes, regardless of their chromosome number, 50.36% in *L. luteus* ($2n = 52$; 2C = 2.37 Gb) and 49.63% in *L. angustifolius* ($2n = 40$; 2C = 1.85 Gb), whereas the two lupins with small genomes and fairly similar chromosome numbers exhibited very contrasted proportions of repeats in their genomes, 23.27% in *L. micranthus* ($2n = 52$; 2C = 1.15 Gb) *versus* 41.10% in *L. albus* ($2n = 50$; 2C = 1.13 Gb). Therefore, the proportion of the repetitive compartment in the smooth-seeded Mediterranean lupins is overall neither correlated to chromosome numbers nor to GS, although large genomes are associated with a strong accumulation of repeated sequences (but not only, regarding the example of *L. albus*).

12.4.2 Gypsy and Copia Retrotransposons Significantly Contribute to Genome Size Variation

The repetitive compartment is mainly composed of transposable elements (~43 to ~85%) in the lupins surveyed and they significantly contribute to the variation of their genome size. Moreover, the overwhelming majority of TEs is composed of Class I *gypsy* and *copia* LTR retrotransposons (ranging from 93.9% of TEs in *L. micranthus* to 97% in *L. luteus*), which *in fine* are the main repeat fractions involved in GS differences (but see later). Together the other Class I (such as LINEs) and Class II elements (DNA transposons) only represented a minor fraction of the lupin genomes (less than 1.6%). This is in general accordance with estimates from other angiosperms, albeit some taxa exhibited a much higher proportion of Class II elements (11–16.5%), such as in *A. thaliana*, *G. max*, wheat, and rice (Hawkins et al. 2006; Oliver et al. 2013).

The analyses based on random amplified RT domains and on Illumina HiSeq sequence data sets revealed in both lupin and Genistoid genomes a wide diversity of shared *copia* and *gypsy* LTR retrotransposons families. The thorough evaluation of LTR retrotransposon elements (via the RepeatExplorer strategy) highlighted the occurrence of a typical general profile of *copia* and *gypsy* families and subfamilies in the smooth-seeded Mediterranean lupin genomes, each species displaying its specific profile characterized by its own relative proportions of these elements. Additionally, a remarkable difference in the *gypsy/copia* ratio was observed among these species, regardless of their genome size as well as of their phylogenetic relationships, which is well exemplified by the prevalence of *copia* elements (~1.4 times more than *gypsy*) in *L. micranthus* and *L. angustifolius* and conversely by the over-accumulation of *gypsy* elements (2.6–3.2 times more than *copia*) in *L. albus* and *L. luteus*. It is noteworthy that few individual *gypsy* (Chromovirus and Athila) and *copia* (Maximus-SIRE) families alone have been remarkably accumulated in the lupins and hence strongly contributed in shaping their LTR retrotransposon profiles and in their GS differences, as shown in *L. luteus* (26.74% of Athila + Chromovirus), *L. micranthus* (18.57% of Chromovirus), and *L. angustifolius* (12.2% of Maximus/SIRE + Chromovirus).

12.4.3 Evolutionary Considerations on the Dynamics of Transposable Elements in Lupins

Altogether the above observations provide interesting insights on the dynamics of the repetitive sequences in lupin genomes, particularly of their major component, LTR retrotransposon elements. Overall, the same types of elements have been retrieved in both lupins and Genistoids (Mahé 2009), which supports their ancient origin from the common ancestor of the Genistoid alliance (and earlier). Nevertheless, it is obvious that the lupin genomes experienced divergent evolutionary dynamics, as demonstrated by the remarkable variability of the species-specific profiles of elements observed among the few representatives of the closely related Old World lupins investigated. Some LTR retrotransposon families appear to have actively proliferated and accumulated in some species (e.g., *Athila*, *Chomovirus*, *Maximus-SIRE* elements, or even *Ogre/Tat*) while they have been maintained at a low level in others. Most other families remained poorly represented throughout species. This strongly suggests that different processes and mechanisms regulating amplification, proliferation, and clearance of these repeats (Lippman et al. 2004; Ma and Bennetzen 2004; Hawkins et al. 2006, 2009; Slotkin and Martienssen 2007; Lisch 2009; Yaakov and Kashkush 2012) have differentially operated in these species over the last ~10 Myr of their diversification. This was also shown in other plant systems (e.g.: Hawkins et al. 2006; Charles et al. 2008; Hu et al. 2011; Estep et al. 2013; Piednoël et al. 2013).

Accordingly, phylogenetic analyses of the most conserved RT sequences (which presumably represent the most recent and potentially yet functional LTR retrotransposons) provided substantial clues which support recent (after species divergence, likely <8–10 Myr) and independent amplifications and accumulations (bursts) of the major *gypsy* and *copia* elements (*Athila*, *Chomovirus*, *Maximus-SIRE*, and even *Ogre/Tat*) in the lupin genomes. The other less common retrotransposons (such as *Angela*, *TAR*, *Tork*, and *Ale*), which seem still potentially able to proliferate in *L. angustifolius*, would represent families that either have low transposition rates or that have been specifically subjected to rapid purging processes following their expansion (Ma and Bennetzen 2004; Bennetzen 2005; Hu et al. 2011; Renny-Byfield et al. 2014; Vu et al. 2015). This leaves open the way to different evolutionary trajectories for the later families. Moreover, some weakly represented *copia* families, such as *AleII/Retrofit* and *Ivana/Oryco*, seem to have lost their ability to transpose. The yet recognizable but degenerated RTs found for these elements would likely represent the witnesses of ancient transposition events experienced by these families, which are ultimately prone to be erased from the DNA repetitive compartment of the smooth-seeded Mediterranean lupins. Another important evolutionary insight derived from the phylogenetic analysis of conserved RTs is that, not only various LTR retrotransposons' families or subfamilies have been differentially accumulated among the different lupin species but also that particular lineages of these families or subfamilies have been differentially amplified within each species, leading to the emergence of species-specific lineages of elements. For example, the major repeats in *L. luteus* essentially result from the recent proliferation of three species-specific *gypsy* lines (one from the *Athila* family, and two from the *Tekay* subfamily). Similarly, the prominent fraction of *gypsy* elements in *L. albus* results from the massive amplification of another specific lineage of the *Tekay* subfamily. Also, there are some evidence of likely recent lineage-specific amplification of *Maximus-SIRE* and *Athila* elements in *L. angustifolius*. Besides, a quick screening (results not shown) of available raw transcriptomic data sets from roots of *L. albus*, *L. luteus*, and *L. mariae-josephae* (Keller et al. 2018) provided some clues indicating a transcriptional activity for various TEs (including for some weakly represented families and ClassII elements). However, deeper investigations of more complete transcriptomic data sets are needed before making any reliable conclusion.

12.4.4 Tandem Repeats May also Greatly Contribute to Genome Obesity and Dynamics in Lupins

In the Mediterranean lupin genomes, the proportion of tandem repeats (excluding nrDNA) remarkably varies from 3 to 6% in *L. albus*, *L. luteus*, and *L. micranthus*, to 26% in *L. angustifolius*. In contrast to the three former lupins and to the general trend in plants (Oliveira et al. 2006; Barghini et al. 2014; Heitkam et al. 2015; Satović et al. 2018), the proportion of tandem repeats is not only tremendous but also is higher than that of transposable elements and represents more than half of the repetitive compartment in *L. angustifolius*. Also, it is noteworthy that even a low proportion of TRs may constitute a substantial fraction, equivalent to ~125 Mb in the large genome of *L. luteus*, for example. Among the best-represented SSRs in the smooth-seeded Mediterranean lupins, three were almost always detected in the genomes (AGGATn, GATGAGn, and GTTTAGGn). This is in agreement with the so-called "library hypothesis" evolution model which predicts that closely related species inherit from a common ancestor a same pool of satellites that are then independently amplified or lost in genomes (Fry and Salser 1977; Oliveira et al. 2006; Plohl et al. 2012; Garrido-Ramos 2017). Accordingly, our results revealed different SSR patterns which reflect the differential evolutionary dynamics experienced by these repeats in the lupin genomes. This is particularly well illustrated by the TR profile of *L. angustifolius*. In the latter species, the microsatellites *k*-mer <10 bp) have been much more accumulated (16.66%) than TRs with *k*-mers >10 bp (2.26%) in the genome, compared to its close Mediterranean relatives and to the lower frequencies reported for most other plants surveyed in the literature (Oliveira et al. 2006; Barghini et al. 2014; Heitkam et al. 2015; Satović et al. 2018). Even more striking, only one SSR (the 6-mer $GATGAG^n$) has been highly amplified and accumulated in *L. angustifolius* (estimated at 15.24% of the genome), whereas it is maintained at less than 0.6% in *L. luteus* and *L. micranthus*, and seem to have been erased from *L. albus*. Such contrasted frequencies of particular SSRs among genomes could be partially explained by divergences in the DNA repair system, as suggested by Oliveira et al. (2006).

Alternatively, while SSRs are yet mostly ubiquitous in the smooth-seeded Mediterranean lupins (regardless of their various proportions), few distinct families of minisatellites and satellites have been each differentially and specifically amplified in either *L. albus*, *L. luteus*, or *L. micranthus*. This suggests that they most likely result from dynamic and complex molecular processes and mechanisms that operated in the repetitive compartment following the diversification of the smooth-seeded Mediterranean lupins, which yielded species-specific satellite families (see: Garrido-Ramos 2015; Ávila Robledillo et al. 2018). It has been suggested that differences in satellites' types and abundance would play a role in speciation through the establishment of reproductive barriers between species, as demonstrated in *Drosophila* (Ferree and Barbash 2009). It is likely that the dramatic expansion of some satellites (alone and/or in conjunction with transposable elements) contributed to isolation and speciation processes among the Mediterranean lupins, as could be suggested by the striking divergent evolutionary dynamics observed following the separation of the closely related *L. luteus* ($2n = 52$; which preferentially accumulated a specific minisatellite and *gypsy* elements) and *L. angustifolius* ($2n = 42$; which rather accumulated a remarkable amount of a particular hexamer SSR and *copia* elements). Additionally, these species-specific satellites represent an important basis for the development of cytogenetic markers to identify chromosomes, and to help understanding genome organization in lupins.

Another interesting observation highlighted from the screening of the available reference genome of *L. angustifolius* (NLL cv. Tanjil) is that all satellites sensu lato detected in our NLL accession (IPG2) were retrieved throughout all the twenty pseudochromosomes. Two different distribution patterns were observed. On one side, almost all the 5-, 6-, and 7-mer SSRs observed in IPG2 are localized outside of the coding sequences in the gene-poor regions, with (GATGAG)n confirmed as the major SSR in this species. On the other side, the tandem repeats found in the coding regions are almost all SSRs with short monomers (*k*-mer < 4 bp), of which the SSR $(CTT)^n$ is the most abundant and widespread throughout the pseudochromosomes. Such prevalence of trinucleotide SSRs in the coding regions indicate that the other types with larger k-mers, which have a greater likelihood to induce frameshift mutations, are subjected to a counter-selection (Metzgar et al. 2000; Toth 2000). The screening of the NLL cv. Tanjil genome identified 1143 genes which contain a (CTT)n SSR with n equal to or greater than 4, which raises important questions to be addressed in order to evaluate their molecular, functional, and evolutionary impact.

12.5 Conclusion and Perspectives

This chapter represents the first study on the repetitive compartment in lupin genomes, using low-depth high-throughput sequencing, reads clustering and annotation. The detailed analyses performed in four smooth-seeded Mediterranean lupins revealed a wide diversity of repeat types and allowed identification of the most abundant categories shaping their genomes. In particular, only few *gypsy* (*Tekay*, *Athila*, *Ogre*) and *copia* (*Maximus-SIRE*) LTR retrotransposon families make up the prominent fraction of the repeats, which significantly contributes to genome size variation among species, regardless of their chromosome numbers and phylogenetic relationships. Interestingly, the results revealed that not only retrotransposons but also tandem repeats, such as microsatellites, may greatly contribute to genome obesity and dynamics in lupins, as demonstrated in *L. angustifolius*. Additionally, it has been shown that differential lineage-specific accumulation of transposable elements and/or tandem repeats occurred in lupins, which strongly supports that different processes and mechanisms regulating amplification, proliferation, and clearance of repeats have differentially operated within the same genus and among closely related Mediterranean species over the last ~10–12 Myr.

Further extension of such evaluation to representatives of the different lupin clades circumscribed in the genus will undoubtedly provide a more accurate and enhanced overview of the repetitive components and their evolutionary dynamics following diversification, evolution, and adaptation to diverse environmental conditions in both the Old and the New World. Additionally, the annotated raw material generated by this work represents a valuable basis to start building a repeats database specifically dedicated to the genus: (i) to accompany and facilitate assembly and annotation of novel lupin genomes; and (ii) to develop potentially useful genetic (e.g., microsatellites) and cytogenetic markers (e.g., specific minisatellites, satellites, and TEs). This will help understanding structure, organization, repeats distribution, and localization, variability, and evolution of the genomic landscape of lupins, and will enable comparative analysis with other legumes. Furthermore, the development of such database of repeats, using and combining genomic resources from both rapid low-depth high-throughput sequencing of various taxa and deep WGS of targeted species or accessions of particular interest, are of great importance to investigate and evaluate their structural, functional, and evolutionary impact on genes, such as, for example, those responsive for important agronomical, adaptive, and defense features.

Acknowledgements We are grateful to INEE-CNRS (France) and to the University of Rennes for their support to this work as part of the research program of the International Associated Laboratory "Ecological Genomics of Polyploidy" involving the University of Rennes (France) and the Iowa State University (Ames, USA). We thank Prof. Barbara Naganowska (Institut of Plant Genetics/PAS, Poznan, Poland) for kindly providing *L. angustifolius* seeds (IPG2 accession).

References

Ainouche A, Bayer RJ, Misset M-T (2004) Molecular phylogeny, diversification and character evolution in *Lupinus* (Fabaceae) with special attention to Mediterranean and African lupines. Plant Syst Evol 246(3–4):211–222

Aïnouche A, Bayer RJ (1999) Phylogenetic relationships in *Lupinus* (Fabaceae: Papilionoideae) based on internal transcribed spacer sequences (ITS) of nuclear ribosomal DNA. Am J Bot 86(4):590–607

Alix K, Heslop-harrison JS (2004) The diversity of retroelements in diploid and allotetraploid *Brassica* species. Plant Mol Biol 54:895–909

Altschul SF, Gish W, Miller W, Myers EW, Lipman DJ (1990) Basic local alignment search tool. J Mol Biol 215:403–410

Atnaf M, Yao N, Martina K, Dagne K, Wegary D, Tesfaye K (2017) Molecular genetic diversity and population structure of Ethiopian white lupin landraces: implications for breeding and conservation. PLoS ONE 12(11):e0188696

Ávila Robledillo L, Koblížková A, Novák P, Böttinger K, Vrbová I, Neumann P, Schubert I, Macas J (2018) Satellite DNA in *Vicia faba* is characterized by remarkable diversity in its sequence composition, association with centromeres, and replication timing. Sci Rep 8:5838

Axtell MJ (2013) Classification and comparison of small RNAs from plants. Annu Rev Plant Biol 64:137–159

Bao W, Kojima KK, Kohany O (2015) Repbase update, a database of repetitive elements in eukaryotic genomes. Mob DNA 6

Barghini E, Natali L, Cossu RM, Giordani T, Pindo M, Cattonaro F, Scalabrin S, Velasco R, Morgante M, Cavallini A (2014) The peculiar landscape of repetitive sequences in the olive (*Olea europaea* L.) genome. Genome Biol Evol 6:776–791

Bennett MD (2005) Nuclear DNA amounts in angiosperms: progress, problems and prospects. Ann Bot 95:45–90

Bennetzen JL (2000) Transposable element contributions to plant gene and genome evolution. Plant Mol Biol 42:251–269

Bennetzen JL (2002) Mechanisms and rates of genome expansion and contraction in flowering plants. Genetica 115:29–36

Bennetzen JL (2005) Transposable elements, gene creation and genome rearrangement in flowering plants. Curr Opin Genet Dev 15:621–627

Bennetzen JL, Wang H (2014) The contributions of transposable elements to the structure, function, and evolution of plant genomes. Annu Rev Plant Biol 65:505–530

Benson G (1999) Tandem repeats finder: a program to analyze DNA sequences. Nucleic Acids Res 27:573–580

Biémont C, Vieira C (2006) Genetics: Junk DNA as an evolutionary force. Nature 443:521–524

Biscotti MA, Olmo E, Heslop-Harrison JS (2015) Repetitive DNA in eukaryotic genomes. Chromosome Res 23(3):415–420

Cabello-Hurtado F, Keller J, Ley J, Sanchez-Lucas R, Jorrín-Novo JV, Aïnouche A (2016) Proteomics for exploiting diversity of lupin seed storage proteins and their use as nutraceuticals for health and welfare. J Proteomics 143:57–68

Camacho C, Coulouris G, Avagyan V, Ma N, Papadopoulos J, Bealer K, Madden TL (2009) BLAST+: architecture and applications. BMC Bioinform 10:421

Castel SE, Martienssen RA (2013) RNA interference in the nucleus: roles for small RNAs in transcription, epigenetics and beyond. Nat Rev Genet 14:100–112

Castresana J (2000) Selection of conserved blocks from multiple alignments for their use in phylogenetic analysis. Mol Biol Evol 17:540–552

Charles M, Belcram H, Just J, Huneau C, Viollet A, Couloux A, Segurens B, Carter M, Huteau V, Coriton O et al (2008) Dynamics and differential proliferation of transposable elements during the evolution of the B and A genomes of wheat. Genetics 180:1071–1086

Chénais B, Caruso A, Hiard S, Casse N (2012) The impact of transposable elements on eukaryotic genomes: from genome size increase to genetic adaptation to stressful environments. Gene 509:7–15

Chevreux B, Wetter T, Suhai S (1999) Genome sequence assembly using trace signals and additional sequence information. Comput Sci Biol: Proc German Conf Bioinform 99:45–56

Chu C, Nielsen R, Wu Y, Antoniewski C (2016) REPdenovo: Inferring de novo repeat motifs from short sequence reads. PLOS ONE 11 (3):e0150719

Conterato IF, Schifino-Wittmann MT (2006) New chromosome numbers, meiotic behaviour and pollen fertility in American taxa of *Lupinus* (Leguminosae):

contributions to taxonomic and evolutionary studies. Bot J Linn Soc 150:229–240

Devos KM (2002) Genome size reduction through illegitimate recombination counteracts genome expansion in *Arabidopsis*. Genome Res 12:1075–1079

Dolezel J, Bartos J, Voglmayr H, Greilhuber J (2003) Nuclear DNA content and genome size of trout and human. Cytometry 51A:127–128

Doolittle WF, Sapienza C (1980) Selfish genes, the phenotype paradigm and genome evolution. Nature 284:601–603

Eastwood RJ, Drummond CS, Schifino-Wittmann MT, Hughes CE (2008) Diversity and evolutionary history of lupins–insights from new phylogenies. In: Lupins for health and wealth: 12th international lupin conference, vol 10

Estep MC, DeBarry JD, Bennetzen JL (2013) The dynamics of LTR retrotransposon accumulation across 25 million years of panicoid grass evolution. Heredity 110:194–204

Ferree PM, Barbash DA (2009) Species-specific heterochromatin prevents mitotic chromosome segregation to cause hybrid lethality in *Drosophila*. PLoS Biol 7: e1000234 (MAF Noor, Ed.)

Flavell AJ, Dunbar E, Anderson R, Pearce SR, Hartley R, Kumar A (1992) *Ty1–copia* group retrotransposons are ubiquitous and heterogeneous in higher plants. Nucleic Acids Res 20:3639–3644

Flutre T, Duprat E, Feuillet C, Quesneville H (2011) Considering transposable element diversification in *de novo* annotation approaches. PLoS ONE 6 (Y Xu, Ed.)

Fry K, Salser W (1977) Nucleotide sequences of HS-a satellite DNA from Kangaroo Rat *Dipodomys ordii* and characterization of similar sequences in other rodents. Cell 12:1069–1084

Garrido-Ramos MA (2015) Satellite DNA in plants: more than just rubbish. Cytogenet Genome Res 146:153–170

Garrido-Ramos M (2017) Satellite DNA: an evolving topic. Genes 8:230

Gladstones JS, Atkins CA, Hamblin J (eds) (1998) Lupins as crop plants: biology, production, and utilization. CAB International, Wallingford, Oxon, UK ; New York, NY, USA

Grandbastien M-A, Audeon C, Bonnivard E, Casacuberta JM, Chalhoub B, Costa A-PP, Le QH, Melayah D, Petit M, Poncet C et al (2005) Stress activation and genomic impact of *Tnt1* retrotransposons in Solanaceae. Cytogenet Genome Res 110:229–241

Gregory TR (2005) The C-value Enigma in plants and animals: a review of parallels and an appeal for partnership. Ann Bot 95:133–146

Greilhuber J, Borsch T, Müller K, Worberg A, Porembski S, Barthlott W (2006) Smallest angiosperm genomes found in lentibulariaceae, with chromosomes of bacterial size. Plant Biol 8:770–777

Hane JK, Ming Y, Kamphuis LG, Nelson MN, Garg G, Atkins CA, Bayer PE, Bravo A, Bringans S, Cannon S et al (2017) A comprehensive draft genome sequence for lupin (*Lupinus angustifolius*), an emerging health food: insights into plant-microbe interactions and legume evolution. Plant Biotechnol J 15:318–330

Hawkins JS, Kim H, Nason JD, Wing RA, Wendel JF (2006) Differential lineage-specific amplification of transposable elements is responsible for genome size variation in *Gossypium*. Genome Res 16:1252–1261

Hawkins JS, Proulx SR, Rapp RA, Wendel JF (2009) Rapid DNA loss as a counterbalance to genome expansion through retrotransposon proliferation in plants. Proc Natl Acad Sci 106:17811–17816

Heitkam T, Petrasch S, Zakrzewski F, Kögler A, Wenke T, Wanke S, Schmidt T (2015) Next-generation sequencing reveals differentially amplified tandem repeats as a major genome component of Northern Europe's oldest *Camellia japonica*. Chromosome Res 23:791–806

Hoang DT, Chernomor O, von Haeseler A, Quang Minh B, Sy VL (2017) Ufboot 2: improving the ultrafast bootstrap approximation. Mol Biol Evol 32:518–522

Hosaka A, Kakutani T (2018) Transposable elements, genome evolution and transgenerational epigenetic variation. Curr Opin Genet Dev 49:43–48

Hřibová E, Neumann P, Matsumoto T, Roux N, Macas J, Doležel J (2010) Repetitive part of the banana (*Musa acuminata*) genome investigated by low-depth 454 sequencing. BMC Plant Biol 10

Hu TT, Pattyn P, Bakker EG, Cao J, Cheng J-F, Clark RM, Fahlgren N, Fawcett JA, Grimwood J, Gundlach H et al (2011) The *Arabidopsis lyrata* genome sequence and the basis of rapid genome size change. Nat Genet 43:476–481

Hughes C, Eastwood R (2006) Island radiation on a continental scale: Exceptional rates of plant diversification after uplift of the Andes. Proc Natl Acad Sci 103 (27):10334–10339. https://doi.org/10.1073/pnas.0601928103

Jiang N, Bao Z, Zhang X, Hirochika H, Eddy SR, McCouch SR, Wessler SR (2003) An active DNA transposon family in rice. Nature 421:163–167

Jiang N, Feschotte C, Zhang X, Wessler SR (2004) Using rice to understand the origin and amplification of miniature inverted repeat transposable elements (MITEs). Curr Opin Plant Biol 7:115–119

Kalendar R, Tanskanen J, Immonen S, Nevo E, Schulman AH (2000) Genome evolution of wild barley (*Hordeum spontaneum*) by BARE-1 retrotransposon dynamics in response to sharp microclimatic divergence. Proc Natl Acad Sci 97:6603–6607

Kalyaanamoorthy S, Minh BQ, Wong TKF, von Haeseler A, Jermiin LS (2017) ModelFinder: fast model selection for accurate phylogenetic estimates. Nat Methods 14:587–589

Kamphuis LG, Hane JK, Nelson MN, Gao L, Atkins CA, Singh KB (2015) Transcriptome sequencing of different narrow-leafed lupin tissue types provides a comprehensive uni-gene assembly and extensive gene-based molecular markers. Plant Biotechnol J 13:14–25

Kashkush K, Feldman M, Levy AA (2003) Transcriptional activation of retrotransposons alters the expression of adjacent genes in wheat. Nat Genet 33:102–106

Katoh K, Standley DM (2013) MAFFT multiple sequence alignment software version 7: improvements in performance and usability. Mol Biol Evol 30:772–780

Keller J, Imperial J, Ruiz-Argüeso T, Privet K, Lima O, Michon-Coudouel S, Biget M, Salmon A, Aïnouche A, Cabello-Hurtado F (2018) RNA sequencing and analysis of three *Lupinus* nodulomes provide new insights into specific host-symbiont relationships with compatible and incompatible *Bradyrhizobium* strains. Plant Sci 266:102–116

Kroc M, Koczyk G, Święcicki W, Kilian A, Nelson MN (2014) New evidence of ancestral polyploidy in the Genistoid legume *Lupinus angustifolius* L. (narrow-leafed lupin). Theor Appl Genet 127:1237–1249

Krzywinski M, Schein J, Birol I, Connors J, Gascoyne R, Horsman D, Jones SJ, Marra MA (2009) Circos: an information aesthetic for comparative genomics. Genome Res 19:1639–1645

Kumar A, Bennetzen JL (1999) Plant retrotransposons. Annu Rev Genet 33:479–532

Leitch AR, Leitch IJ (2008) Genomic plasticity and the diversity of polyploid plants. Science 320:481–483

Lerat E (2010) Identifying repeats and transposable elements in sequenced genomes: how to find your way through the dense forest of programs. Heredity 104:520–533

Levinson G, Gutman G (1987) Slipped-strand mispairing: a major mechanism for DNA sequence evolution. Mol Biol Evol 4:203–221

Li Y-C (2004) Microsatellites within genes: structure, function, and evolution. Mol Biol Evol 21:991–1007

Lim KG, Kwoh CK, Hsu LY, Wirawan A (2013) Review of tandem repeat search tools: a systematic approach to evaluating algorithmic performance. Brief Bioinform 14:67–81

Lippman Z, Gendrel A-V, Black M, Vaughn MW, Dedhia N, Richard McCombie W, Lavine K, Mittal V, May B, Kasschau KD et al (2004) Role of transposable elements in heterochromatin and epigenetic control. Nature 430:471–476

Lisch D (2009) Epigenetic regulation of transposable elements in plants. Annu Rev Plant Biol 60:43–66

Lisch D (2013) How important are transposons for plant evolution? Nat Rev Genet 14:49–61

Liu B, Wendel JF (2000) Retrotransposon activation followed by rapid repression in introgressed rice plants. Genome 43:874–880

Lönnig W-E, Saedler H (1997) Plant transposons: contributors to evolution? Gene 205:245–253

Lower SS, McGurk MP, Clark AG, Barbash DA (2018) Satellite DNA evolution: old ideas, new approaches. Curr Opin Genet Dev 49:70–78

Lynch VJ, Nnamani MC, Kapusta A, Brayer K, Plaza SL, Mazur EC, Emera D, Sheikh SZ, Grützner F, Bauersachs S et al (2015) Ancient transposable elements transformed the uterine regulatory landscape and transcriptome during the evolution of mammalian pregnancy. Cell Rep 10:551–561

Ma J, Bennetzen JL (2004) Rapid recent growth and divergence of rice nuclear genomes. Proc Natl Acad Sci 101:12404–12410

Macas J, Neumann P, Navrátilová A (2007) Repetitive DNA in the pea (*Pisum sativum* L.) genome: comprehensive characterization using 454 sequencing and comparison to soybean and *Medicago truncatula*. BMC Genomics 8

Mahé F (2009) Phylogénie, éléments transposables et évolution de la taille des génomes chez les lupins

Mahé F, Pascual H, Coriton O, Huteau V, Navarro Perris A, Misset M-T, Aïnouche A (2011) New data and phylogenetic placement of the enigmatic Old World lupin: *Lupinus mariae-josephi* H Pascual. Genet Resour Crop Evol 58:101–114

Mayer KFX, Martis M, Hedley PE, Šimková H, Liu H, Morris JA, Steuernagel B, Taudien S, Roessner S, Gundlach H et al (2011) Unlocking the barley genome by chromosomal and comparative genomics. Plant Cell 23:1249–1263

McClintock B (1948) Mutable loci in maize. Carnegie Institution of Washington Year Book, 47, 155–169

Metzgar D, Bytof J, Wills C (2000) Selection against frameshift mutations limits microsatellite expansion in coding DNA. Genome Res 9

Morgante M, Brunner S, Pea G, Fengler K, Zuccolo A, Rafalski A (2005) Gene duplication and exon shuffling by helitron-like transposons generate intraspecies diversity in maize. Nat Genet 37:997–1002

Naganowska B (2003) Nuclear DNA content variation and species relationships in the genus *Lupinus* (Fabaceae). Ann Bot 92:349–355

Naganowska B, Wolko B, Śliwińska E, Kaczmarek Z, Schifino-Wittmann MT (2005) 2C DNA variation and relationships among New World species of the genus *Lupinus* (Fabaceae). Plant Syst Evol 256:147–157

Nguyen L-T, Schmidt HA, von Haeseler A, Minh BQ (2015) IQ-TREE: a fast and effective stochastic algorithm for estimating maximum-likelihood phylogenies. Mol Biol Evol 32:268–274

Novák P, Neumann P, Macas J (2010) Graph-based clustering and characterization of repetitive sequences in next-generation sequencing data. BMC Bioinform 11

Novák P, Ávila Robledillo L, Koblížková A, Vrbová I, Neumann P, Macas J (2017) TAREAN: a computational tool for identification and characterization of satellite DNA from unassembled short reads. Nucleic Acids Res 45

Novák P, Neumann P, Pech J, Steinhaisl J, Macas J (2013) RepeatExplorer: a Galaxy-based web server for genome-wide characterization of eukaryotic repetitive elements from next-generation sequence reads. Bioinformatics 29:792–793

O'Rourke JA, Yang SS, Miller SS, Bucciarelli B, Liu J, Rydeen A, Bozsoki Z, Uhde-Stone C, Tu ZJ, Allan D et al (2013) An RNA-Seq transcriptome analysis of orthophosphate-deficient white lupin reveals novel insights into phosphorus acclimation in plants. Plant Physiol 161:705–724

Oliveira EJ, Pádua JG, Zucchi MI, Vencovsky R, Vieira MLC (2006) Origin, evolution and genome distribution of microsatellites. Genet Mol Biol 29:294–307

Oliver KR, McComb JA, Greene WK (2013) Transposable elements: powerful contributors to angiosperm evolution and diversity. Genome Biol Evol 5:1886–1901

Orgel LE, Crick FH, Sapienza C (1980) Selfish DNA. Nature 288:645–646

Parra-González LB, Aravena-Abarzúa GA, Navarro-Navarro CS, Udall J, Maughan J, Peterson LM, Salvo-Garrido HE, Maureira-Butler IJ (2012) Yellow lupin (*Lupinus luteus* L.) transcriptome sequencing: molecular marker development and comparative studies. BMC Genomics 13:425

Pellicer J, Hidalgo O, Dodsworth S, Leitch I (2018) Genome size diversity and its impact on the evolution of land plants. Genes 9:88

Petes TD (1980) Unequal meiotic recombination within tandem arrays of yeast ribosomal DNA genes. Cell 19:765–774

Piednoël M, Carrete-Vega G, Renner SS (2013) Characterization of the LTR retrotransposon repertoire of a plant clade of six diploid and one tetraploid species. Plant J 75:699–709

Piegu B, Guyot R, Picault N, Roulin A, Saniyal A, Kim H, Collura K, Brar DS, Jackson S, Wing RA et al (2006) Doubling genome size without polyploidization: dynamics of retrotransposition-driven genomic expansions in *Oryza australiensis*, a wild relative of rice. Genome Res 16:1262–1269

Plohl M, Mestrovic N, Mravinac B (2012) Satellite DNA evolution. In: Garrido-Ramos MA (ed) Genome dynamics. S. KARGER AG, Basel, pp 126–152

Priyam A, Woodcroft BJ, Rai V, Munagala A, Moghul I, Ter F, Gibbins MA, Moon H, Leonard G, Rumpf W, Wurm Y (2015) Sequenceserver: a modern graphical user interface for custom BLAST databases. biorxiv. https://doi.org/10.1101/033142

Quesneville H, Bergman CM, Andrieu O, Autard D, Nouaud D, Ashburner M, Anxolabehere D (2005) Combined evidence annotation of transposable elements in genome sequences. PLoS Comput Biol 1:166–175

Raman R, Cowley RB, Raman HD, Luckett DJ (2014) Analyses using SSR and DArT molecular markers reveal that Ethiopian accessions of white lupin (*Lupinus albus* L.) represent a unique gene pool. Open J Genet 4:87–98

Renny-Byfield S, Wendel JF (2014) Doubling down on genomes: polyploidy and crop plants. Am J Bot 101:1711–1725

Renny-Byfield S, Chester M, Kovarik A, Le Comber SC, Grandbastien M-A, Deloger M, Nichols RA, Macas J, Novak P, Chase MW et al (2011) Next generation sequencing reveals genome downsizing in allotetraploid *Nicotiana tabacum*, predominantly through the elimination of paternally derived repetitive DNAs. Mol Biol Evol 28:2843–2854

Renny-Byfield S, Gallagher JP, Grover CE, Szadkowski E, Page JT, Udall JA, Wang X, Paterson AH, Wendel JF (2014) Ancient gene duplicates in *Gossypium* (Cotton) exhibit near-complete expression divergence. Genome Biol Evol 6:559–571

Ruiz-Ruano FJ, López-León MD, Cabrero J, Camacho JPM (2016) High-throughput analysis of the satellitome illuminates satellite DNA evolution. Sci Rep 6

Satović E, Vojvoda Zeljko T, Plohl M (2018) Characteristics and evolution of satellite DNA sequences in bivalve mollusks. Eur Zool J 85:94–103

Schmuths H (2004) Genome size variation among accessions of *Arabidopsis thaliana*. Ann Bot 93:317–321

Sequencing Project IRG (2005) The map-based sequence of the rice genome. Nature 436:793–800

Shi J, Huang S, Fu D, Yu J, Wang X, Hua W, Liu S, Liu G, Wang H (2013) Evolutionary dynamics of microsatellite distribution in plants: insight from the comparison of sequenced *Brassica*, *Arabidopsis* and other angiosperm species. PLoS ONE 8:e59988 (BA Vinatzer, Ed.)

Sievers F, Wilm A, Dineen D, Gibson TJ, Karplus K, Li W, Lopez R, McWilliam H, Remmert M, Soding J et al (2014) Fast, scalable generation of high-quality protein multiple sequence alignments using clustal omega. Mol Syst Biol 7:539–539

Slotkin RK, Martienssen R (2007) Transposable elements and the epigenetic regulation of the genome. Nat Rev Genet 8:272–285

Soltis DE, Albert VA, Leebens-Mack J, Bell CD, Paterson AH, Zheng C, Sankoff D, dePamphilis CW, Wall PK, Soltis PS (2009) Polyploidy and angiosperm diversification. Am J Bot 96:336–348

Staton SE, Burke JM (2015) Transposome: Annotation of transposable element families from unassembled sequence reads. Bioinform, https://doi.org/10.1093/bioinformatics/btv05

Streelman JT, Kocher TD (2002) Microsatellite variation associated with prolactin expression and growth of salt-challenged tilapia. Physiol Genomics 9:1–4

Thomas CA (1971) The genetic organization of chromosomes. Annu Rev Genet 5:237–256

Toth G (2000) Microsatellites in different eukaryotic genomes: survey and analysis. Genome Res 10:967–981

Treangen TJ, Salzberg SL (2011) Repetitive DNA and next-generation sequencing: computational challenges and solutions. Nat Rev Genet 13:36–46

Usai G, Mascagni F, Natali L, Giordani T, Cavallini A (2017) Comparative genome-wide analysis of repetitive DNA in the genus *Populus* L. Tree Genet Genomes 13:96

Vicient CM, Suoniemi A, Anamthawat-Jónsson K, Tanskanen J, Beharav A, Nevo E, Schulman AH (1999) Retrotransposon BARE-1 and its role in genome evolution in the genus *Hordeum*. Plant Cell 11:17

Vu GTH, Schmutzer T, Bull F, Cao HX, Fuchs J, Tran TD, Jovtchev G, Pistrick K, Stein N, Pecinka A et al (2015) Comparative genome analysis reveals divergent genome size evolution in a carnivorous plant genus. Plant Genome 8(3):1–14

Wajid B, Serpedin E (2012) Review of general algorithmic features for genome assemblers for next generation sequencers. Genomics Proteomics Bioinform 10:58–73

Wendel JF, Jackson SA, Meyers BC, Wing RA (2016) Evolution of plant genome architecture. Genome Biol 17:37

Wessler SR (2006) Transposable elements and the evolution of eukaryotic genomes. Proc Natl Acad Sci 103:17600–17601

Wicker T, Sabot F, Hua-Van A, Bennetzen JL, Capy P, Chalhoub B, Flavell A, Leroy P, Morgante M, Panaud O et al (2007) A unified classification system for eukaryotic transposable elements. Nat Rev Genet 8:973–982

Wicker T, Gundlach H, Spannagl M, Uauy C, Borrill P, Ramírez-González RH, De Oliveira R, Mayer KFX, Paux E, Choulet F (2018) Impact of transposable elements on genome structure and evolution in bread wheat. Genome Biol 19(1):103

Wolko B, Weeden NF (1989) Estimation of *Lupinus* genome polyploidy on the basis of isozymic loci number. Genet Pol 30:165–171

Wu DD, Ruban A, Fuchs J, Macas J, Novak P, Vaio M, Zhou YH, Houben A (2019) Nondisjunction and unequal spindle organization accompany the drive of *Aegilops speltoides* B chromosomes. New Phytol 223:1340–1352

Yaakov B, Kashkush K (2012) Mobilization of Stowaway-like MITEs in newly formed allohexaploid wheat species. Plant Mol Biol 80:419–427

Correction to: Genetic Diversity in Narrow-Leafed Lupin Breeding After the Domestication Bottleneck

Wallace A. Cowling

Correction to:
Chapter 1 in: K. B. Singh et al. (eds.), The Lupin Genome, Compendium of Plant Genomes,
https://doi.org/10.1007/978-3-030-21270-4_1

The original version of this book was inadvertently published with the below-mentioned error:

Incorrect Figure 1.1 had been placed in Chapter 1.

The erratum chapter and the book have been updated.

The updated version of this chapter can be found at
https://doi.org/10.1007/978-3-030-21270-4_1

K. B. Singh et al. (eds.), *The Lupin Genome*, Compendium of Plant Genomes,
https://doi.org/10.1007/978-3-030-21270-4_13

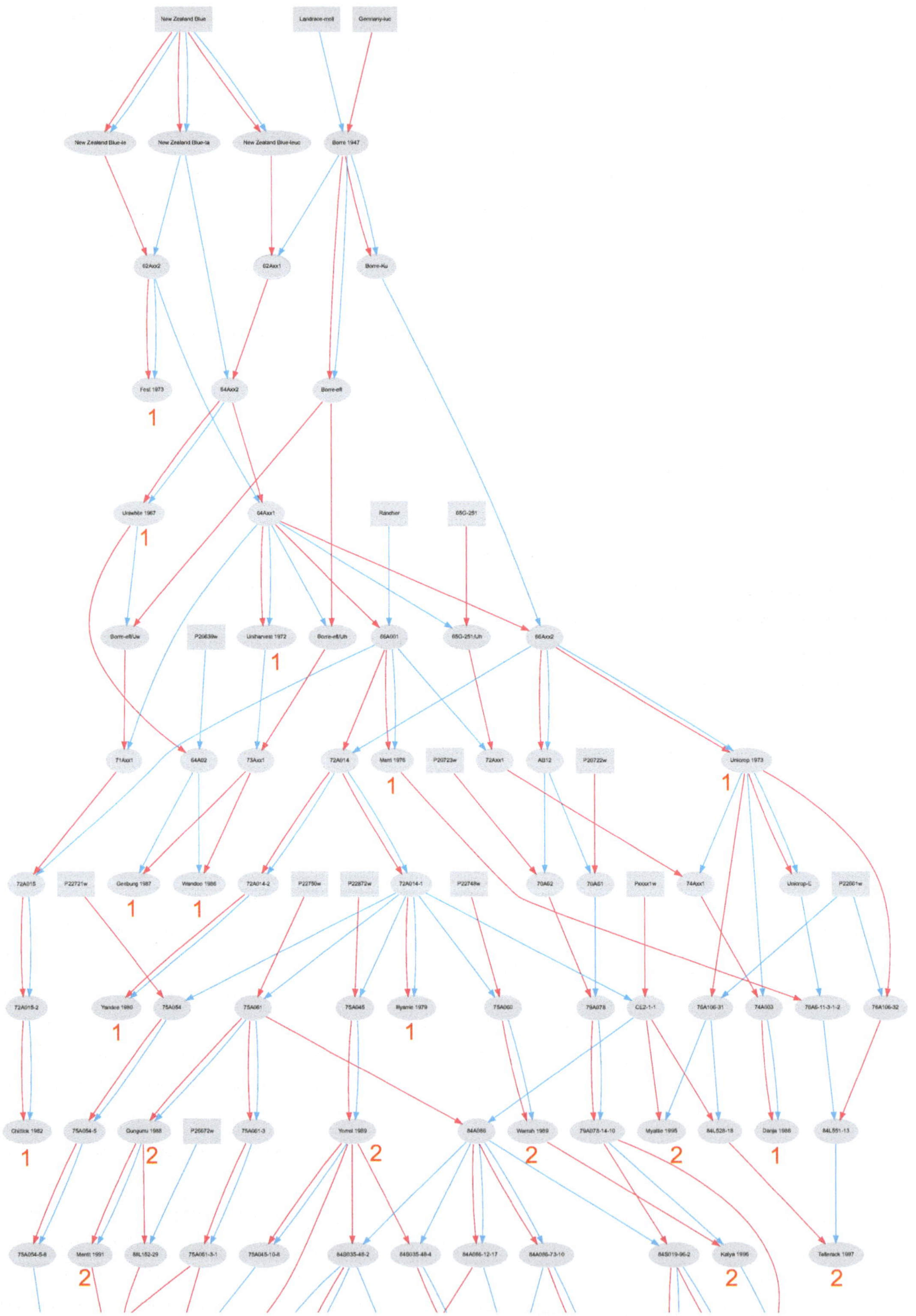

Fig. 1.1 Pedigree diagram of Australian narrow-leafed lupin varieties over four Phases of variety release, indicated by numerals 1, 2, 3, and 4 below the variety name. Female parents are indicated by red lines and male parents as blue lines. Selfing is indicated by both red and blue lines connecting from the parent

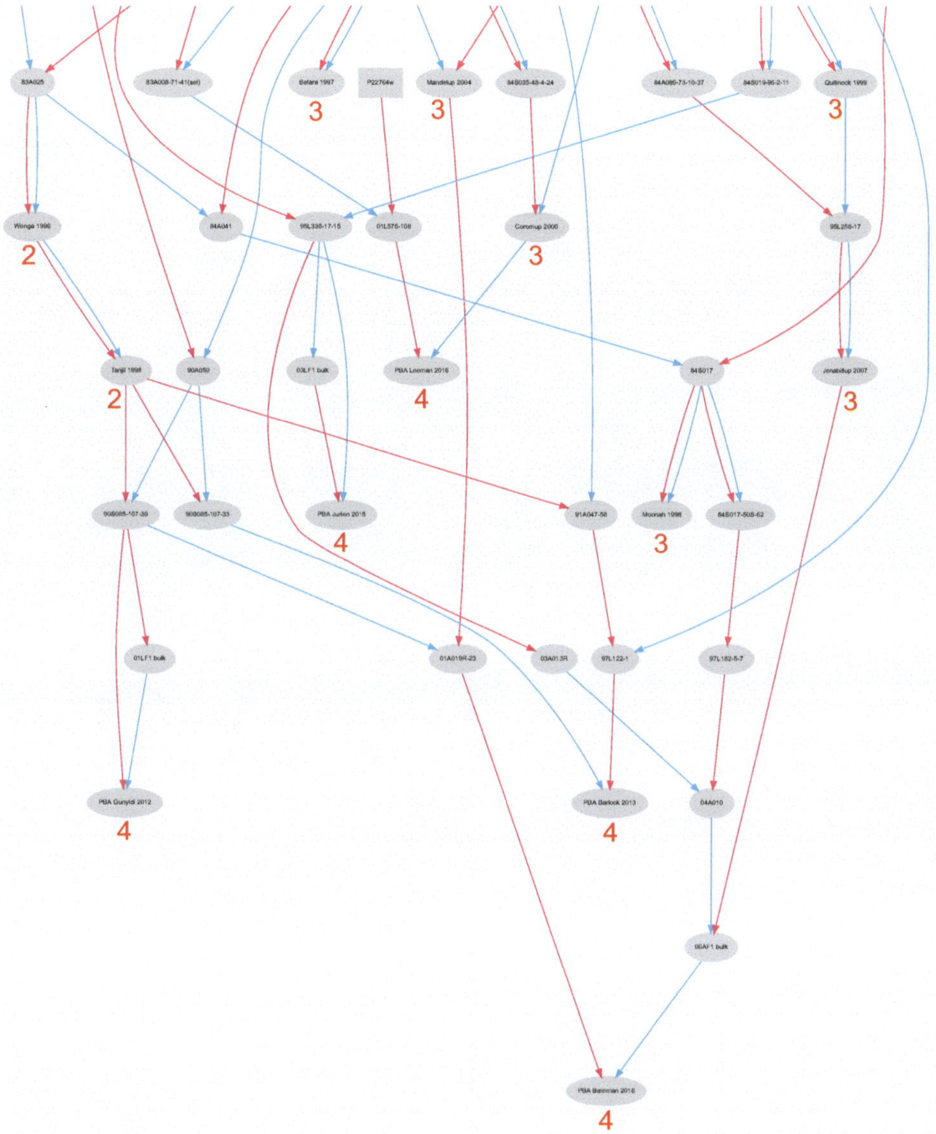

Fig. 1.1 (continued)

The manufacturer's authorised representative in the EU is Springer
Nature Customer Service Centre GmbH, Europaplatz 3, 69115 Heidelberg,
Germany. If you have any concerns regarding our products, please
contact ProductSafety@springernature.com

Printed and bound by CPI Group (UK) Ltd, Croydon, CR0 4YY
30/09/2026
02225651-0020